AF327085

Karl-Rudolf Koch

Introduction to Bayesian Statistics

Second Edition

Karl-Rudolf Koch

Introduction to Bayesian Statistics

Second, updated and enlarged Edition

With 17 Figures

 Springer

Professor Dr.-Ing.,
Dr.-Ing. E.h. mult. Karl-Rudolf Koch (em.)
University of Bonn
Institute of Theoretical Geodesy
Nussallee 17
53115 Bonn

E-mail: koch@geod.uni-bonn.de

Library of Congress Control Number: 2007929992

ISBN   978-3-540-72723-1 Springer Berlin Heidelberg New York

ISBN (1. Aufl) 978-3-540-66670-7   Einführung in Bayes-Statistik

Springer is a part of Springer Science+Business Media
springer.com

Cover design: deblik, Berlin
Production: Almas Schimmel
Typesetting: Camera-ready by Author

Printed on acid-free paper   30/3180/as 5 4 3 2 1 0

Preface to the Second Edition

This is the second and translated edition of the German book "Einführung in die Bayes-Statistik, Springer-Verlag, Berlin Heidelberg New York, 2000". It has been completely revised and numerous new developments are pointed out together with the relevant literature. The Chapter 5.2.4 is extended by the stochastic trace estimation for variance components. The new Chapter 5.2.6 presents the estimation of the regularization parameter of type Tykhonov regularization for inverse problems as the ratio of two variance components. The reconstruction and the smoothing of digital three-dimensional images is demonstrated in the new Chapter 5.3. The Chapter 6.2.1 on importance sampling for the Monte Carlo integration is rewritten to solve a more general integral. This chapter contains also the derivation of the SIR (sampling-importance-resampling) algorithm as an alternative to the rejection method for generating random samples. Markov Chain Monte Carlo methods are now frequently applied in Bayesian statistics. The first of these methods, the Metropolis algorithm, is therefore presented in the new Chapter 6.3.1. The kernel method is introduced in Chapter 6.3.3, to estimate density functions for unknown parameters, and used for the example of Chapter 6.3.6. As a special application of the Gibbs sampler, finally, the computation and propagation of large covariance matrices is derived in the new Chapter 6.3.5.

I want to express my gratitude to Mrs. Brigitte Gundlich, Dr.-Ing., and to Mr. Boris Kargoll, Dipl.-Ing., for their suggestions to improve the book. I also would like to mention the good cooperation with Dr. Chris Bendall of Springer-Verlag.

Bonn, March 2007   Karl-Rudolf Koch

Preface to the First German Edition

This book is intended to serve as an introduction to Bayesian statistics which is founded on Bayes' theorem. By means of this theorem it is possible to estimate unknown parameters, to establish confidence regions for the unknown parameters and to test hypotheses for the parameters. This simple approach cannot be taken by traditional statistics, since it does not start from Bayes' theorem. In this respect Bayesian statistics has an essential advantage over traditional statistics.

The book addresses readers who face the task of statistical inference on unknown parameters of complex systems, i.e. who have to estimate unknown parameters, to establish confidence regions and to test hypotheses for these parameters. An effective use of the book merely requires a basic background in analysis and linear algebra. However, a short introduction to one-dimensional random variables with their probability distributions is followed by introducing multidimensional random variables so that the knowledge of one-dimensional statistics will be helpful. It also will be of an advantage for the reader to be familiar with the issues of estimating parameters, although the methods here are illustrated with many examples.

Bayesian statistics extends the notion of probability by defining the probability for statements or propositions, whereas traditional statistics generally restricts itself to the probability of random events resulting from random experiments. By logical and consistent reasoning three laws can be derived for the probability of statements from which all further laws of probability may be deduced. This will be explained in Chapter 2. This chapter also contains the derivation of Bayes' theorem and of the probability distributions for random variables. Thereafter, the univariate and multivariate distributions required further along in the book are collected though without derivation. Prior density functions for Bayes' theorem are discussed at the end of the chapter.

Chapter 3 shows how Bayes' theorem can lead to estimating unknown parameters, to establishing confidence regions and to testing hypotheses for the parameters. These methods are then applied in the linear model covered in Chapter 4. Cases are considered where the variance factor contained in the covariance matrix of the observations is either known or unknown, where informative or noninformative priors are available and where the linear model is of full rank or not of full rank. Estimation of parameters robust with respect to outliers and the Kalman filter are also derived.

Special models and methods are given in Chapter 5, including the model of prediction and filtering, the linear model with unknown variance and covariance components, the problem of pattern recognition and the segmentation of digital images. In addition, Bayesian networks are developed for decisions in systems with uncertainties. They are, for instance, applied for the automatic interpretation of digital images.

If it is not possible to analytically solve the integrals for estimating parameters, for establishing confidence regions and for testing hypotheses, then numerical techniques have to be used. The two most important ones are the Monte Carlo integration and the Markoff Chain Monte Carlo methods. They are presented in Chapter 6.

Illustrative examples have been variously added. The end of each is indicated by the symbol Δ, and the examples are numbered within a chapter if necessary.

For estimating parameters in linear models traditional statistics can rely on methods, which are simpler than the ones of Bayesian statistics. They are used here to derive necessary results. Thus, the techniques of traditional statistics and of Bayesian statistics are not treated separately, as is often the case such as in two of the author's books "Parameter Estimation and Hypothesis Testing in Linear Models, 2nd Ed., Springer-Verlag, Berlin Heidelberg New York, 1999" and "Bayesian Inference with Geodetic Applications, Springer-Verlag, Berlin Heidelberg New York, 1990". By applying Bayesian statistics with additions from traditional statistics it is tried here to derive as simply and as clearly as possible methods for the statistical inference on parameters.

Discussions with colleagues provided valuable suggestions that I am grateful for. My appreciation is also forwarded to those students of our university who contributed ideas for improving this book. Equally, I would like to express my gratitude to my colleagues and staff of the Institute of Theoretical Geodesy who assisted in preparing it. My special thanks go to Mrs. Brigitte Gundlich, Dipl.-Ing., for various suggestions concerning this book and to Mrs. Ingrid Wahl for typesetting and formatting the text. Finally, I would like to thank the publisher for valuable input.

Bonn, August 1999   Karl-Rudolf Koch

Contents

1 Introduction

Bayesian statistics has the advantage, in comparison to traditional statistics, which is not founded on Bayes' theorem, of being easily established and derived. Intuitively, methods become apparent which in traditional statistics give the impression of arbitrary computational rules. Furthermore, problems related to testing hypotheses or estimating confidence regions for unknown parameters can be readily tackled by Bayesian statistics. The reason is that by use of Bayes' theorem one obtains probability density functions for the unknown parameters. These density functions allow for the estimation of unknown parameters, the testing of hypotheses and the computation of confidence regions. Therefore, application of Bayesian statistics has been spreading widely in recent times.

Traditional statistics introduces probabilities for random events which result from random experiments. Probability is interpreted as the relative frequency with which an event occurs given many repeated trials. This notion of probability has to be generalized for Bayesian statistics, since probability density functions are introduced for the unknown parameters, as already mentioned above. These parameters may represent constants which do not result from random experiments. Probability is therefore not only associated with random events but more generally with statements or propositions, which refer in case of the unknown parameters to the values of the parameters. Probability is therefore not only interpreted as frequency, but it represents in addition the plausibility of statements. The state of knowledge about a proposition is expressed by the probability. The rules of probability follow from logical and consistent reasoning.

Since unknown parameters are characterized by probability density functions, the method of testing hypotheses for the unknown parameters besides their estimation can be directly derived and readily established by Bayesian statistics. Intuitively apparent is also the computation of confidence regions for the unknown parameters based on their probability density functions. Whereas in traditional statistics the estimate of confidence regions follows from hypothesis testing which in turn uses test statistics, which are not readily derived.

The advantage of traditional statistics lies with simple methods for estimating parameters in linear models. These procedures are covered here in detail to augment the Bayesian methods. As will be shown, Bayesian statistics and traditional statistics give identical results for linear models. For this important application Bayesian statistics contains the results of traditional statistics. Since Bayesian statistics is simpler to apply, it is presented here as a meaningful generalization of traditional statistics.

2 Probability

The foundation of statistics is built on the theory of probability. Plausibility and uncertainty, respectively, are expressed by probability. In traditional statistics probability is associated with random events, i.e. with results of random experiments. For instance, the probability is expressed that a face with a six turns up when throwing a die. Bayesian statistics is not restricted to defining probabilities for the results of random experiments, but allows also for probabilities of statements or propositions. The statements may refer to random events, but they are much more general. Since probability expresses a plausibility, probability is understood as a measure of plausibility of a statement.

2.1 Rules of Probability

The rules given in the following are formulated for conditional probabilities. Conditional probabilities are well suited to express empirical knowledge. This is necessary, for instance, if decisions are to be made in systems with uncertainties, as will be explained in Chapter 5.5. Three rules are sufficient to establish the theory of probability.

2.1.1 Deductive and Plausible Reasoning

Starting from a cause we want to deduce the consequences. The formalism of *deductive reasoning* is described by mathematical logic. It only knows the states true or false. Deductive logic is thus well suited for mathematical proofs.

Often, after observing certain effects one would like to deduce the underlying causes. Uncertainties may arise from having insufficient information. Instead of deductive reasoning one is therefore faced with *plausible* or *inductive reasoning*. By deductive reasoning one derives consequences or effects from causes, while plausible reasoning allows to deduce possible causes from effects. The effects are registered by observations or the collection of data. Analyzing these data may lead to the possible causes.

2.1.2 Statement Calculus

A *statement* of mathematical logic, for instance, a sentence in the English language, is either true or false. Statements will be denoted by large letters $A, B, \ldots$ and will be called *statement variables*. They only take the values true (T) or false (F). They are linked by connectivities which are defined by truth tables, see for instance HAMILTON (1988, p.4). In the following we need the *conjunction* $A \wedge B$ of the statement variables A and B which has the truth table

$$
\begin{array}{cc|c}
A & B & A \wedge B \\
\hline
T & T & T \\
T & F & F \\
F & T & F \\
F & F & F
\end{array}
\tag{2.1}
$$

The conjunction is also called the *product* of the statement variables. It corresponds in the English language to "and". The conjunction $A \wedge B$ is denoted in the following by

$$
AB
\tag{2.2}
$$

in agreement with the common notation of probability theory.

The *disjunction* $A \vee B$ of the statement variables A and B which produces the truth table

$$
\begin{array}{cc|c}
A & B & A \vee B \\
\hline
T & T & T \\
T & F & T \\
F & T & T \\
F & F & F
\end{array}
\tag{2.3}
$$

is also called the *sum* of A and B. It corresponds in English to "or". It will be denoted by

$$
A + B
\tag{2.4}
$$

in the sequel.

The *negation* $\neg A$ of the statement A is described by the truth table

$$
\begin{array}{c|c}
A & \neg A \\
\hline
T & F \\
F & T
\end{array}
\tag{2.5}
$$

and is denoted by

$$
\bar{A}
\tag{2.6}
$$

in the following.

Expressions involving statement variables and connectivities are called *statement forms* which obey certain laws, see for instance HAMILTON (1988, p.11) and NOVIKOV (1973, p.23). In the following we need the commutative laws

$$
A + B = B + A \quad \text{and} \quad AB = BA \, ,
\tag{2.7}
$$

the associative laws

$$(A + B) + C = A + (B + C) \quad \text{and} \quad (AB)C = A(BC) \,, \tag{2.8}$$

the distributive laws

$$A(B + C) = AB + AC \quad \text{and} \quad A + (BC) = (A + B)(A + C) \tag{2.9}$$

and DE MORGAN's laws

$$\overline{(A + B)} = \bar{A}\bar{B} \quad \text{and} \quad \overline{AB} = \bar{A} + \bar{B} \tag{2.10}$$

where the equal signs denote logical equivalences.

The set of statement forms fulfilling the laws mentioned above is called statement algebra. It is as the set algebra a Boolean algebra, see for instance WHITESITT (1969, p.53). The laws given above may therefore be verified also by Venn diagrams.

2.1.3   Conditional Probability

A statement or a proposition depends in general on the question, whether a further statement is true. One writes $A|B$ to denote the situation that A is true under the condition that B is true. A and B are statement variables and may represent statement forms. The *probability* of $A|B$, also called *conditional probability*, is denoted by

$$P(A|B) \,. \tag{2.11}$$

It gives a measure for the plausibility of the statement $A|B$ or in general a measure for the uncertainty of the plausible reasoning mentioned in Chapter 2.1.1.

Example 1: We look at the probability of a burglary under the condition that the alarm system has been triggered.   △

Conditional probabilities are well suited to express empirical knowledge. The statement B points to available knowledge and $A|B$ to the statement A in the context specified by B. By $P(A|B)$ the probability is expressed with which available knowledge is relevant for further knowledge. This representation allows to structure knowledge and to consider the change of knowledge. Decisions under uncertainties can therefore be reached in case of changing information. This will be explained in more detail in Chapter 5.5 dealing with Bayesian networks.

Traditional statistics introduces the probabilities for random events of random experiments. Since these experiments fulfill certain conditions and certain information exists about these experiments, the probabilities of traditional statistics may be also formulated by conditional probabilities, if the statement B in (2.11) represents the conditions and the information.

Example 2: The probability that a face with a three turns up, when throwing a symmetrical die, is formulated according to (2.11) as the probability of a three under the condition of a symmetrical die.   △

Traditional statistics also knows the conditional probability, as will be mentioned in connection with (2.26).

2.1.4 Product Rule and Sum Rule of Probability

The quantitative laws, which are fulfilled by the probability, may be derived solely by logical and consistent reasoning. This was shown by COX (1946). He introduces a certain degree of plausibility for the statement $A|B$, i.e. for the statement that A is true given that B is true. JAYNES (2003) formulates three basic requirements for the plausibility:

1. Degrees of plausibility are represented by real numbers.

2. The qualitative correspondence with common sense is asked for.

3. The reasoning has to be consistent.

A relation is derived between the plausibility of the product AB and the plausibility of the statement A and the statement B given that the proposition C is true. The probability is introduced as a function of the plausibility. Using this approach COX (1946) and with additions JAYNES (2003), see also LOREDO (1990) and SIVIA (1996), obtain by extensive derivations, which need not to be given here, the *product rule* of probability

$$P(AB|C) = P(A|C)P(B|AC) = P(B|C)P(A|BC) \qquad (2.12)$$

with

$$P(S|C) = 1 \qquad (2.13)$$

where $P(S|C)$ denotes the probability of the *sure statement*, i.e. the statement S is with certainty true given that C is true. The statement C contains additional information or background information about the context in which statements A and B are being made.

From the relation between the plausibility of the statement A and the plausibility of its negation $\bar{A}$ under the condition C the *sum rule* of probability follows

$$P(A|C) + P(\bar{A}|C) = 1 \; . \qquad (2.14)$$

Example: Let an experiment result either in a success or a failure. Given the background information C about this experiment, let the statement A denote the success whose probability shall be $P(A|C) = p$. Then, because of (2.6), $\bar{A}$ stands for failure whose probability follows from (2.14) by $P(\bar{A}|C) = 1 - p$.   △

If $S|C$ in (2.13) denotes the sure statement, then $\bar{S}|C$ is the *impossible statement*, i.e. $\bar{S}$ is according to (2.5) with certainty false given that C is true. The probability of this impossible statement follows from (2.13) and (2.14) by

$$P(\bar{S}|C) = 0 \,. \tag{2.15}$$

Thus, the probability $P(A|C)$ is a real number between zero and one

$$0 \leq P(A|C) \leq 1 \,. \tag{2.16}$$

It should be mentioned here that the three rules (2.12) to (2.14) are sufficient to derive the following laws of probability which are needed in Bayesian statistics. These three rules only are sufficient for the further development of the theory of probability. They are derived, as explained at the beginning of this chapter, by logical and consistent reasoning.

2.1.5   Generalized Sum Rule

The probability of the sum $A + B$ of the statements A and B under the condition of the true statement C shall be derived. By (2.10) and by repeated application of (2.12) and (2.14) we obtain

$$
\begin{aligned}
P(A + B|C) &= P(\overline{\bar{A}\bar{B}}|C) = 1 - P(\bar{A}\bar{B}|C) = 1 - P(\bar{A}|C)P(\bar{B}|\bar{A}C) \\
&= 1 - P(\bar{A}|C)[1 - P(B|\bar{A}C)] = P(A|C) + P(\bar{A}B|C) \\
&= P(A|C) + P(B|C)P(\bar{A}|BC) = P(A|C) + P(B|C)[1 - P(A|BC)] \,.
\end{aligned}
$$

The *generalized sum rule* therefore reads

$$P(A + B|C) = P(A|C) + P(B|C) - P(AB|C) \,. \tag{2.17}$$

If $B = \bar{A}$ is substituted here, the statement $A + \bar{A}$ takes the truth value T and $A\bar{A}$ the truth value F according to (2.1), (2.3) and (2.5) so that $A + \bar{A}|C$ represents the sure statement and $A\bar{A}|C$ the impossible statement. The sum rule (2.14) therefore follows with (2.13) and (2.15) from (2.17). Thus indeed, (2.17) generalizes (2.14).

Let the statements A and B in (2.17) now be *mutually exclusive*. It means that the condition C requires that A and B cannot simultaneously take the truth value T. The product AB therefore obtains from (2.1) the truth value F. Then, according to (2.15)

$$P(AB|C) = 0 \,. \tag{2.18}$$

Example 1: Under the condition C of the experiment of throwing a die, let the statement A refer to the event that a two shows up and the statement B to the concurrent event that a three appears. Since the two statements A and B cannot be true simultaneously, they are mutually exclusive. Δ

We get with (2.18) instead of (2.17) the generalized sum rule for the two mutually exclusive statements A and B, that is

$$P(A + B|C) = P(A|C) + P(B|C) \,. \tag{2.19}$$

This rule shall now be generalized to the case of n mutually exclusive statements $A_1, A_2, \ldots, A_n$. Hence, (2.18) gives

$$P(A_i A_j|C) = 0 \quad \text{for} \quad i \neq j,\ i, j \in \{1, \ldots, n\}\,, \tag{2.20}$$

and we obtain for the special case $n = 3$ with (2.17) and (2.19)

$$\begin{aligned}
P(A_1 + A_2 + A_3|C) &= P(A_1 + A_2|C) + P(A_3|C) - P((A_1 + A_2)A_3|C) \\
&= P(A_1|C) + P(A_2|C) + P(A_3|C)
\end{aligned}$$

because of

$$P((A_1 + A_2)A_3|C) = P(A_1 A_3|C) + P(A_2 A_3|C) = 0$$

by virtue of (2.9) and (2.20). Correspondingly we find

$$P(A_1 + A_2 + \ldots + A_n|C) = P(A_1|C) + P(A_2|C) + \ldots + P(A_n|C) \,. \tag{2.21}$$

If the statements $A_1, A_2, \ldots, A_n$ are not only mutually exclusive but also *exhaustive* which means that the background information C stipulates that one and only one statement must be true and if one is true the remaining statements must be false, then we obtain with (2.13) and (2.15) from (2.21)

$$P(A_1 + A_2 + \ldots + A_n|C) = \sum_{i=1}^{n} P(A_i|C) = 1 \,. \tag{2.22}$$

Example 2: Let $A_1, A_2, \ldots, A_6$ be the statements of throwing a one, a two, and so on, or a six given the information C of a symmetrical die. These statements are mutually exclusive, as explained by Example 1 to (2.18). They are also exhaustive. With (2.22) therefore follows

$$P(A_1 + A_2 + \ldots + A_6|C) = \sum_{i=1}^{6} P(A_i|C) = 1 \,.$$

$$\Delta$$

To assign numerical values to the probabilities $P(A_i|C)$ in (2.22), it is assumed that the probabilities are equal, and it follows

$$P(A_i|C) = \frac{1}{n} \quad \text{for} \quad i \in \{1, 2, \ldots, n\} \,. \tag{2.23}$$

JAYNES (2003, p.40) shows that this result may be derived not only by intuition as done here but also by logical reasoning.

Let A under the condition C now denote the statement that is true in n_A cases for which (2.23) holds, then we obtain with (2.21)

$$P(A|C) = \frac{n_A}{n} \ . \tag{2.24}$$

This rule corresponds to the *classical definition of probability*. It says that if an experiment can result in n mutually exclusive and equally likely outcomes and if n_A of these outcomes are connected with the event A, then the probability of the event A is given by n_A/n. Furthermore, the definition of the *relative frequency* of the event A follows from (2.24), if n_A denotes the number of outcomes of the event A and n the number of trials for the experiment.

Example 3: Given the condition C of a symmetrical die the probability is $2/6 = 1/3$ to throw a two or a three according to the classical definition (2.24) of probability. $\quad\Delta$

Example 4: A card is taken from a deck of 52 cards under the condition C that no card is marked. What is the probability that it will be an ace or a diamond? If A denotes the statement of drawing a diamond and B the one of drawing an ace, $P(A|C) = 13/52$ and $P(B|C) = 4/52$ follow from (2.24). The probability of drawing the ace of diamonds is $P(AB|C) = 1/52$. Using (2.17) the probability of an ace or diamond is then $P(A + B|C) = 13/52 + 4/52 - 1/52 = 4/13$. $\quad\Delta$

Example 5: Let the condition C be true that an urn contains 15 red and 5 black balls of equal size and weight. Two balls are drawn without being replaced. What is the probability that the first ball is red and the second one black? Let A be the statement to draw a red ball and B the statement to draw a black one. With (2.24) we obtain $P(A|C) = 15/20 = 3/4$. The probability $P(B|AC)$ of drawing a black ball under the condition that a red one has been drawn is $P(B|AC) = 5/19$ according to (2.24). The probability of drawing without replacement a red ball and then a black one is therefore $P(AB|C) = (3/4)(5/19) = 15/76$ according to the product rule (2.12). $\quad\Delta$

Example 6: The grey value g of a picture element, also called pixel, of a digital image takes on the values $0 \leq g \leq 255$. If 100 pixels of a digital image with 512×512 pixels have the gray value $g = 0$, then the relative frequency of this value equals $100/512^2$ according to (2.24). The distribution of the relative frequencies of the gray values $g = 0, g = 1, \ldots, g = 255$ is called a histogram. $\quad\Delta$

2.1.6 Axioms of Probability

Probabilities of random events are introduced by axioms for the probability theory of traditional statistics, see for instance KOCH (1999, p.78). Starting from the set S of elementary events of a random experiment, a special system Z of subsets of S known as σ-algebra is introduced to define the random events. Z contains as elements subsets of S and in addition as elements the empty set and the set S itself. Z is closed under complements and countable unions. Let A with $A \in Z$ be a random event, then the following axioms are presupposed,

Axiom 1: A real number $P(A) \geq 0$ is assigned to every event A of Z. $P(A)$ is called the probability of A.

Axiom 2: The probability of the sure event is equal to one, $P(S) = 1$.

Axiom 3: If $A_1, A_2, \ldots$ is a sequence of a finite or infinite but countable number of events of Z which are mutually exclusive, that is $A_i \cap A_j = \emptyset$ for $i \neq j$, then

$$P(A_1 \cup A_2 \cup \ldots) = P(A_1) + P(A_2) + \ldots \ . \qquad (2.25)$$

The axioms introduce the probability as a measure for the sets which are the elements of the system Z of random events. Since Z is a σ-algebra, it may contain a finite or infinite number of elements, whereas the rules given in Chapter 2.1.4 and 2.1.5 are valid only for a finite number of statements.

If the system Z of random events contains a finite number of elements, the σ-algebra becomes a set algebra and therefore a Boolean algebra, as already mentioned at the end of Chapter 2.1.2. Axiom 1 is then equivalent to the requirement 1 of Chapter 2.1.4, which was formulated with respect to the plausibility. Axiom 2 is identical with (2.13) and Axiom 3 with (2.21), if the condition C in (2.13) and (2.21) is not considered. We may proceed to an infinite number of statements, if a well defined limiting process exists. This is a limitation of the generality, but is is compensated by the fact that the probabilities (2.12) to (2.14) have been derived as rules by consistent and logical reasoning. This is of particular interest for the product rule (2.12). It is equivalent in the form

$$P(A|BC) = \frac{P(AB|C)}{P(B|C)} \quad \text{with} \quad P(B|C) > 0 \ , \qquad (2.26)$$

if the condition C is not considered, to the definition of the conditional probability of traditional statistics. This definition is often interpreted by relative frequencies which in contrast to a derivation is less obvious.

For the foundation of Bayesian statistics it is not necessary to derive the rules of probability only for a finite number of statements. One may, as is shown for instance by BERNARDO and SMITH (1994, p.105), introduce by additional requirements a σ-algebra for the set of statements whose probabilities are sought. The probability is then defined not only for the sum of a finite number of statements but also for a countable infinite number of statements. This method will not be applied here. Instead we will restrict ourselves to an intuitive approach to Bayesian statistics. The theory of probability is therefore based on the rules (2.12), (2.13) and (2.14).

2.1.7 Chain Rule and Independence

The probability of the product of n statements is expressed by the chain rule of probability. We obtain for the product of three statements A_1, A_2 and A_3 under the condition C with the product rule (2.12)

$$P(A_1 A_2 A_3 | C) = P(A_3 | A_1 A_2 C) P(A_1 A_2 | C)$$

and by a renewed application of the product rule

$$P(A_1 A_2 A_3 | C) = P(A_3 | A_1 A_2 C) P(A_2 | A_1 C) P(A_1 | C) .$$

With this result and the product rule follows

$$P(A_1 A_2 A_3 A_4 | C) = P(A_4 | A_1 A_2 A_3 C) P(A_3 | A_1 A_2 C) P(A_2 | A_1 C) P(A_1 | C)$$

or for the product of n statements $A_1, A_2, \ldots, A_n$ the *chain rule* of probability

$$\begin{aligned}
P(A_1 A_2 \ldots A_n | C) = {} & P(A_n | A_1 A_2 \ldots A_{n-1} C) \\
& P(A_{n-1} | A_1 A_2 \ldots A_{n-2} C) \ldots P(A_2 | A_1 C) P(A_1 | C) .
\end{aligned} \quad (2.27)$$

We obtain for the product of the statements A_1 to A_{n-k-1} by the chain rule

$$\begin{aligned}
P(A_1 A_2 \ldots A_{n-k-1} | C) = {} & P(A_{n-k-1} | A_1 A_2 \ldots A_{n-k-2} C) \\
& P(A_{n-k-2} | A_1 A_2 \ldots A_{n-k-3} C) \ldots P(A_2 | A_1 C) P(A_1 | C) .
\end{aligned}$$

If this result is substituted in (2.27), we find

$$\begin{aligned}
P(A_1 A_2 \ldots A_n | C) = {} & P(A_n | A_1 A_2 \ldots A_{n-1} C) \ldots \\
& P(A_{n-k} | A_1 A_2 \ldots A_{n-k-1} C) P(A_1 A_2 \ldots A_{n-k-1} | C) .
\end{aligned} \quad (2.28)$$

In addition, we get by the product rule (2.12)

$$\begin{aligned}
P(A_1 A_2 \ldots A_n | C) = {} & P(A_1 A_2 \ldots A_{n-k-1} | C) \\
& P(A_{n-k} A_{n-k+1} \ldots A_n | A_1 A_2 \ldots A_{n-k-1} C) .
\end{aligned} \quad (2.29)$$

By substituting this result in (2.28) the *alternative chain rule* follows

$$\begin{aligned}
& P(A_{n-k} A_{n-k+1} \ldots A_n | A_1 A_2 \ldots A_{n-k-1} C) \\
& = P(A_n | A_1 A_2 \ldots A_{n-1} C) \ldots P(A_{n-k} | A_1 A_2 \ldots A_{n-k-1} C) .
\end{aligned} \quad (2.30)$$

The product rule and the chain rule simplify in case of independent statements. The two statements A and B are said to be *conditionally independent* or shortly expressed *independent*, if and only if under the condition C

$$P(A | BC) = P(A | C) . \quad (2.31)$$

If two statements A and B are independent, then the probability of the statement A given the condition of the product BC is therefore equal to the probability of the statement A given the condition C only. If conversely (2.31) holds, the two statements A and B are independent.

Example 1: Let the statement B given the condition C of a symmetrical die refer to the result of the first throw of a die and the statement A to the result of a second throw. The statements A and B are independent, since the probability of the result A of the second throw given the condition C and the condition that the first throw results in B is independent of this result B so that (2.31) holds. $\triangle$

The computation of probabilities in Bayesian networks presented in Chapter 5.5 is based on the chain rule (2.27) together with (2.31).

If (2.31) holds, we obtain instead of the product rule (2.12) the product rule of two independent statements

$$P(AB|C) = P(A|C)P(B|C) \tag{2.32}$$

and for n independent statements A_1 to A_n instead of the chain rule (2.27) the product rule of independent statements

$$P(A_1 A_2 \ldots A_n|C) = P(A_1|C)P(A_2|C) \ldots P(A_n|C) \, . \tag{2.33}$$

Example 2: Let the condition C denote the trial to repeat an experiment n times. Let the repetitions be independent and let each experiment result either in a success or a failure. Let the statement A denote the success with probability $P(A|C) = p$. The probability of the failure $\bar{A}$ then follows from the sum rule (2.14) with $P(\bar{A}|C) = 1-p$. Let n trials result first in x successes A and then in $n-x$ failures $\bar{A}$. The probability of this sequence follows with (2.33) by

$$P(AA \ldots A\bar{A}\bar{A} \ldots \bar{A}|C) = p^x(1-p)^{n-x} \, ,$$

since the individual trials are independent. This result leads to the binomial distribution presented in Chapter 2.2.3. $\triangle$

2.1.8 Bayes' Theorem

The probability of the statement AB given C and the probability of the statement $A\bar{B}$ given C follow from the product rule (2.12). Thus, we obtain after adding the probabilities

$$P(AB|C) + P(A\bar{B}|C) = [P(B|AC) + P(\bar{B}|AC)]P(A|C) \, . \tag{2.34}$$

The sum rule (2.14) leads to

$$P(B|AC) + P(\bar{B}|AC) = 1$$

and therefore

$$P(A|C) = P(AB|C) + P(A\bar{B}|C) \,. \tag{2.35}$$

If instead of AB and $A\bar{B}$ the statements $AB_1, AB_2, \ldots, AB_n$ under the condition C are given, we find in analogy to (2.34)

$$P(AB_1|C) + P(AB_2|C) + \ldots + P(AB_n|C)$$
$$= [P(B_1|AC) + P(B_2|AC) + \ldots + P(B_n|AC)]P(A|C) \,.$$

If $B_1, \ldots, B_n$ given C are mutually exclusive and exhaustive statements, we find with (2.22) the generalization of (2.35)

$$P(A|C) = \sum_{i=1}^{n} P(AB_i|C) \tag{2.36}$$

or with (2.12)

$$P(A|C) = \sum_{i=1}^{n} P(B_i|C)P(A|B_iC) \,. \tag{2.37}$$

These two results are remarkable, because the probability of the statement A given C is obtained by summing the probabilities of the statements in connection with B_i. Examples are found in the following examples for Bayes' theorem.

If the product rule (2.12) is solved for $P(A|BC)$, *Bayes' theorem* is obtained

$$P(A|BC) = \frac{P(A|C)P(B|AC)}{P(B|C)} \,. \tag{2.38}$$

In common applications of Bayes' theorem A denotes the statement about an unknown phenomenon. B represents the statement which contains information about the unknown phenomenon and C the statement for background information. $P(A|C)$ is denoted as *prior probability*, $P(A|BC)$ as *posterior probability* and $P(B|AC)$ as *likelihood*. The prior probability of the statement concerning the phenomenon, before information has been gathered, is modified by the likelihood, that is by the probability of the information given the statement about the phenomenon. This leads to the posterior probability of the statement about the unknown phenomenon under the condition that the information is available. The probability $P(B|C)$ in the denominator of Bayes' theorem may be interpreted as normalization constant which will be shown by (2.40).

The bibliography of Thomas Bayes, creator of Bayes' theorem, and references for the publications of Bayes' theorem may be found, for instance, in PRESS (1989, p.15 and 173).

If mutually exclusive and exhaustive statements $A_1, A_2, \ldots, A_n$ are given, we obtain with (2.37) for the denominator of (2.38)

$$P(B|C) = \sum_{j=1}^{n} P(A_j|C)P(B|A_jC) \tag{2.39}$$

and Bayes' theorem (2.38) takes on the form

$$P(A_i|BC) = P(A_i|C)P(B|A_iC)/c \quad \text{for} \quad i \in \{1, \ldots, n\} \tag{2.40}$$

with

$$c = \sum_{j=1}^{n} P(A_j|C)P(B|A_jC) \, . \tag{2.41}$$

Thus, the constant c acts as a normalization constant because of

$$\sum_{i=1}^{n} P(A_i|BC) = 1 \tag{2.42}$$

in agreement with (2.22).

The normalization constant (2.40) is frequently omitted, in which case Bayes' theorem (2.40) is represented by

$$P(A_i|BC) \propto P(A_i|C)P(B|A_iC) \tag{2.43}$$

where $\propto$ denotes proportionality. Hence,

$$\text{posterior probability} \propto \text{prior probability} \times \text{likelihood} \, .$$

Example 1: Three machines M_1, M_2, M_3 share the production of an object with portions 50%, 30% and 20%. The defective objects are registered, they amount to 2% for machine M_1, 5% for M_2 and 6% for M_3. An object is taken out of the production and it is assessed to be defective. What is the probability that it has been produced by machine M_1?

Let A_i with $i \in \{1, 2, 3\}$ be the statement that an object randomly chosen from the production stems from machine M_i. Then according to (2.24), given the condition C of the production the prior probabilities of these statements are $P(A_1|C) = 0.5$, $P(A_2|C) = 0.3$ and $P(A_3|C) = 0.2$. Let statement B denote the defective object. Based on the registrations the probabilities $P(B|A_1C) = 0.02$, $P(B|A_2C) = 0.05$ and $P(B|A_3C) = 0.06$ follow from (2.24). The probability $P(B|C)$ of a defective object of the production amounts with (2.39) to

$$P(B|C) = 0.5 \times 0.02 + 0.3 \times 0.05 + 0.2 \times 0.06 = 0.037$$

or to 3.7%. The posterior probability $P(A_1|BC)$ that any defective object stems from machine M_1 follows with Bayes' theorem (2.40) to be

$$P(A_1|BC) = 0.5 \times 0.02/0.037 = 0.270 \ .$$

By registering the defective objects the prior probability of 50% is reduced to the posterior probability of 27% that any defective object is produced by machine M_1.   △

Example 2: By a simple medical test it shall be verified, whether a person is infected by a certain virus. It is known that 0.3% of a certain group of the population is infected by this virus. In addition, it is known that 95% of the infected persons react positive to the simple test but also 0.5% of the healthy persons. This was determined by elaborate investigations. What is the probability that a person which reacts positive to the simple test is actually infected by the virus?

Let A be the statement that a person to be checked is infected by the virus and $\bar{A}$ according to (2.6) the statement that it is not infected. Under the condition C of the background information on the test procedure the prior probabilities of these two statements are according to (2.14) and (2.24) $P(A|C) = 0.003$ and $P(\bar{A}|C) = 0.997$. Furthermore, let B be the statement that the simple test has reacted. The probabilities $P(B|AC) = 0.950$ and $P(B|\bar{A}C) = 0.005$ then follow from (2.24). The probability $P(B|C)$ of a positive reaction is obtained with (2.39) by

$$P(B|C) = 0.003 \times 0.950 + 0.997 \times 0.005 = 0.007835 \ .$$

The posterior probability $P(A|BC)$ that a person showing a positive reaction is infected follows from Bayes' theorem (2.40) with

$$P(A|BC) = 0.003 \times 0.950/0.007835 = 0.364 \ .$$

For a positive reaction of the test the probability of an infection by the virus increases from the prior probability of 0.3% to the posterior probability of 36.4%.

The probability shall also be computed for the event that a person is infected by the virus, if the test reacts negative. $\bar{B}$ according to (2.6) is the statement of a negative reaction. With (2.14) we obtain $P(\bar{B}|AC) = 0.050, P(\bar{B}|\bar{A}C) = 0.995$ and $P(\bar{B}|C) = 0.992165$. Bayes' theorem (2.40) then gives the very small probability of

$$P(A|\bar{B}C) = 0.003 \times 0.050/0.992165 = 0.00015$$

or with (2.14) the very large probability

$$P(\bar{A}|\bar{B}C) = 0.99985$$

of being healthy in case of a negative test result. This probability must not be derived with (2.14) from the posterior probability $P(A|BC)$ because of

$$P(\bar{A}|\bar{B}C) \neq 1 - P(A|BC) \ .$$

                                                                                  △

2.1.9   Recursive Application of Bayes' Theorem

If the information on the unknown phenomenon A is given by the product $B_1 B_2 \ldots B_n$ of the statements $B_1, B_2, \ldots, B_n$, we find with Bayes' theorem (2.43)

$$P(A|B_1 B_2 \ldots B_n C) \propto P(A|C) P(B_1 B_2 \ldots B_n | AC)$$

and in case of independent statements from (2.33)

$$P(A|B_1 B_2 \ldots B_n C) \propto P(A|C) P(B_1|AC) P(B_2|AC) \ldots P(B_n|AC) . \quad (2.44)$$

Thus in case of independent information, Bayes' theorem may be applied recursively. The information B_1 gives with (2.43)

$$P(A|B_1 C) \propto P(A|C) P(B_1|AC) .$$

This posterior probability is applied as prior probability to analyze the information B_2, and we obtain

$$P(A|B_1 B_2 C) \propto P(A|B_1 C) P(B_2|AC) .$$

Analyzing B_3 leads to

$$P(A|B_1 B_2 B_3 C) \propto P(A|B_1 B_2 C) P(B_3|AC) .$$

If one proceeds accordingly up to information B_k, the recursive application of Bayes' theorem gives

$$P(A|B_1 B_2 \ldots B_k C) \propto P(A|B_1 B_2 \ldots B_{k-1} C) P(B_k|AC)$$
$$\text{for }\ k \in \{2, \ldots, n\} \quad (2.45)$$

with

$$P(A|B_1 C) \propto P(A|C) P(B_1|AC) .$$

This result agrees with (2.44). By analyzing the information B_1 to B_n the state of knowledge A about the unknown phenomenon is successively updated. This is equivalent to the process of learning by the gain of additional information.

2.2   Distributions

So far, the statements have been kept very general. In the following they shall refer to the numerical values of variables, i.e. to real numbers. The statements may refer to the values of any variables, not only to the random variables of traditional statistics whose values result from random experiments. Nevertheless, the name *random variable* is retained in order not to deviate from the terminology of traditional statistics.

Random variables frequently applied in the sequel are the *unknown parameters* which describe unknown phenomena. They represent in general fixed quantities, for instance, the unknown coordinates of a point at the rigid surface of the earth. The statements refer to the values of the fixed quantities. The unknown parameters are treated in detail in Chapter 2.2.8. Random variables are also often given as measurements, observations or data. They follow from measuring experiments or in general from random experiments whose results are registered digitally. Another source of data are surveys with numerical results. The measurements or observations are carried out and the data are collected to gain information on the unknown parameters. The analysis of the data is explained in Chapters 3 to 5.

It will be shown in Chapters 2.2.1 to 2.2.8 that the rules obtained for the probabilities of statements and Bayes' theorem hold analogously for the probability density functions of random variables, which are derived by the statements concerning their values. To get these results the rules of probability derived so far are sufficient. As will be shown, summing the probability density functions in case of discrete random variables has only to be replaced by an integration in case of continuous random variables.

2.2.1 Discrete Distribution

The statements shall first refer to the discrete values of a variable so that the *discrete random variable* X with the discrete values $x_i \in \mathbb{R}$ for $i \in \{1, \ldots, m\}$ is obtained. The probability $P(X = x_i|C)$ that X takes on the value x_i given the statement C, which contains background information, is denoted by a small letter in agreement with the notation for the continuous random variables to be presented in the following chapter

$$p(x_i|C) = P(X = x_i|C) \quad \text{for} \quad i \in \{1, \ldots, m\} . \tag{2.46}$$

We call $p(x_i|C)$ the *discrete probability density function* or abbreviated *discrete density function* or also *discrete probability distribution* or shortly *discrete distribution* for the discrete random variable X.

The statements referring to the values x_i of the discrete random variable X are mutually exclusive according to (2.18). Since with $i \in \{1, \ldots, m\}$ all values x_i are denoted which the random variable X can take, the statements for all values x_i are also exhaustive. We therefore get with (2.16) and (2.22)

$$p(x_i|C) \geq 0 \quad \text{and} \quad \sum_{i=1}^{m} p(x_i|C) = 1 . \tag{2.47}$$

The discrete density function $p(x_i|C)$ for the discrete random variable X has to satisfy the conditions (2.47). They hold for a random variable with a finite number of values x_i. If a countable infinite number of values x_i is present, one concludes in analogy to (2.47)

$$p(x_i|C) \geq 0 \quad \text{and} \quad \sum_{i=1}^{\infty} p(x_i|C) = 1 . \tag{2.48}$$

The probability $P(X < x_i|C)$ of the statement $X < x_i|C$, which is a function of x_i given the information C, is called the *probability distribution function* or shortly *distribution function $F(x_i)$*

$$F(x_i) = P(X < x_i|C) . \tag{2.49}$$

The statement referring to the values x_i are mutually exclusive, as already mentioned in connection with (2.47). Thus, we find with (2.21)

$$F(x_i) = P(X < x_i|C) = p(x_1|C) + p(x_2|C) + \ldots + p(x_{i-1}|C)$$

or

$$F(x_i) = \sum_{j<i} p(x_j|C) . \tag{2.50}$$

Because of $p(x_i|C) \geq 0$ it is obvious that the distribution function is a monotonically increasing function. Since $X < -\infty$ represents an impossible statement, we obtain with (2.15) and (2.47) for the distribution function

$$F(-\infty) = 0, \ F(\infty) = 1 \quad \text{and} \quad F(x_i) \leq F(x_j) \quad \text{for} \quad x_i < x_j . \tag{2.51}$$

The most important example of a discrete distribution, the binomial distribution, is presented in Chapter 2.2.3.

2.2.2 Continuous Distribution

Let X now be a *continuous random variable* with the values $x \in \mathbb{R}$ in the interval $-\infty < x < \infty$. The probability $P(X < x|C)$ of the statement $X < x|C$, which depends on x given the information C, is called again the *probability distribution function* or shortly *distribution function $F(x)$*

$$F(x) = P(X < x|C) . \tag{2.52}$$

Let the distribution function $F(x)$ be continuous and continuously differentiable. Given these assumptions the probability shall be determined that the random variable X takes on values in the interval $a \leq X < b$. The following three statements are considered

$$X < a, \ X < b \quad \text{and} \quad a \leq X < b .$$

The statement $X < b$ results as the sum of the statements $X < a$ and $a \leq X < b$. Since the two latter statements are mutually exclusive, we find by the sum rule (2.19)

$$P(X < b|C) = P(X < a|C) + P(a \leq X < b|C)$$

or

$$P(a \leq X < b|C) = P(X < b|C) - P(X < a|C)$$

and with the distribution function (2.52)

$$P(a \leq X < b|C) = F(b) - F(a) \ .$$

The fundamental theorem of calculus gives, see for instance BLATTER (1974, II, p.15),

$$F(b) - F(a) = \int_{-\infty}^{b} p(x|C)dx - \int_{-\infty}^{a} p(x|C)dx = \int_{a}^{b} p(x|C)dx \quad (2.53)$$

with

$$dF(x)/dx = p(x|C) \ . \tag{2.54}$$

We call $p(x|C)$ the *continuous probability density function* or abbreviated *continuous density function*, also *continuous probability distribution* or shortly *continuous distribution* for the random variable X. The distribution for a one-dimensional continuous random variable X is also called *univariate distribution*.

The distribution function $F(x)$ from (2.52) follows therefore with the density function $p(x|C)$ according to (2.53) as an area function by

$$F(x) = \int_{-\infty}^{x} p(t|C)dt \tag{2.55}$$

where t denotes the variable of integration. The distribution function $F(x)$ of a continuous random variable is obtained by an integration of the continuous density function $p(x|C)$, while the distribution function $F(x_i)$ of the discrete random variable follows with (2.50) by a summation of the discrete density function $p(x_j|C)$. The integral (2.55) may therefore be interpreted as a limit of the sum (2.50).

Because of (2.53) we obtain $P(a \leq X < b|C) = P(a < X < b|C)$. Thus, we will only work with open intervals $a < X < b$ in the following. For the interval $x < X < x + dx$ we find with (2.53)

$$P(x < X < x + dx|C) = p(x|C)dx \ . \tag{2.56}$$

The values x of the random variable X are defined by the interval $-\infty < x < \infty$ so that $X < \infty$ represents an exhaustive statement. Therefore, it follows from (2.22)

$$F(\infty) = P(X < \infty|C) = 1 \ .$$

Because of (2.16) we have $F(x) \geq 0$ which according to (2.55) is only fulfilled, if $p(x|C) \geq 0$. Thus, the two conditions are obtained which the density function $p(x|C)$ for the continuous random variable X has to fulfill

$$p(x|C) \geq 0 \quad \text{and} \quad \int_{-\infty}^{\infty} p(x|C)dx = 1 \ . \tag{2.57}$$

For the distribution function $F(x)$ we find with (2.15), since $F(-\infty)$ represents the probability of the impossible statement,

$$F(-\infty) = 0, \ F(\infty) = 1 \quad \text{and} \quad F(x_i) \leq F(x_j) \quad \text{for} \quad x_i < x_j \ . \tag{2.58}$$

The statement $X < x_j$ follows from the sum of the two statements $X < x_i$ and $x_i \leq X < x_j$. The latter statements are mutually exclusive and $P(x_i \leq X < x_j|C) \geq 0$ holds, therefore $P(X < x_i|C) \leq P(X < x_j|C)$.

Example: The random variable X has the *uniform distribution* with parameters a and b, if its density function $p(x|a,b)$ is given by

$$p(x|a,b) = \begin{cases} \dfrac{1}{b-a} & \text{for} \quad a \leq x \leq b \\ 0 & \text{for} \quad x < a \quad \text{and} \quad x > b \ . \end{cases} \tag{2.59}$$

As the density function $p(x|a,b)$ is constant in the interval $a \leq x \leq b$, one speaks of a uniform distribution.

The distribution function $F(x;a,b)$ of the uniform distribution is computed with (2.55) by

$$F(x;a,b) = \int_{a}^{x} \frac{dt}{b-a} \quad \text{for} \quad a \leq x \leq b \ .$$

We therefore obtain

$$F(x;a,b) = \begin{cases} 0 & \text{for} \quad x < a \\ \frac{x-a}{b-a} & \text{for} \quad a \leq x \leq b \\ 1 & \text{for} \quad x > b \ . \end{cases} \tag{2.60}$$

The density function (2.59) satisfies both conditions (2.57) because of $p(x|a,b) \geq 0$ and $F(\infty;a,b) = 1$. $\quad\quad\quad\quad\quad\quad\quad\quad\quad\quad\quad\quad\quad\quad\quad\quad\quad\quad\quad\quad$ Δ

Additional examples for univariate distributions will be presented in Chapter 2.4.

2.2.3   Binomial Distribution

A discrete random variable X has the *binomial distribution* with parameters n and p, if its density function $p(x|n,p)$ is given by

$$p(x|n,p) = \binom{n}{x} p^x (1-p)^{n-x}$$

$$\text{for} \quad x \in \{0,1,\ldots,n\} \quad \text{and} \quad 0 < p < 1 \ . \tag{2.61}$$

The values for the parameters n and p may vary, but a pair of values for n and p determines in each case the binomial distribution.

The binomial distribution expresses the probability that in n independent trials of an experiment x successes occur. Each experiment results either in a success or a failure, and the success has the probability p. In Example 2 for (2.33) the probability $p^x(1-p)^{n-x}$ was determined which follows from n successive trials with x successes first and then $n-x$ failures. The first x trials do not have to end in x successes, because there are $\binom{n}{x}$ possibilities that x successes may occur in n trials, see for instance KOCH (1999, p.36). With (2.21) we therefore determine the probability of x successes among n trials by $\binom{n}{x}p^x(1-p)^{n-x}$.

The density function (2.61) fulfills the two conditions (2.47), since with $p > 0$ and $(1-p) > 0$ we find $p(x|n,p) > 0$. Furthermore, the binomial series leads to

$$1 = (p + (1-p))^n = \sum_{x=0}^{n}\binom{n}{x}p^x(1-p)^{n-x} = \sum_{x=0}^{n}p(x|n,p) \ . \qquad (2.62)$$

As will be derived in Example 1 to (2.138) and in Example 2 to (2.141), the expected value $E(X)$ of the random variable X with the density function (2.61) is given by

$$E(X) = np \qquad (2.63)$$

and its variance $V(X)$ by

$$V(X) = np(1-p) \ . \qquad (2.64)$$

Example: What is the probability that in a production of 4 objects x objects with $x \in \{0,1,2,3,4\}$ are defective, if the probability that a certain object is defective is given by $p = 0.3$ and if the productions of the single objects are idependent? Using (2.61) we find

$$p(x|n = 4, \ p = 0.3) = \binom{4}{x}0.3^x \times 0.7^{4-x} \quad \text{for} \quad x \in \{0,1,2,3,4\}$$

and therefore

$$p(0|\ldots) = 0.240, \ p(1|\ldots) = 0.412, \ p(2|\ldots) = 0.264,$$
$$p(3|\ldots) = 0.076, \ p(4|\ldots) = 0.008 \ .$$

By applying the distribution function (2.50) we may, for instance, compute the probability $P(X < 2|C)$ that less than two products are defective by $P(X < 2|C) = 0.652$. Δ

If the number of repetitions of an experiment goes to infinity and the probability of the occurrence of a success approaches zero, the Poisson distribution follows from the binomial distribution, see for instance KOCH (1999, p.87).

2.2.4　Multidimensional Discrete and Continuous Distributions

Statements for which probabilities are defined shall now refer to the discrete values of n variables so that the *n-dimensional discrete random variable* $X_1, \ldots, X_n$ is obtained. Each random variable X_k with $k \in \{1, \ldots, n\}$ of the n-dimensional random variable $X_1, \ldots, X_n$ may take on the m_k discrete values $x_{k1}, \ldots, x_{km_k} \in \mathbb{R}$. We introduce the probability that given the condition C the random variables X_1 to X_n take on the given values $x_{1j_1}, \ldots, x_{nj_n}$ which means according to (2.46)

$$p(x_{1j_1}, \ldots, x_{nj_n}|C) = P(X_1 = x_{1j_1}, \ldots, X_n = x_{nj_n}|C)$$
$$\text{with} \quad j_k \in \{1, \ldots, m_k\},\ k \in \{1, \ldots, n\} \ . \quad (2.65)$$

We call $p(x_{1j_1}, \ldots, x_{nj_n}|C)$ the *n-dimensional discrete probability density function* or shortly *discrete density function* or *discrete multivariate distribution* for the n-dimensional discrete random variable $X_1, \ldots, X_n$.

We look at all values x_{kj_k} of the random variable X_k with $k \in \{1, \ldots, n\}$ so that in analogy to (2.47) and (2.48) the conditions follow which a discrete density function $p(x_{1j_1}, \ldots, x_{nj_n}|C)$ must satisfy

$$p(x_{1j_1}, \ldots, x_{nj_n}|C) \geq 0 \quad \text{and} \quad \sum_{j_1=1}^{m_1} \cdots \sum_{j_n=1}^{m_n} p(x_{1j_1}, \ldots, x_{nj_n}|C) = 1 \quad (2.66)$$

or for a countable infinite number of values x_{kj_k}

$$\sum_{j_1=1}^{\infty} \cdots \sum_{j_n=1}^{\infty} p(x_{1j_1}, \ldots, x_{nj_n}|C) = 1 \ . \quad (2.67)$$

The distribution function $F(x_{1j_1}, \ldots, x_{nj_n})$ for the n-dimensional discrete random variable $X_1, \ldots, X_n$ is defined in analogy to (2.49) by

$$F(x_{1j_1}, \ldots, x_{nj_n}) = P(X_1 < x_{1j_1}, \ldots, X_n < x_{nj_n}|C) \ . \quad (2.68)$$

It is computed as in (2.50) by

$$F(x_{1j_1}, \ldots, x_{nj_n}) = \sum_{k_1<j_1} \cdots \sum_{k_n<j_n} p(x_{1k_1}, \ldots, x_{nk_n}|C) \ . \quad (2.69)$$

An *n-dimensional continuous random variable* $X_1, \ldots, X_n$ takes on the values $x_1, \ldots, x_n \in \mathbb{R}$ in the intervals $-\infty < x_k < \infty$ with $k \in \{1, \ldots, n\}$. The distribution function $F(x_1, \ldots, x_n)$ for this random variable is defined as in (2.52) by

$$F(x_1, \ldots, x_n) = P(X_1 < x_1, \ldots, X_n < x_n|C) \ . \quad (2.70)$$

It represents corresponding to (2.53) the probability that the random variables X_k take on values in the given intervals $x_{ku} < X_k < x_{ko}$ for $k \in \{1, \ldots, n\}$

$$P(x_{1u} < X_1 < x_{1o}, \ldots, x_{nu} < X_n < x_{no}|C)$$
$$= \int_{x_{nu}}^{x_{no}} \ldots \int_{x_{1u}}^{x_{1o}} p(x_1, \ldots, x_n|C)dx_1 \ldots dx_n \quad (2.71)$$

with

$$\partial^n F(x_1, \ldots, x_n)/\partial x_1 \ldots \partial x_n = p(x_1, \ldots, x_n|C) \ . \quad (2.72)$$

We call $p(x_1, \ldots, x_n|C)$ the *n-dimensional continuous probability density function* or abbreviated *continuous density function* or *multivariate distribution* for the n-dimensional continuous random variable $X_1, \ldots, X_n$.

The distribution function $F(x_1, \ldots, x_n)$ is obtained with (2.70) by the density function $p(x_1, \ldots, x_n|C)$

$$F(x_1, \ldots, x_n) = \int_{-\infty}^{x_n} \ldots \int_{-\infty}^{x_1} p(t_1, \ldots, t_n|C)dt_1 \ldots dt_n \quad (2.73)$$

where $t_1, \ldots, t_n$ denote the variables of integration. The conditions which a density function $p(x_1, \ldots, x_n|C)$ has to fulfill follows in analogy to (2.57) with

$$p(x_1, \ldots, x_n|C) \geq 0 \quad \text{and}$$
$$\int_{-\infty}^{\infty} \ldots \int_{-\infty}^{\infty} p(x_1, \ldots, x_n|C)dx_1 \ldots dx_n = 1 \ . \quad (2.74)$$

The n-dimensional discrete or continuous random variable $X_1, \ldots, X_n$ will be often denoted in the following by the $n \times 1$ discrete or continuous random vector $\boldsymbol{x}$ with

$$\boldsymbol{x} = |X_1, \ldots, X_n|' \ . \quad (2.75)$$

The values which the discrete random variables of the random vector $\boldsymbol{x}$ take on will be also denoted by the $n \times 1$ vector $\boldsymbol{x}$ with

$$\boldsymbol{x} = |x_{1j_1}, \ldots, x_{nj_n}|', \ j_k \in \{1, \ldots, m_k\}, \ k \in \{1, \ldots, n\} \quad (2.76)$$

or in a more compact writing, if x_k symbolizes one of the m_k values $x_{k1}, \ldots, x_{km_k}$ for $k \in \{1, \ldots, n\}$, with

$$\boldsymbol{x} = |x_1, \ldots, x_n|' \ . \quad (2.77)$$

The values of the random vector of a continuous random variable are also collected in the vector $\boldsymbol{x}$ with

$$\boldsymbol{x} = |x_1, \ldots, x_n|', \ -\infty < x_k < \infty, \ k \in \{1, \ldots, n\} \ . \quad (2.78)$$

The reason for not distinguishing in the sequel between the vector $\boldsymbol{x}$ of random variables and the vector $\boldsymbol{x}$ of values of the random variables follows from the notation of vectors and matrices which labels the vectors by small letters and the matrices by capital letters. If the distinction is necessary, it will be explained by additional comments.

The n-dimensional discrete or continuous density function of the discrete or continuous random vector $\boldsymbol{x}$ follows with (2.76), (2.77) or (2.78) instead of (2.65) or (2.72) by

$$p(\boldsymbol{x}|C) \ . \tag{2.79}$$

Examples for multivariate distributions can be found in Chapter 2.5.

2.2.5   Marginal Distribution

Let X_1, X_2 be a two-dimensional discrete random variable with the m_1 values x_{1j_1} for X_1 and $j_1 \in \{1, \ldots, m_1\}$ and with the m_2 values x_{2j_2} for X_2 with $j_2 \in \{1, \ldots, m_2\}$. If given the condition C the statement A in (2.36) refers to a value of X_1 and the statement B_i to the ith value of X_2, we get

$$P(X = x_{1j_1}|C) = \sum_{j_2=1}^{m_2} P(X_1 = x_{1j_1}, X_2 = x_{2j_2}|C) \tag{2.80}$$

or with (2.65)

$$p(x_{1j_1}|C) = \sum_{j_2=1}^{m_2} p(x_{1j_1}, x_{2j_2}|C) \ . \tag{2.81}$$

By summation of the two-dimensional density function $p(x_{1j_1}, x_{2j_2}|C)$ for X_1, X_2 over the values of the random variable X_2, the density function $p(x_{1j_1}|C)$ follows for the random variable X_1. It is called the *marginal density function* or *marginal distribution* for X_1.

Since the statements A and B_i in (2.36) may refer to several discrete random variables, we obtain by starting from the n-dimensional discrete density function for $X_1, \ldots, X_i, X_{i+1}, \ldots, X_n$

$$p(x_{1j_1}, \ldots, x_{ij_i}, x_{i+1,j_{i+1}}, \ldots, x_{nj_n}|C)$$

the marginal density function $p(x_{1j_1}, \ldots, x_{ij_i}|C)$ for the random variables $X_1, \ldots, X_i$ by summing over the values of the remaining random variables $X_{i+1}, \ldots, X_n$, thus

$$p(x_{1j_1}, \ldots, x_{ij_i}|C)$$
$$= \sum_{j_{i+1}=1}^{m_{i+1}} \cdots \sum_{j_n=1}^{m_n} p(x_{1j_1}, \ldots, x_{ij_i}, x_{i+1,j_{i+1}}, \ldots, x_{nj_n}|C) \ . \tag{2.82}$$

In a more compact notation we obtain with (2.77)

$$p(x_1,\ldots,x_i|C) = \sum_{x_{i+1}}\cdots\sum_{x_n} p(x_1,\ldots,x_i,x_{i+1},\ldots,x_n|C) \tag{2.83}$$

or with (2.79) and

$$\boldsymbol{x}_1 = |x_1,\ldots,x_i|' \quad\text{and}\quad \boldsymbol{x}_2 = |x_{i+1},\ldots,x_n|' \tag{2.84}$$

finally

$$p(\boldsymbol{x}_1|C) = \sum_{\boldsymbol{x}_2} p(\boldsymbol{x}_1,\boldsymbol{x}_2|C) \; . \tag{2.85}$$

The *marginal distribution function*, that is the distribution function which belongs to the marginal density (2.82), is determined in analogy to (2.69) by

$$\begin{aligned}
&F(x_{1j_1},\ldots,x_{ij_i},x_{i+1,m_{i+1}},\ldots,x_{nm_n})\\
&= P(X_1 < x_{1j_1},\ldots,X_i < x_{ij_i}|C)\\
&= \sum_{k_1<j_1}\cdots\sum_{k_i<j_i}\sum_{k_{i+1}=1}^{m_{i+1}}\cdots\sum_{k_n=1}^{m_n} p(x_{1k_1},\ldots,x_{ik_i},x_{i+1,k_{i+1}},\ldots,x_{nk_n}|C) \; .
\end{aligned}$$
$$\tag{2.86}$$

If an n-dimensional continuous random variable $X_1,\ldots,X_n$ is given, the marginal distribution function for $X_1,\ldots,X_i$ is obtained instead of (2.86) in analogy to (2.70) up to (2.73) by

$$\begin{aligned}
&F(x_1,\ldots,x_i,\infty,\ldots,\infty) = P(X_1 < x_1,\ldots,X_i < x_i|C)\\
&= \int_{-\infty}^{\infty}\cdots\int_{-\infty}^{\infty}\int_{-\infty}^{x_i}\cdots\int_{-\infty}^{x_1} p(t_1,\ldots,t_i,t_{i+1},\ldots,t_n|C)dt_1\ldots dt_n
\end{aligned}$$
$$\tag{2.87}$$

with the variables $t_1,\ldots,t_n$ of integration. The marginal density function $p(x_1,\ldots,x_i|C)$ for $X_1,\ldots,X_i$ results from

$$\partial^i F(x_1,\ldots,x_i,\infty,\ldots,\infty)/\partial x_1\ldots\partial x_i = p(x_1,\ldots,x_i|C) \; , \tag{2.88}$$

hence

$$\begin{aligned}
&p(x_1,\ldots,x_i|C)\\
&= \int_{-\infty}^{\infty}\cdots\int_{-\infty}^{\infty} p(x_1,\ldots,x_i,x_{i+1},\ldots,x_n|C)dx_{i+1}\ldots dx_n \; .
\end{aligned} \tag{2.89}$$

The marginal density function $p(x_1,\ldots,x_i|C)$ therefore follows by integrating over the values of the random variables $X_{i+1},\ldots,X_n$. We introduce with (2.78)

$$\boldsymbol{x}_1 = |x_1,\ldots,x_i|' \quad\text{and}\quad \boldsymbol{x}_2 = |x_{i+1},\ldots,x_n|' \tag{2.90}$$

and get in a more compact notation

$$p(\boldsymbol{x}_1|C) = \int_{\mathcal{X}_2} p(\boldsymbol{x}_1, \boldsymbol{x}_2|C)d\boldsymbol{x}_2 \tag{2.91}$$

where $\mathcal{X}_2$ denotes the domain for integrating $\boldsymbol{x}_2$.

2.2.6   Conditional Distribution

The statement AB in the product rule (2.12) shall now refer to any value of a two-dimensional discrete random variable X_1, X_2. We obtain

$$P(X_1 = x_{1j_1}, X_2 = x_{2j_2}|C)$$
$$= P(X_2 = x_{2j_2}|C)P(X_1 = x_{1j_1}|X_2 = x_{2j_2}, C) \tag{2.92}$$

and with (2.65)

$$p(x_{1j_1}|x_{2j_2}, C) = \frac{p(x_{1j_1}, x_{2j_2}|C)}{p(x_{2j_2}|C)} . \tag{2.93}$$

We call $p(x_{1j_1}|x_{2j_2}, C)$ the *conditional discrete density function* or *conditional discrete distribution* for X_1 under the conditions that X_2 takes on the value x_{2j_2} and that the background information C is available. The conditional distribution for X_1 is therefore found by dividing the joint distribution for X_1 and X_2 by the marginal distribution for X_2.

Since the statement AB in the product rule (2.12) may also refer to the values of several discrete random variables, we obtain the conditional discrete density function for the random variables $X_1, \ldots, X_i$ of the discrete n-dimensional random variable $X_1, \ldots, X_n$ under the condition of given values for $X_{i+1}, \ldots, X_n$ by

$$p(x_{1j_1}, \ldots, x_{ij_i}|x_{i+1,j_{i+1}}, \ldots, x_{nj_n}, C)$$
$$= \frac{p(x_{1j_1}, \ldots, x_{ij_i}, x_{i+1,j_{i+1}}, \ldots, x_{nj_n}|C)}{p(x_{i+1,j_{i+1}}, \ldots, x_{nj_n}|C)} . \tag{2.94}$$

To find the conditional continuous density function for a two-dimensional continuous random variable X_1, X_2 corresponding to (2.93), one has to be aware that $P(X_2 = x_2|C) = 0$ holds for the continuous random variable X_2 because of (2.53). The statement AB in the product rule (2.12) must therefore be formulated such that we obtain for the continuous random variable X_1, X_2

$$P(X_1 < x_1, x_2 < X_2 < x_2 + \Delta x_2|C)$$
$$= P(x_2 < X_2 < x_2 + \Delta x_2|C)P(X_1 < x_1|x_2 < X_2 < x_2 + \Delta x_2, C) .$$

This leads with (2.71), (2.87), with the density function $p(x_1, x_2|C)$ for X_1 and X_2 and with the variables t_1 and t_2 of integration to

$$P(X_1 < x_1|x_2 < X_2 < x_2 + \Delta x_2, C)$$

$$= \frac{\int_{x_2}^{x_2+\Delta x_2} \int_{-\infty}^{x_1} p(t_1, t_2|C) dt_1 dt_2}{\int_{x_2}^{x_2+\Delta x_2} \int_{-\infty}^{\infty} p(t_1, t_2|C) dt_1 dt_2} \ . \tag{2.95}$$

The probability $P(X_1 < x_1|X_2 = x_2, C)$, which corresponds to the *conditional distribution function* $F(x_1|x_2)$, needs to be determined. It follows from (2.95) by the limit process $\Delta x_2 \to 0$. With the marginal density $p(x_2|C) = \int_{-\infty}^{\infty} p(t_1, t_2|C) dt_1$ from (2.89) we find

$$F(x_1|x_2) = \lim_{\Delta x_2 \to 0} P(X_1 < x_1|x_2 < X_2 < x_2 + \Delta x_2, C)$$

$$= \frac{\int_{-\infty}^{x_1} p(t_1, t_2|C) dt_1 \Delta x_2}{p(x_2|C)\Delta x_2} \ . \tag{2.96}$$

By differentiating with respect to x_1 in analogy to (2.72) the *conditional continuous density function* $p(x_1|x_2, C)$ for X_1 is obtained under the conditions that the value x_2 of X_2 and that C are given

$$p(x_1|x_2, C) = \frac{p(x_1, x_2|C)}{p(x_2|C)} \ . \tag{2.97}$$

Starting from the n-dimensional continuous random variable $X_1, \ldots, X_n$ with the density function $p(x_1, \ldots, x_n|C)$, the conditional continuous density function for the random variables $X_1, \ldots, X_i$ given the values $x_{i+1}, \ldots, x_n$ for $X_{i+1}, \ldots, X_n$ is obtained analogously by

$$p(x_1, \ldots, x_i|x_{i+1}, \ldots, x_n, C) = \frac{p(x_1, \ldots, x_i, x_{i+1}, \ldots, x_n|C)}{p(x_{i+1}, \ldots, x_n|C)} \ . \tag{2.98}$$

If corresponding to (2.75) the discrete or continuous random variables are arranged in the discrete or continuous random vectors

$$\boldsymbol{x}_1 = |X_1, \ldots, X_i|' \quad \text{and} \quad \boldsymbol{x}_2 = |X_{i+1}, \ldots, X_n|' \tag{2.99}$$

and as in (2.76) the values of the discrete random variables in

$$\boldsymbol{x}_1 = |x_{1j_1}, \ldots, x_{ij_i}|' \quad \text{and} \quad \boldsymbol{x}_2 = |x_{i+1,j_{i+1}}, \ldots, x_{nj_n}|' \tag{2.100}$$

or corresponding to (2.77) and (2.78) the values of the discrete or continuous random variables in

$$\boldsymbol{x}_1 = |x_1, \ldots, x_i|' \quad \text{and} \quad \boldsymbol{x}_2 = |x_{i+1}, \ldots, x_n|' \ , \tag{2.101}$$

then we get instead of (2.94) and (2.98) the conditional discrete or continuous density function for the discrete or continuous random vector $\boldsymbol{x}_1$ given the values for $\boldsymbol{x}_2$ by

$$p(\boldsymbol{x}_1|\boldsymbol{x}_2, C) = \frac{p(\boldsymbol{x}_1, \boldsymbol{x}_2|C)}{p(\boldsymbol{x}_2|C)} \; . \tag{2.102}$$

The conditional discrete density function (2.94) or (2.102) has to fulfill corresponding to (2.66) or (2.67) the two conditions

$$p(x_{1j_1}, \ldots, x_{ij_i}|x_{i+1,j_{i+1}}, \ldots, x_{nj_n}, C) \geq 0$$

and

$$\sum_{j_1=1}^{m_1} \cdots \sum_{j_i=1}^{m_i} p(x_{1j_1}, \ldots, x_{ij_i}|x_{i+1,j_{i+1}}, \ldots, x_{nj_n}, C) = 1 \tag{2.103}$$

or

$$\sum_{j_1=1}^{\infty} \cdots \sum_{j_i=1}^{\infty} p(x_{1j_1}, \ldots, x_{ij_i}|x_{i+1,j_{i+1}}, \ldots, x_{nj_n}, C) = 1 \tag{2.104}$$

for a countable infinite number of values of the discrete random variables X_1 to X_i. This follows from the fact that by summing up the numerator on the right-hand side of (2.94) the denominator is obtained because of (2.82) as

$$\sum_{j_1=1}^{m_1} \cdots \sum_{j_i=1}^{m_i} p(x_{1j_1}, \ldots, x_{ij_i}, x_{i+1,j_{i+1}}, \ldots, x_{nj_n}|C)$$

$$= p(x_{i+1,j_{i+1}}, \ldots, x_{nj_n}|C) \; .$$

Correspondingly, the conditional continuous density function (2.98) or (2.102) satisfies

$$p(x_1, \ldots, x_i|x_{i+1}, \ldots, x_n, C) \geq 0 \tag{2.105}$$

and

$$\int_{-\infty}^{\infty} \cdots \int_{-\infty}^{\infty} p(x_1, \ldots, x_i|x_{i+1}, \ldots, x_n, C)dx_1 \ldots dx_i = 1 \; . \tag{2.106}$$

2.2.7 Independent Random Variables and Chain Rule

The concept of conditional independency (2.31) of statements A and B shall now be transferred to random variables. Starting from the n-dimensional discrete random variable $X_1, \ldots, X_i, X_{i+1}, \ldots, X_n$ the statement A shall refer to the random variables $X_1, \ldots, X_i$ and the statement B to the random variables $X_{i+1}, \ldots, X_n$. The random variables $X_1, \ldots, X_i$ and the random variables $X_{i+1}, \ldots, X_n$ are *conditionally independent* or shortly expressed *independent*, if and only if under the condition C

$$p(x_{1j_1}, \ldots, x_{ij_i} | x_{i+1,j_{i+1}}, \ldots, x_{nj_n}, C) = p(x_{1j_1}, \ldots, x_{ij_i} | C) \ . \qquad (2.107)$$

If this expression is substituted on the left-hand side of (2.94), we find

$$p(x_{1j_1}, \ldots, x_{ij_i}, x_{i+1,j_{i+1}}, \ldots, x_{nj_n} | C)$$
$$= p(x_{1j_1}, \ldots, x_{ij_i} | C) p(x_{i+1,j_{i+1}}, \ldots, x_{nj_n} | C) \ . \qquad (2.108)$$

Thus, the random variables $X_1, \ldots, X_i$ are independent of the random variables $X_{i+1}, \ldots, X_n$, if and only if the density function for the n-dimensional random variable $X_1, \ldots, X_n$ can be factorized into the marginal density functions for $X_1, \ldots, X_i$ and $X_{i+1}, \ldots, X_n$.

The factorization (2.108) of the density function for the n-dimensional discrete random variable $X_1, \ldots, X_n$ into the two marginal density functions follows also from the product rule (2.32) of the two independent statements A and B, if A refers to the random variables $X_1, \ldots, X_i$ and B to the random variables $X_{i+1}, \ldots, X_n$.

By derivations corresponding to (2.95) up to (2.98) we conclude from (2.107) that the random variables $X_1, \ldots, X_i$ and $X_{i+1}, \ldots, X_n$ of the n-dimensional continuous random variable $X_1, \ldots, X_n$ are independent, if and only if under the condition C the relation

$$p(x_1, \ldots, x_i | x_{i+1}, \ldots, x_n, C) = p(x_1, \ldots, x_i | C) \qquad (2.109)$$

holds. By substituting this result on the left-hand side of (2.98) the factorization of the density function for the continuous random variable $X_1, \ldots, X_n$ corresponding to (2.108) follows

$$p(x_1, \ldots, x_i, x_{i+1}, \ldots, x_n | C) = p(x_1, \ldots, x_i | C) p(x_{i+1}, \ldots, x_n | C) \ . \qquad (2.110)$$

After introducing the discrete or continuous random vectors $\boldsymbol{x}_1$ and $\boldsymbol{x}_2$ defined by (2.99) and their values (2.100) or (2.101), we obtain instead of (2.108) or (2.110)

$$p(\boldsymbol{x}_1, \boldsymbol{x}_2 | C) = p(\boldsymbol{x}_1 | C) p(\boldsymbol{x}_2 | C) \ . \qquad (2.111)$$

The n statements $A_1, A_2, \ldots, A_n$ of the chain rule (2.27) shall now refer to the values of the n-dimensional discrete random variable $X_1, \ldots, X_n$. We therefore find with (2.65) the chain rule for a discrete density function

$$p(x_{1j_1}, x_{2j_2}, \ldots, x_{nj_n} | C) = p(x_{nj_n} | x_{1j_1}, x_{2j_2}, \ldots, x_{n-1,j_{n-1}}, C)$$
$$p(x_{n-1,j_{n-1}} | x_{1j_1}, x_{2j_2}, \ldots, x_{n-2,j_{n-2}}, C) \ldots p(x_{2j_2} | x_{1j_1}, C) p(x_{1j_1} | C) \ .$$
$$(2.112)$$

The density function $p(x_{ij_i}|C)$ takes on because of (2.65) the m_i values

$$p(x_{i1}|C), p(x_{i2}|C), \ldots, p(x_{im_i}|C) \;, \tag{2.113}$$

the density function $p(x_{ij_i}|x_{kj_k}, C)$ the $m_i \times m_k$ values

$$
\begin{aligned}
&p(x_{i1}|x_{k1}, C), p(x_{i2}|x_{k1}, C), \ldots, p(x_{im_i}|x_{k1}, C) \\
&p(x_{i1}|x_{k2}, C), p(x_{i2}|x_{k2}, C), \ldots, p(x_{im_i}|x_{k2}, C) \\
&\cdots\cdots\cdots\cdots\cdots\cdots\cdots\cdots\cdots\cdots\cdots\cdots\cdots \\
&p(x_{i1}|x_{km_k}, C), p(x_{i2}|x_{km_k}, C), \ldots, p(x_{im_i}|x_{km_k}, C) \;,
\end{aligned} \tag{2.114}
$$

the density function $p(x_{ij_i}|x_{kj_k}, x_{lj_l}, C)$ the $m_i \times m_k \times m_l$ values

$$
\begin{aligned}
&p(x_{i1}|x_{k1}, x_{l1}, C), p(x_{i2}|x_{k1}, x_{l1}, C), \ldots, p(x_{im_i}|x_{k1}, x_{l1}, C) \\
&p(x_{i1}|x_{k2}, x_{l1}, C), p(x_{i2}|x_{k2}, x_{l1}, C), \ldots, p(x_{im_i}|x_{k2}, x_{l1}, C) \\
&\cdots\cdots\cdots\cdots\cdots\cdots\cdots\cdots\cdots\cdots\cdots\cdots\cdots \\
&p(x_{i1}|x_{km_k}, x_{l1}, C), p(x_{i2}|x_{km_k}, x_{l1}, C), \ldots, p(x_{im_i}|x_{km_k}, x_{l1}, C) \\
&\cdots\cdots\cdots\cdots\cdots\cdots\cdots\cdots\cdots\cdots\cdots\cdots\cdots \\
&p(x_{i1}|x_{k1}, x_{lm_l}, C), p(x_{i2}|x_{k1}, x_{lm_l}, C), \ldots, p(x_{im_i}|x_{k1}, x_{lm_l}, C) \\
&p(x_{i1}|x_{k2}, x_{lm_l}, C), p(x_{i2}|x_{k2}, x_{lm_l}, C), \ldots, p(x_{im_i}|x_{k2}, x_{lm_l}, C) \\
&\cdots\cdots\cdots\cdots\cdots\cdots\cdots\cdots\cdots\cdots\cdots\cdots\cdots \\
&p(x_{i1}|x_{km_k}, x_{lm_l}, C), p(x_{i2}|x_{km_k}, x_{lm_l}, C), \ldots, p(x_{im_i}|x_{km_k}, x_{lm_l}, C)
\end{aligned} \tag{2.115}
$$

and so on. The more random variables appear in the condition, the greater is the number of density values.

We may write with (2.77) the chain rule (2.112) in a more compact form

$$
\begin{aligned}
p(x_1, x_2, \ldots, x_n|C) &= p(x_n|x_1, x_2, \ldots, x_{n-1}, C) \\
&\quad p(x_{n-1}|x_1, x_2, \ldots, x_{n-2}, C) \ldots p(x_2|x_1, C)p(x_1|C) \;.
\end{aligned} \tag{2.116}
$$

In case of independency of random variables the chain rule simplifies. If X_i and X_k are independent, we obtain with (2.31) or (2.107) for $i > k$

$$
\begin{aligned}
&p(x_i|x_1, x_2, \ldots, x_{k-1}, x_k, x_{k+1}, \ldots, x_{i-1}, C) \\
&\quad = p(x_i|x_1, x_2, \ldots, x_{k-1}, x_{k+1}, \ldots, x_{i-1}, C) \;.
\end{aligned} \tag{2.117}
$$

Thus, the random variable X_k disappears in the enumeration of the random variables whose values enter the density function as conditions. An application can be found in Chapter 5.5.2.

Corresponding to the derivations, which lead from (2.95) to (2.98), the relations (2.116) and (2.117) are also valid for the continuous density functions of continuous random variables. We therefore obtain for the discrete or continuous random vectors $\boldsymbol{x}_1, \ldots, \boldsymbol{x}_n$ from (2.116)

$$p(\boldsymbol{x}_1, \boldsymbol{x}_2, \ldots, \boldsymbol{x}_n | C) = p(\boldsymbol{x}_n | \boldsymbol{x}_1, \boldsymbol{x}_2, \ldots, \boldsymbol{x}_{n-1}, C)$$
$$p(\boldsymbol{x}_{n-1} | \boldsymbol{x}_1, \boldsymbol{x}_2, \ldots, \boldsymbol{x}_{n-2}, C) \ldots p(\boldsymbol{x}_2 | \boldsymbol{x}_1, C) p(\boldsymbol{x}_1 | C) \ . \quad (2.118)$$

If the random vector $\boldsymbol{x}_i$ is independent from $\boldsymbol{x}_j$, we obtain in analogy to (2.117)

$$p(\boldsymbol{x}_i | \boldsymbol{x}_j, \boldsymbol{x}_k, C) = p(\boldsymbol{x}_i | \boldsymbol{x}_k, C) \ . \quad (2.119)$$

If independent vectors exist among the random vectors $\boldsymbol{x}_1, \ldots, \boldsymbol{x}_n$, the chain rule (2.118) simplifies because of (2.119).

2.2.8 Generalized Bayes' Theorem

Bayes' theorem (2.38), which has been derived for the probability of statements, shall now be generalized such that it is valid for the density functions of discrete or continuous random variables.

If $\boldsymbol{x}$ and $\boldsymbol{y}$ are discrete or continuous random variables, we obtain with (2.102)

$$p(\boldsymbol{x} | \boldsymbol{y}, C) = \frac{p(\boldsymbol{x}, \boldsymbol{y} | C)}{p(\boldsymbol{y} | C)}$$

under the condition that values of the random vector $\boldsymbol{y}$ are given, which because of (2.76) to (2.78) are also denoted by $\boldsymbol{y}$. Furthermore, we have

$$p(\boldsymbol{y} | \boldsymbol{x}, C) = \frac{p(\boldsymbol{x}, \boldsymbol{y} | C)}{p(\boldsymbol{x} | C)} \ . \quad (2.120)$$

If these two equations are solved for $p(\boldsymbol{x}, \boldsymbol{y} | C)$ and the resulting expressions are equated, the *generalized Bayes' theorem* is found in a form corresponding to (2.38)

$$p(\boldsymbol{x} | \boldsymbol{y}, C) = \frac{p(\boldsymbol{x} | C) p(\boldsymbol{y} | \boldsymbol{x}, C)}{p(\boldsymbol{y} | C)} \ . \quad (2.121)$$

Since the vector $\boldsymbol{y}$ contains fixed values, $p(\boldsymbol{y} | C)$ is constant. Bayes' theorem is therefore often applied in the form corresponding to (2.43)

$$p(\boldsymbol{x} | \boldsymbol{y}, C) \propto p(\boldsymbol{x} | C) p(\boldsymbol{y} | \boldsymbol{x}, C) \ . \quad (2.122)$$

The components of the discrete or continuous random vector $\boldsymbol{x}$ are now identified with unknown parameters which were already mentioned in Chapter 2.2. The binomial distribution (2.61) possesses, for instance, the parameters n and p. They may take on different values, but a pair of values determines the binomial distribution. In the following example p is an unknown parameter. In general, unknown parameters are understood to be quantities which describe unknown phenomena. The values of the parameters are unknown. To estimate them, measurements, observations or data have to be taken which contain information about the unknown parameters. This was already indicated in Chapter 2.2.

The vector of values of the random vector $\boldsymbol{x}$ of unknown parameters is also called $\boldsymbol{x}$ because of (2.76) to (2.78). The set of vectors $\boldsymbol{x}$, that is the set of vectors which contains all possible values for the parameters, is called the *parameter space* $\mathcal{X}$, hence we have $\boldsymbol{x} \in \mathcal{X}$ in (2.121) or (2.122). The values $\boldsymbol{y}$ of the discrete or continuous random vector $\boldsymbol{y}$ represent given data. The density function $p(\boldsymbol{x}|C)$ given the background information C contains information about the parameters $\boldsymbol{x}$ before the data $\boldsymbol{y}$ have been taken. One calls $p(\boldsymbol{x}|C)$ therefore *prior density function* or *prior distribution* for the parameters $\boldsymbol{x}$. It contains the prior information about the unknown parameters. By taking into account the observations $\boldsymbol{y}$ the density function $p(\boldsymbol{x}|\boldsymbol{y}, C)$ follows. It is called *posterior density function* or *posterior distribution* for the parameters $\boldsymbol{x}$. Via the density function $p(\boldsymbol{y}|\boldsymbol{x}, C)$ the information available in the data $\boldsymbol{y}$ reaches the parameters $\boldsymbol{x}$. Since the data $\boldsymbol{y}$ are given, this density function is not interpreted as a function of the data $\boldsymbol{y}$ but as a function of the parameters $\boldsymbol{x}$, and it is called the *likelihood function*. Thus,

$$\text{posterior density function} \propto \text{prior density function} \times \text{likelihood function} .$$

The data modify the prior density function by the likelihood function and lead to the posterior density function for the unknown parameters.

Example 1: Under the condition C that a box contains m balls of equal shape and weight, among which k red and $m - k$ black balls are present, the statement A refers to drawing a red ball. Its probability is according to (2.24)

$$P(A|C) = \frac{k}{m} = p .$$

This probability is equal to the ratio p of the number of red balls to the total number of balls. The experiment C is augmented such that the drawn ball is put back and the box is shaken so that the new draw of a ball is independent from the result of the first draw. The probability, to get x red balls after n draws with replacements, then follows from the binomial distribution (2.61) by

$$p(x|n, p, C) = \binom{n}{x} p^x (1 - p)^{n-x} .$$

Let the proportion p of red balls to the total number of balls in the box now be the unknown parameter to be determined by n draws with replacements. The number x of drawn red balls is registered. The binomial distribution given above is then interpreted as likelihood function $p(n, x|p, C)$ for the unknown parameter p with the data n and x in Bayes' theorem (2.122)

$$p(n, x | p, C) = \binom{n}{x} p^x (1 - p)^{n-x} \ .$$ (2.123)

The unknown proportion p takes because of (2.16) values in the interval $0 \leq p \leq 1$. No prior information exists regarding the possible values of p. The uniform distribution (2.59) is therefore chosen as prior density function $p(p|C)$ for the unknown parameter p

$$p(p|0, 1, C) = \begin{cases} 1 & \text{for} \quad 0 \leq p \leq 1 \\ 0 & \text{for} \quad p < 0 \quad \text{and} \quad p > 1 \ . \end{cases}$$

The posterior density function $p(p|n, x, C)$ for the unknown parameter p then follows from Bayes' theorem (2.122) with

$$p(p|n, x, C) \propto p^x (1 - p)^{n-x} \quad \text{for} \quad 0 \leq p \leq 1$$ (2.124)

where the term $\binom{n}{x}$ of the binomial distribution is constant and therefore need not be considered. If this density function is compared with the density function (2.178) of the beta distribution, one recognizes that the unknown parameter p possesses as posterior density function

$$p(p|n, x, C) = \frac{\Gamma(n + 2)}{\Gamma(x + 1)\Gamma(n - x + 1)} \, p^x (1 - p)^{n-x}$$ (2.125)

which is the density function of the beta distribution.

If we draw, for instance, $x = 4$ red balls in $n = 10$ trials, we obtain with (2.173) the posterior density function for p

$$p(p|10, 4, C) = 2\,310 p^4 (1 - p)^6 \ .$$

The graph of this posterior density function is shown in Figure 2.1. For the 10 draws which result in 4 red balls the prior density function for p with identical

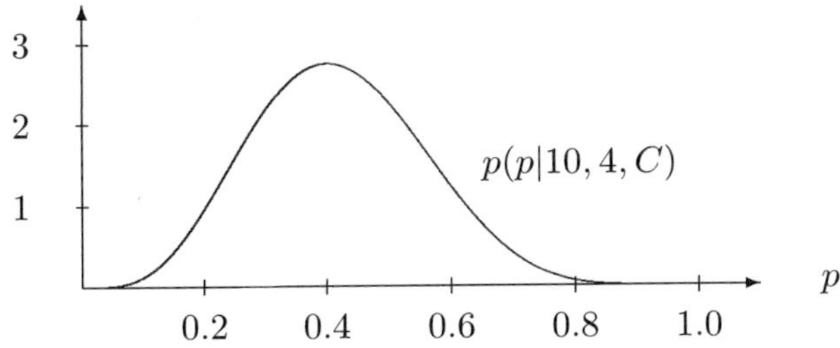

Figure 2.1: Posterior Density Function for p

values of one is modified to the posterior density function for p of Figure 2.1 with the maximum at $p = 0.4$.

As will be shown by (3.30), an estimate $\hat{p}_M$ of p may be determined such that with $\hat{p}_M$ the posterior density function for p becomes maximal. The posterior density function is therefore differentiated with respect to p, and the derivative is set equal to zero. We get

$$\frac{d(p(p|n,x,C))}{dp} \propto xp^{x-1}(1-p)^{n-x} - (n-x)p^x(1-p)^{n-x-1}$$

and

$$x\hat{p}_M^{-1} = (n-x)(1-\hat{p}_M)^{-1}$$

and finally the estimate $\hat{p}_M$ of p

$$\hat{p}_M = \frac{x}{n} \; . \tag{2.126}$$

An identical result follows, if in (2.63) the expected value $E(X)$ is replaced by the measurement x and the parameter p by its estimate $\hat{p}_M$. With the values $n = 10$ and $x = 4$ of Figure 2.1 the estimate $\hat{p}_M$ follows with $\hat{p}_M = 0.4$. For these values $p(p|10,4,C)$ attains its maximum.   △

As will be shown in Chapter 3, knowledge of the posterior density function suffices to estimate the unknown parameters, to test hypotheses for the unknown parameters or to establish regions, within which the values of the parameters lie with a given probability. For these problems Bayes' theorem (2.122) presents a solution.

The vector $\boldsymbol{x}$ of unknown parameters is defined in Bayes' theorem as a random vector with which a prior density and a posterior density function is associated. This approach is contrary to the one of traditional statistics which is not based on Bayes' theorem and which defines the vector of unknown parameters generally as a vector of constants. But this does not mean that the vector $\boldsymbol{x}$ of parameters in Bayes' theorem (2.122) may not represent constants such as the coordinates of a point at the rigid surface of the earth. By the prior and posterior density function for the unknown parameters the probability is determined that the values of the parameters lie within certain regions. The probability expresses, as explained in Chapter 2.1.4, the plausibility of these statements. The probability does not need to be interpreted as frequency of random experiments which may be inappropriate, as the unknown parameters do not in general result from random experiments. The probability serves the purpose to express the plausibility of values of the parameters. The parameters may therefore represent constant quantities or variable quantities, too, which vary for instance with time. By Bayes' theorem we obtain a posterior density function which characterizes the values of the parameters which exist when recording the data.

The density function in the denominator of (2.121) may be expressed by a marginal density function. We obtain with (2.85) and (2.120) for a discrete random vector $\boldsymbol{y}$

$$p(\boldsymbol{y}|C) = \sum_{\boldsymbol{x}\in\mathcal{X}} p(\boldsymbol{x},\boldsymbol{y}|C) = \sum_{\boldsymbol{x}\in\mathcal{X}} p(\boldsymbol{x}|C)p(\boldsymbol{y}|\boldsymbol{x},C) \qquad (2.127)$$

and with (2.91) and (2.120) for a continuous random vector $\boldsymbol{y}$

$$p(\boldsymbol{y}|C) = \int_{\mathcal{X}} p(\boldsymbol{x},\boldsymbol{y}|C)d\boldsymbol{x} = \int_{\mathcal{X}} p(\boldsymbol{x}|C)p(\boldsymbol{y}|\boldsymbol{x},C)d\boldsymbol{x} \qquad (2.128)$$

where the parameter space $\mathcal{X}$ denotes the domain over which $\boldsymbol{x}$ has to be summed or integrated. Bayes' theorem (2.121) then follows in a form corresponding to (2.40) by

$$p(\boldsymbol{x}|\boldsymbol{y},C) = p(\boldsymbol{x}|C)p(\boldsymbol{y}|\boldsymbol{x},C)/c \qquad (2.129)$$

with the constant c from (2.127) for a discrete random vector $\boldsymbol{y}$

$$c = \sum_{\boldsymbol{x}\in\mathcal{X}} p(\boldsymbol{x}|C)p(\boldsymbol{y}|\boldsymbol{x},C) \qquad (2.130)$$

or from (2.128) for a continuous random vector $\boldsymbol{y}$

$$c = \int_{\mathcal{X}} p(\boldsymbol{x}|C)p(\boldsymbol{y}|\boldsymbol{x},C)d\boldsymbol{x} \; . \qquad (2.131)$$

Thus, it becomes obvious that c acts as a normalization constant which must be introduced to satisfy (2.103), (2.104) or (2.106).

If instead of the vector $\boldsymbol{y}$ of data the vectors $\boldsymbol{y}_1, \boldsymbol{y}_2, \ldots, \boldsymbol{y}_n$ are given, we obtain instead of (2.122)

$$p(\boldsymbol{x}|\boldsymbol{y}_1, \boldsymbol{y}_2, \ldots, \boldsymbol{y}_n, C) \propto p(\boldsymbol{x}|C)p(\boldsymbol{y}_1, \boldsymbol{y}_2, \ldots, \boldsymbol{y}_n|\boldsymbol{x},C) \; . \qquad (2.132)$$

Let the vector $\boldsymbol{y}_i$ of data be independent of the vector $\boldsymbol{y}_j$ for $i \neq j$ and $i,j \in \{1,\ldots,n\}$, then we get with (2.111) instead of (2.132)

$$p(\boldsymbol{x}|\boldsymbol{y}_1, \boldsymbol{y}_2, \ldots, \boldsymbol{y}_n, C) \propto p(\boldsymbol{x}|C)p(\boldsymbol{y}_1|\boldsymbol{x},C)p(\boldsymbol{y}_2|\boldsymbol{x},C)\ldots p(\boldsymbol{y}_n|\boldsymbol{x},C) \; . \qquad (2.133)$$

For independent data Bayes' theorem may therefore be applied recursively. We find with the data $\boldsymbol{y}_1$ from (2.122)

$$p(\boldsymbol{x}|\boldsymbol{y}_1, C) \propto p(\boldsymbol{x}|C)p(\boldsymbol{y}_1|\boldsymbol{x},C) \; .$$

This posterior density function is introduced as prior density for the analysis of $\boldsymbol{y}_2$, thus

$$p(\boldsymbol{x}|\boldsymbol{y}_1, \boldsymbol{y}_2, C) \propto p(\boldsymbol{x}|\boldsymbol{y}_1, C)p(\boldsymbol{y}_2|\boldsymbol{x},C) \; .$$

If one proceeds in this manner up to the data $\boldsymbol{y}_k$, the recursive application of Bayes' theorem follows corresponding to (2.45) by

$$p(\boldsymbol{x}|\boldsymbol{y}_1,\boldsymbol{y}_2,\ldots,\boldsymbol{y}_k,C) \propto p(\boldsymbol{x}|\boldsymbol{y}_1,\boldsymbol{y}_2,\ldots,\boldsymbol{y}_{k-1},C)p(\boldsymbol{y}_k|\boldsymbol{x},C)$$
$$\text{for}\quad k \in \{2,\ldots,n\} \quad (2.134)$$

with

$$p(\boldsymbol{x}|\boldsymbol{y}_1,C) \propto p(\boldsymbol{x}|C)p(\boldsymbol{y}_1|\boldsymbol{x},C) \; .$$

This result agrees with (2.133). By analyzing the observations $\boldsymbol{y}_1$ to $\boldsymbol{y}_n$ the knowledge about the unknown parameters $\boldsymbol{x}$ is sequentially updated.

Example 2: In Example 1 to Bayes' theorem (2.122) x red balls were drawn in n trials. They shall now be denoted by x_1 and n_1, thus $x_1 = 4$ and $n_1 = 10$. Again, we draw and obtain x_2 red balls in n_2 trials, namely $x_2 = 6$ and $n_2 = 20$. Because of the setup of the experiment the data n_1 and x_1 are independent of n_2 and x_2. The posterior density function for p based on the additional draws may therefore be derived by the recursive application of Bayes' theorem by introducing the posterior density function obtained from the data n_1 and x_1 as prior density function for the analysis of the data n_2 and x_2. We find with (2.134)

$$p(p|n_1,x_1,n_2,x_2,C) \propto p(p|n_1,x_1,C)p(n_2,x_2|p,C) \qquad (2.135)$$

where the prior density function $p(p|n_1,x_1,C)$ is identical with the posterior density (2.124), if n and x are substituted by n_1 and x_1 and if the likelihood function again follows from (2.123). Thus, we get instead of (2.135)

$$p(p|n_1,x_1,n_2,x_2,C) \propto p^{x_1}(1-p)^{n_1-x_1}p^{x_2}(1-p)^{n_2-x_2}$$

or by a comparison with (2.178) the density function of the beta distribution

$$p(p|n_1,x_1,n_2,x_2,C)$$
$$= \frac{\Gamma(n_1+n_2+2)}{\Gamma(x_1+x_2+1)\Gamma(n_1+n_2-(x_1+x_2)+1)} \, p^{x_1+x_2}(1-p)^{n_1+n_2-(x_1+x_2)} \; .$$
$$(2.136)$$

Instead of deriving recursively the posterior density function (2.136) by analyzing first the data n_1 and x_1 and then the data n_2 and x_2, the data may be also evaluated jointly. We replace in (2.125) just n by $n_1 + n_2$ and x by $x_1 + x_2$ and obtain immediately the posterior density function (2.136).

For the data $n_1 = 10, x_1 = 4, n_2 = 20, x_2 = 6$ we find because of (2.173)

$$p(p|10,4,20,6,C) = 931\,395\,465\,p^{10}(1-p)^{20} \; .$$

The posterior density function $p(p|10,4,20,6,C)$ is shown together with the prior density function $p(p|10,4,C)$ for the unknown parameter p in Figure 2.2. As can be recognized, the posterior density function is considerably more concentrated around its maximum value than the prior density function because of the additional observations. The maximum is also shifted.

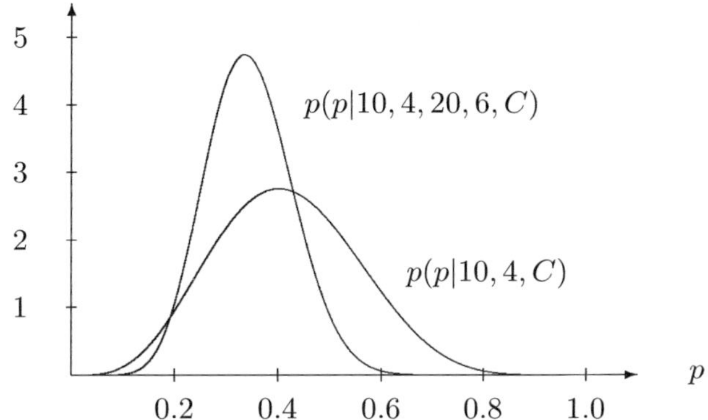

Figure 2.2: Prior and Posterior Density Function for p

The estimate $\hat{p}_M$ of p follows from (2.126) by

$$\hat{p}_M = \frac{x_1 + x_2}{n_1 + n_2} \tag{2.137}$$

which is $\hat{p}_M = 0.333$ for the given data. $\qquad\qquad\triangle$

An additional example for the recursive application of Bayes' theorem is presented in Chapter 4.2.7.

2.3 Expected Value, Variance and Covariance

2.3.1 Expected Value

The *expected value* or the *expectation* denoted by μ or $E(X)$ of a discrete random variable X with the density function $p(x_i|C)$ for $i \in \{1, \ldots, m\}$ is defined by

$$\mu = E(X) = \sum_{i=1}^{m} x_i p(x_i|C) \tag{2.138}$$

and for $i \in \{1, \ldots, \infty\}$ in analogy to (2.138) by

$$\mu = E(X) = \sum_{i=1}^{\infty} x_i p(x_i|C) \ . \tag{2.139}$$

Because of the condition (2.47) that the summation of the density function $p(x_i|C)$ gives one, the expected value can be expressed according to (4.20) as a weighted arithmetic mean with the density values $p(x_i|C)$ as weights

$$\mu = E(X) = \sum_{i=1}^{m} x_i p(x_i|C) / \sum_{i=1}^{m} p(x_i|C) \ .$$

Thus, the expected value $E(X)$ of a random variable X may be interpreted as a mean value.

The expected value $E(X)$ is computed by the density function $p(x_i|C)$. It therefore depends on the condition concerning the information C. This could be expressed by the notation $E(X|C)$ which will be dispensed with in the following for the sake of simplification. Later in Chapters 2.6.3 and 4 it will be necessary to introduce this notation.

Example 1: We obtain from (2.138) the expected value $E(X)$ of the random variable X with the binomial distribution (2.61) by

$$E(X) = \sum_{x=0}^{n} x \binom{n}{x} p^x (1-p)^{n-x} = \sum_{x=1}^{n} \frac{n!}{(x-1)!(n-x)!} p^x (1-p)^{n-x}$$

$$= np \sum_{x=1}^{n} \frac{(n-1)!}{(x-1)!(n-x)!} p^{(x-1)} (1-p)^{n-x} \; .$$

By substituting $j = x - 1$ we find with the binomial series

$$E(X) = np \sum_{j=0}^{n-1} \frac{(n-1)!}{j!(n-j-1)!} p^j (1-p)^{n-j-1}$$

$$= np[p + (1-p)]^{n-1}$$

$$= np \; . \hspace{5cm} \triangle$$

The expected value (2.138) is a special case of the definition of the expected value μ_i or $E(X_i)$ of the random variable X_i of the $n \times 1$ discrete random vector $\boldsymbol{x} = |X_1, \ldots, X_n|'$ with the density function $p(x_{1j_1}, \ldots, x_{nj_n}|C)$ with $j_k \in \{1, \ldots, m_k\}$ and $k \in \{1, \ldots, n\}$ from (2.65)

$$\mu_i = E(X_i) = \sum_{j_1=1}^{m_1} \cdots \sum_{j_n=1}^{m_n} x_{ij_i} \, p(x_{1j_1}, \ldots, x_{nj_n}|C) \; . \tag{2.140}$$

Correspondingly, the expected value $E(f(X_i))$ of the function $f(X_i)$ of the random variable X_i follows with

$$E(f(X_i)) = \sum_{j_1=1}^{m_1} \cdots \sum_{j_n=1}^{m_n} f(x_{ij_i}) p(x_{1j_1}, \ldots, x_{nj_n}|C) \; . \tag{2.141}$$

Example 2: To derive the variance $V(X)$ of a random variable X, the expected value $E(X^2)$ needs to be computed, as will be shown with (2.147). For a random variable X with the binomial distribution (2.61) one obtains

$$E(X^2) = \sum_{x=0}^{n} x^2 \binom{n}{x} p^x (1-p)^{n-x} = \sum_{x=1}^{n} \frac{xn!}{(x-1)!(n-x)!} p^x (1-p)^{n-x}$$

$$= \sum_{x=1}^{n} \left[\frac{(x-1)n!}{(x-1)!(n-x)!} + \frac{n!}{(x-1)!(n-x)!} \right] p^x (1-p)^{n-x} \; .$$

As was shown in Example 1 to (2.138), summing the second term gives np, hence

$$E(X^2) = \sum_{x=2}^{n} \frac{n!}{(x-2)!(n-x)!} p^x (1-p)^{n-x} + np$$

$$= n(n-1)p^2 \sum_{x=2}^{n} \frac{(n-2)!}{(x-2)!(n-x)!} p^{x-2}(1-p)^{n-x} + np$$

$$= n(n-1)p^2 + np .$$

This finally leads with (2.147) to

$$V(X) = E(X^2) - [E(X)]^2 = np - np^2$$

$$= np(1-p) .$$
$\triangle$

The expected value μ or $E(X)$ of the continuous random variable X with the density function $p(x|C)$ is in analogy to (2.138) defined by

$$\mu = E(X) = \int_{-\infty}^{\infty} x p(x|C) dx . \tag{2.142}$$

Example 3: The expected value $E(X)$ of the random variable X with the exponential distribution (2.189) follows from

$$E(X) = \frac{1}{\mu} \int_{0}^{\infty} x e^{-x/\mu} dx = \frac{1}{\mu} \left[e^{-x/\mu}(-\mu x - \mu^2) \right]_{0}^{\infty} = \mu .$$
$\triangle$

Definition (2.142) is a special case of the definition of the expected value μ_i or $E(X_i)$ of the continuous random variable X_i of the $n \times 1$ continuous random vector $\boldsymbol{x} = |X_1, \ldots, X_n|'$ with the density function $p(x_1, \ldots, x_n|C)$

$$\mu_i = E(X_i) = \int_{-\infty}^{\infty} \cdots \int_{-\infty}^{\infty} x_i p(x_1, \ldots, x_n|C) dx_1 \ldots dx_n . \tag{2.143}$$

Correspondingly, the expected value $E(f(X_i))$ of the function $f(X_i)$ of the random variable X_i follows with

$$E(f(X_i)) = \int_{-\infty}^{\infty} \cdots \int_{-\infty}^{\infty} f(x_i) p(x_1, \ldots, x_n|C) dx_1 \ldots dx_n . \tag{2.144}$$

If the marginal distribution $p(x_i|C)$ for the random variable X_i from (2.89) is substituted in (2.143), we obtain

$$\mu_i = E(X_i) = \int_{-\infty}^{\infty} x_i p(x_i|C) dx_i . \tag{2.145}$$

The expected value lies in the center of the marginal density function $p(x_i|C)$, as shown in Figure 2.3. This is true, because the x_i axis can be imagined as a bar with the density $p(x_i|C)$. The mass center x_s of the bar is computed by the laws of mechanics by

$$x_s = \int_{-\infty}^{\infty} x_i p(x_i|C)dx_i \Big/ \int_{-\infty}^{\infty} p(x_i|C)dx_i$$

whence $x_s = E(X_i)$ follows because of (2.74).

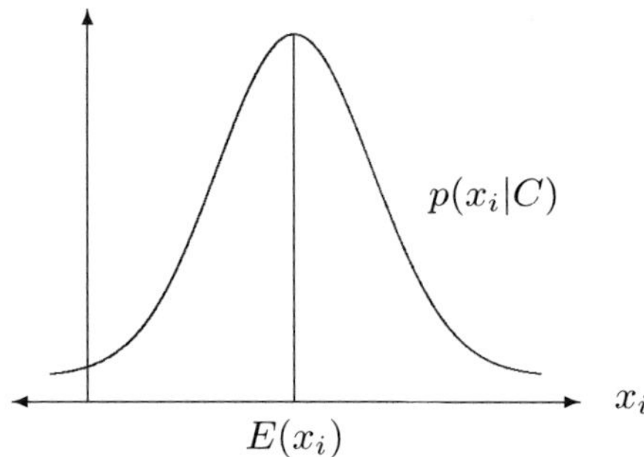

Figure 2.3: Expected Value

For the computation with expected values the following relation holds

$$E(\boldsymbol{Ax} + \boldsymbol{By} + \boldsymbol{c}) = \boldsymbol{A}E(\boldsymbol{x}) + \boldsymbol{B}E(\boldsymbol{y}) + \boldsymbol{c} \qquad (2.146)$$

where $\boldsymbol{A}$ denotes an $m \times n$, $\boldsymbol{B}$ an $m \times o$ matrix, $\boldsymbol{c}$ an $m \times 1$ vector of constants, $\boldsymbol{x} = |X_1, \ldots, X_n|'$ an $n \times 1$ random vector and $\boldsymbol{y} = |Y_1, \ldots, Y_o|'$ a $o \times 1$ random vector. To derive the result (2.146), we define $\boldsymbol{d} = \boldsymbol{Ax}$ with $\boldsymbol{d} = (d_i)$, $\boldsymbol{e} = \boldsymbol{By}$ with $\boldsymbol{e} = (e_i)$, $\boldsymbol{A} = (a_{ij})$, $\boldsymbol{B} = (b_{ij})$ and $\boldsymbol{c} = (c_i)$ and obtain with (2.144)

$$E(d_i + e_i + c_i) = E(\sum_{j=1}^{n} a_{ij}X_j + \sum_{k=1}^{o} b_{ik}Y_k + c_i)$$

$$= \sum_{j=1}^{n} a_{ij}E(X_j) + \sum_{k=1}^{o} b_{ik}E(Y_k) + E(c_i) \ .$$

The constant c_i is introduced by $c_i = f(X_l)$ as a function $f(X_l)$ of a random variable X_l with the density function $p(x_l|C)$. Thus, with (2.57) and (2.144) $E(c_i) = E(f(X_l)) = c_i \int_{-\infty}^{\infty} p(x_l|C)dx_l = c_i$.

Example 4: The expected value $E((X - \mu)^2)$ for the random variable X with the expected value $\mu = E(X)$ shall be computed by (2.146). It leads because of (2.151) to the variance $V(X)$ of X. We get

$$V(X) = E((X - \mu)^2) = E(X^2) - 2\mu E(X) + \mu^2$$
$$= E(X^2) - \mu^2 \ . \qquad (2.147)$$

△

2.3.2 Variance and Covariance

Expected values of random variables are specializations of the multivariate moments of random variables. They will be defined in the following for continuous random variables only. The definitions for discrete random variables follow corresponding to the definition (2.141) of the expected value for the function of a random variable in a discrete random vector. Let $\boldsymbol{x} = |X_1, \ldots, X_n|'$ be the $n \times 1$ continuous random vector with the density function $p(x_1, \ldots, x_n|C)$, then $\mu_{x_1,\ldots,x_n}^{(k)}$ with $k = \sum_{i=1}^{n} k_i$ and $k_i \in \mathbb{N}$ denotes the kth *multivariate moment* of $X_1, \ldots, X_n$

$$
\begin{aligned}
\mu_{x_1,\ldots,x_n}^{(k)} &= E(X_1^{k_1} X_2^{k_2} \ldots X_n^{k_n}) \\
&= \int_{-\infty}^{\infty} \ldots \int_{-\infty}^{\infty} x_1^{k_1} x_2^{k_2} \ldots x_n^{k_n} p(x_1, \ldots, x_n|C) dx_1 \ldots dx_n .
\end{aligned}
\tag{2.148}
$$

It gives with $k_i = 1$ and $k_j = 0$ for $i \neq j$ the expected value μ_i or $E(X_i)$ of the random variable X_i defined in (2.143).

Moments about the expected values μ_i of the random variables X_i are called *central moments*

$$
E((X_1 - \mu_1)^{k_1}(X_2 - \mu_2)^{k_2} \ldots (X_n - \mu_n)^{k_n}) .
\tag{2.149}
$$

The second central moments are of particular interest.

The second central moment σ_{ij} or $C(X_i, X_j)$ of the random variables X_i and X_j is called the *covariance*

$$
\begin{aligned}
\sigma_{ij} &= C(X_i, X_j) = E((X_i - \mu_i)(X_j - \mu_j)) \\
&= \int_{-\infty}^{\infty} \ldots \int_{-\infty}^{\infty} (x_i - \mu_i)(x_j - \mu_j)p(x_1, \ldots, x_n|C) dx_1 \ldots dx_n
\end{aligned}
\tag{2.150}
$$

and the second central moment σ_i^2 or $V(X_i)$ of the random variable X_i the *variance*

$$
\begin{aligned}
\sigma_i^2 &= V(X_i) = \sigma_{ii} = C(X_i, X_i) = E((X_i - \mu_i)^2) \\
&= \int_{-\infty}^{\infty} \ldots \int_{-\infty}^{\infty} (x_i - \mu_i)^2 p(x_1, \ldots, x_n|C) dx_1 \ldots dx_n .
\end{aligned}
\tag{2.151}
$$

Considering (2.74) it is obvious from the definition of the variance that $\sigma_i^2 \geq 0$ holds. The positive square root of σ_i^2 is called the *standard deviation* σ_i, thus $\sigma_i = \sqrt{\sigma_i^2}$.

Example 1: The variance $V(X)$ of the random variable X with the exponential distribution (2.189) follows with $E(X) = \mu$ from Example 3 to (2.142) and with (2.147) by

$$
\begin{aligned}
V(X) &= \frac{1}{\mu} \int_0^{\infty} x^2 e^{-x/\mu} dx - \mu^2 \\
&= \frac{1}{\mu} \left[e^{-x/\mu}(-\mu x^2 - 2\mu^2 x - 2\mu^3) \right]_0^{\infty} - \mu^2 = \mu^2 .
\end{aligned}
$$

$\triangle$

The variance σ_i^2 is a measure of dispersion of the random variable X_i about its expected value μ_i and therefore a measure for the *accuracy* of a random variable which represents a measurement, since we obtain by the marginal density function $p(x_i|C)$ for X_i from (2.89) instead of (2.151)

$$\sigma_i^2 = \int_{-\infty}^{\infty} (x_i - \mu_i)^2 p(x_i|C) dx_i \ . \tag{2.152}$$

If most of the area under the graph of the marginal density function $p(x_i|C)$ sketched in Figure 2.3 lies near the expected value $E(X_i)$, the variance σ_i^2 is small. On the other hand σ_i^2 is large, if the area is spread out. A small variance of a measurement means a high accuracy, a large variance a low accuracy.

The covariance σ_{ij} is a measure of dependency of two random variables X_i and X_j. To explain this we have on the one side

$$\sigma_{ij} = 0 \ , \tag{2.153}$$

if the random variables X_i and X_j are independent.

To show this, σ_{ij} is computed with (2.150) by

$$\begin{aligned}
\sigma_{ij} &= E(X_i X_j) - \mu_j E(X_i) - \mu_i E(X_j) + \mu_i \mu_j \\
&= E(X_i X_j) - E(X_i)E(X_j) \ .
\end{aligned} \tag{2.154}$$

Furthermore, we get with (2.148) and the marginal density function $p(x_i, x_j|C)$ for X_i and X_j from (2.89)

$$E(X_i X_j) = \int_{-\infty}^{\infty} \int_{-\infty}^{\infty} x_i x_j p(x_i, x_j|C) dx_i dx_j \ .$$

Because of the independency of X_i and X_j we obtain with (2.110) the factorization $p(x_i, x_j|C) = p(x_i|C)p(x_j|C)$. Thus, $E(X_i X_j) = E(X_i)E(X_j)$ follows with (2.145) and finally (2.153).

On the other side we obtain, if the covariance σ_{ij} is normalized, the *correlation coefficient* ρ_{ij} of X_i and X_j as

$$\rho_{ij} = \sigma_{ij}/(\sigma_i \sigma_j) \quad \text{for} \quad \sigma_i > 0 \quad \text{and} \quad \sigma_j > 0 \ , \tag{2.155}$$

and the relation, see for instance KOCH (1999, p.98),

$$-1 \le \rho_{ij} \le 1 \ . \tag{2.156}$$

We get $\rho_{ij} = \pm 1$, if and only if there is a linear relation between the random variables X_i and X_j with probability one, that is $P(X_j = cX_i + d|C) = 1$ with c and d being constants.

The variances and covariances of a random vector $\boldsymbol{x} = |X_1, \ldots, X_n|'$ are collected in the *covariance matrix* $D(\boldsymbol{x})$, also called *variance-covariance matrix* or *dispersion matrix*

$$D(\boldsymbol{x}) = (\sigma_{ij}) = (C(X_i, X_j)) = E((\boldsymbol{x} - E(\boldsymbol{x}))(\boldsymbol{x} - E(\boldsymbol{x}))')$$

$$= \begin{vmatrix} \sigma_1^2 & \sigma_{12} & \cdots & \sigma_{1n} \\ \sigma_{21} & \sigma_2^2 & \cdots & \sigma_{2n} \\ \hdotsfor{4} \\ \sigma_{n1} & \sigma_{n2} & \cdots & \sigma_n^2 \end{vmatrix} . \tag{2.157}$$

From (2.150) follows $\sigma_{ij} = \sigma_{ji}$ so that the covariance matrix is symmetric. In addition, it is positive definite or positive semidefinite, see for instance KOCH (1999, p.100).

Let the $n \times 1$ random vector $\boldsymbol{x}$ with the covariance matrix $D(\boldsymbol{x})$ be linearly transformed with the $m \times n$ matrix $\boldsymbol{A}$ and the $m \times 1$ vector $\boldsymbol{b}$ of constants into the $m \times 1$ random vector $\boldsymbol{y}$ by $\boldsymbol{y} = \boldsymbol{A}\boldsymbol{x} + \boldsymbol{b}$. Its $m \times m$ covariance matrix $D(\boldsymbol{y})$ follows with

$$D(\boldsymbol{y}) = D(\boldsymbol{A}\boldsymbol{x} + \boldsymbol{b}) = \boldsymbol{A}D(\boldsymbol{x})\boldsymbol{A}' , \tag{2.158}$$

since we obtain with the definition (2.157) of a covariance matrix $D(\boldsymbol{y}) = E((\boldsymbol{y} - E(\boldsymbol{y}))(\boldsymbol{y} - E(\boldsymbol{y}))')$ and with (2.146) $D(\boldsymbol{y}) = E((\boldsymbol{A}\boldsymbol{x} + \boldsymbol{b} - \boldsymbol{A}E(\boldsymbol{x}) - \boldsymbol{b})(\boldsymbol{A}\boldsymbol{x} - \boldsymbol{A}E(\boldsymbol{x}))') = \boldsymbol{A}E((\boldsymbol{x} - E(\boldsymbol{x}))(\boldsymbol{x} - E(\boldsymbol{x}))')\boldsymbol{A}'$.

In case of nonlinear transformations the matrix $\boldsymbol{A}$ contains as elements the derivatives of the transformed values of the random vector $\boldsymbol{x}$ with respect to the values of $\boldsymbol{x}$, see for instance KOCH (1999, p.100). In such a case (2.158) is called the *law of error propagation*.

Example 2: The 3×1 random vector $\boldsymbol{x}$ with $\boldsymbol{x} = (X_i)$ and with the 3×3 covariance matrix $\boldsymbol{\Sigma}_x$ is linearly transformed by

$$Y_1 = X_1 + X_2 + X_3$$
$$Y_2 = X_1 - X_2 + X_3$$

into the 2×1 random vector $\boldsymbol{y}$ with $\boldsymbol{y} = (Y_i)$. The 2×2 covariance matrix $\boldsymbol{\Sigma}_y$ of $\boldsymbol{y}$ then follows from $\boldsymbol{\Sigma}_y = \boldsymbol{A}\boldsymbol{\Sigma}_x\boldsymbol{A}'$ with

$$\boldsymbol{A} = \begin{vmatrix} 1 & 1 & 1 \\ 1 & -1 & 1 \end{vmatrix} .$$

$\triangle$

Let $D(\boldsymbol{x}) = \boldsymbol{\Sigma}$ be the $n \times n$ positive definite covariance matrix of the $n \times 1$ random vector $\boldsymbol{x} = |X_1, \ldots, X_n|'$. The $n \times n$ matrix $\boldsymbol{P}$

$$\boldsymbol{P} = c\boldsymbol{\Sigma}^{-1} , \tag{2.159}$$

where c denotes a constant, is then called the *weight matrix* and the diagonal element p_{ii} of $\boldsymbol{P} = (p_{ij})$ the *weight* of the random variable X_i. If the random variables are independent, the covariance matrix $\boldsymbol{\Sigma}$ simplifies because of (2.153) to the diagonal matrix $\boldsymbol{\Sigma} = \mathrm{diag}(\sigma_1^2, \ldots, \sigma_n^2)$. The weight p_i of X_i then results from

$$p_i = p_{ii} = c/\sigma_i^2 \; . \tag{2.160}$$

The smaller the variance of the random variable X_i the larger is its weight and the higher is the precision or the accuracy for a random variable which represents a measurement.

If the $m \times 1$ random vector $\boldsymbol{z}$ is decomposed with $\boldsymbol{z} = |\boldsymbol{x}', \boldsymbol{y}'|'$ into the $n \times 1$ random vector $\boldsymbol{x} = (X_i)$ and the $p \times 1$ random vector $\boldsymbol{y} = (Y_j)$ with $m = n + p$, then $C(\boldsymbol{x}, \boldsymbol{y})$ denotes the $n \times p$ covariance matrix of the random vectors $\boldsymbol{x}$ and $\boldsymbol{y}$. Corresponding to the definition (2.150) and (2.157) we have

$$C(\boldsymbol{x}, \boldsymbol{y}) = (C(X_i, Y_j)) = E((\boldsymbol{x} - E(\boldsymbol{x}))(\boldsymbol{y} - E(\boldsymbol{y}))') \; . \tag{2.161}$$

The covariance matrix $D(\boldsymbol{z})$ of the random vector $\boldsymbol{z}$ then follows with $C(\boldsymbol{x}, \boldsymbol{x}) = D(\boldsymbol{x})$ and $C(\boldsymbol{y}, \boldsymbol{y}) = D(\boldsymbol{y})$ from

$$D(\boldsymbol{z}) = D(\left|\begin{array}{c} \boldsymbol{x} \\ \boldsymbol{y} \end{array}\right|) = \left|\begin{array}{cc} D(\boldsymbol{x}) & C(\boldsymbol{x}, \boldsymbol{y}) \\ C(\boldsymbol{y}, \boldsymbol{x}) & D(\boldsymbol{y}) \end{array}\right| \; . \tag{2.162}$$

If $\boldsymbol{x}$ and $\boldsymbol{y}$ are two $n \times 1$ random vectors and if $\boldsymbol{u} = \boldsymbol{x} - \boldsymbol{y}$ is the $n \times 1$ random vector of their difference, then

$$\boldsymbol{u} = \boldsymbol{x} - \boldsymbol{y} = |\boldsymbol{I}, -\boldsymbol{I}| \left|\begin{array}{c} \boldsymbol{x} \\ \boldsymbol{y} \end{array}\right| \; .$$

The covariance matrix $D(\boldsymbol{x} - \boldsymbol{y})$ is therefore obtained from (2.158) and (2.162) by

$$D(\boldsymbol{x} - \boldsymbol{y}) = D(\boldsymbol{x}) - C(\boldsymbol{x}, \boldsymbol{y}) - C(\boldsymbol{y}, \boldsymbol{x}) + D(\boldsymbol{y}) \; . \tag{2.163}$$

If two random vectors $\boldsymbol{x}$ and $\boldsymbol{y}$ are linearly transformed by $\boldsymbol{A}\boldsymbol{x} + \boldsymbol{a}$ and $\boldsymbol{B}\boldsymbol{y} + \boldsymbol{b}$, where the matrices $\boldsymbol{A}$ and $\boldsymbol{B}$ as well as the vectors $\boldsymbol{a}$ and $\boldsymbol{b}$ contain constants, the covariance matrix of the linearly transformed vectors follows with

$$C(\boldsymbol{A}\boldsymbol{x} + \boldsymbol{a}, \boldsymbol{B}\boldsymbol{y} + \boldsymbol{b}) = \boldsymbol{A}C(\boldsymbol{x}, \boldsymbol{y})\boldsymbol{B}' \; , \tag{2.164}$$

since based on the definition (2.161) we obtain with (2.146)

$$C(\boldsymbol{A}\boldsymbol{x} + \boldsymbol{a}, \boldsymbol{B}\boldsymbol{y} + \boldsymbol{b}) = \boldsymbol{A}E((\boldsymbol{x} - E(\boldsymbol{x}))(\boldsymbol{y} - E(\boldsymbol{y}))')\boldsymbol{B}' = \boldsymbol{A}C(\boldsymbol{x}, \boldsymbol{y})\boldsymbol{B}' \; .$$

2.3.3 Expected Value of a Quadratic Form

Let the $n \times 1$ random vector $\boldsymbol{x}$ have the $n \times 1$ vector $E(\boldsymbol{x}) = \boldsymbol{\mu}$ of expected values and the $n \times n$ covariance matrix $D(\boldsymbol{x}) = \boldsymbol{\Sigma}$, then the expected value of the quadratic form $\boldsymbol{x}'\boldsymbol{A}\boldsymbol{x}$ with the symmetric $n \times n$ matrix $\boldsymbol{A}$ is given by

$$E(\boldsymbol{x}'\boldsymbol{A}\boldsymbol{x}) = \mathrm{tr}(\boldsymbol{A}\boldsymbol{\Sigma}) + \boldsymbol{\mu}'\boldsymbol{A}\boldsymbol{\mu} \; . \tag{2.165}$$

This result follows from the fact that the quadratic form is a scalar, thus with (2.154)

$$E(\boldsymbol{x}'\boldsymbol{A}\boldsymbol{x}) = E(\operatorname{tr}(\boldsymbol{x}'\boldsymbol{A}\boldsymbol{x})) = E(\operatorname{tr}(\boldsymbol{A}\boldsymbol{x}\boldsymbol{x}'))$$
$$= \operatorname{tr}(\boldsymbol{A}E(\boldsymbol{x}\boldsymbol{x}')) = \operatorname{tr}(\boldsymbol{A}(\boldsymbol{\Sigma} + \boldsymbol{\mu}\boldsymbol{\mu}'))$$

whence (2.165) follows.

2.4 Univariate Distributions

The simplest univariate distribution for a continuous random variable, the uniform distribution, was already introduced in (2.59). In this chapter additional distributions needed later are shortly presented and some of their properties are mentioned. For their derivations from experiments or from known distributions see for instance BOX and TIAO (1973), JOHNSON and KOTZ (1970), KOCH (1990) and ZELLNER (1971). Formulas for computing the distribution functions and the percentage points of the distributions are found, for instance, in KOCH (1999).

2.4.1 Normal Distribution

A random variable X is said to be *normally* distributed with the parameters μ and σ^2, which is written as $X \sim N(\mu, \sigma^2)$, if its density function $p(x|\mu, \sigma^2)$ is given by

$$p(x|\mu, \sigma^2) = \frac{1}{\sqrt{2\pi}\sigma} e^{-(x-\mu)^2/2\sigma^2} \quad \text{for} \quad -\infty < x < \infty . \tag{2.166}$$

It can be shown that the normal distribution fulfills the two conditions (2.57). If $X \sim N(\mu, \sigma^2)$, then

$$E(X) = \mu \quad \text{and} \quad V(X) = \sigma^2 . \tag{2.167}$$

Thus, the parameter μ of the normal distribution is determined by the expected value and the parameter σ^2 by the variance of the random variable X. For the multivariate normal distribution the corresponding relation (2.196) is valid, which is one of the reasons that the normal distribution is the most important distribution. It is frequently applied, since for an experiment based on measurements, the expected value and the variance are at least approximately known. These quantities determine already the normal distribution.

The *central limit theorem* gives a further reason for the frequent application of the normal distribution. This theorem is also valid for the multivariate normal distribution. It states that for n independent random variables with any distributions the distribution of the sum of these random variables moves under certain but very general conditions asymptotically towards a normal distribution, if n goes to infinity. Very often a random variable resulting from a measurement can be thought of as originating from a sum of many independent random variables with different distributions, for instance, the electro-optical distance measurement, which results from many influences caused by the instrument and the atmosphere. Measurements or observations can therefore generally be assumed as being normally distributed.

Finally, the normal distribution is obtained, if the expected value and the variance of a random variable are known and if one chooses among the distributions the one which except for this information contains the highest measure of uncertainty, i.e. maximum entropy. This will be dealt with in Chapter 2.6.2.

The distribution function $F(x_F; \mu, \sigma^2)$ of the random variable X with $X \sim N(\mu, \sigma^2)$ for the value x_F follows from (2.55) with (2.166)

$$F(x_F; \mu, \sigma^2) = \frac{1}{\sqrt{2\pi}\sigma} \int_{-\infty}^{x_F} e^{-(x-\mu)^2/2\sigma^2} \, dx \; . \tag{2.168}$$

By the transformation of the variable with

$$z = (x - \mu)/\sigma \quad \text{and} \quad dz = dx/\sigma \tag{2.169}$$

this integral is transformed together with $z_F = (x_F - \mu)/\sigma$ into

$$F(z_F; 0, 1) = \frac{1}{\sqrt{2\pi}} \int_{-\infty}^{z_F} e^{-z^2/2} dz \; . \tag{2.170}$$

This is the distribution function of the random variable Z with the *standard normal distribution* $N(0, 1)$, thus $Z \sim N(0, 1)$. Values for $F(z_F; 0, 1)$ are tabulated and there are approximate formulas for (2.170). Furthermore, $F(z_F; 0, 1)$ is found as a function of compilers.

Example: The probabilities $P(\mu - \sigma < X < \mu + \sigma | C)$ and $P(\mu - 3\sigma < X < \mu + 3\sigma | C)$ regarding the random variable X with the normal distribution $X \sim N(\mu, \sigma^2)$ are computed given the information C with (2.52) and (2.168) by

$$P(\mu - \sigma < X < \mu + \sigma | C) = \frac{1}{\sqrt{2\pi}\sigma} \int_{\mu-\sigma}^{\mu+\sigma} e^{-(x-\mu)^2/2\sigma^2} \, dx \; .$$

Transforming the variable according to (2.169) gives because of the symmetry of the normal distribution

$$P(\mu - \sigma < X < \mu + \sigma | C) = \frac{1}{\sqrt{2\pi}} \int_{-1}^{1} e^{-z^2/2} dz$$
$$= F(1; 0, 1) - (1 - F(1; 0, 1))$$

and with $F(1; 0, 1) = 0.8413$

$$P(\mu - \sigma < X < \mu + \sigma | C) = 0.683 \; .$$

Similarly, with $F(3; 0, 1) = 0.9987$ we obtain

$$P(\mu - 3\sigma < X < \mu + 3\sigma | C) = 0.997 \;.$$

▲

The last result establishes the so-called 3σ rule. It says that with a probability of nearly 100% the values of a normally distributed random variable lie within the interval $(\mu - 3\sigma, \mu + 3\sigma)$.

The value x_α for which with (2.170)

$$F(x_\alpha; 0, 1) = \alpha \tag{2.171}$$

holds is called the *lower α-percentage point* of the standard normal distribution and $x_{1-\alpha}$ the *upper α-percentage point*. Tables or approximate formulas can be used for obtaining the percentage points.

The normal distribution has been defined here. It may also be derived as the distribution of experiments, for instance, as distribution of observational errors.

2.4.2 Gamma Distribution

A random variable X has the *gamma distribution* $G(b, p)$ with the real-valued parameters b and p, written as $X \sim G(b, p)$, if its density function is given by

$$p(x|b, p) = \frac{b^p}{\Gamma(p)} x^{p-1} e^{-bx} \quad \text{for} \quad b > 0 \,, \; p > 0 \,, \; 0 < x < \infty \tag{2.172}$$

and by $p(x|b, p) = 0$ for the remaining values of X with $\Gamma(p)$ being the gamma function. The density function of the gamma distribution is unequal to zero only for positive values of X. As will be shown later, it is therefore used as the distribution of the reciprocal value of a variance. The distribution of a variance therefore follows from the inverted gamma distribution, see the following chapter. The gamma distribution satisfies the two conditions (2.57). For $p \in \mathbb{N}$ and $p > 0$ we have

$$\Gamma(p) = (p - 1)! \tag{2.173}$$

and

$$\Gamma(p + \frac{1}{2}) = \frac{(2p - 1)(2p - 3)\ldots 5 \times 3 \times 1}{2^p} \sqrt{\pi} \;. \tag{2.174}$$

Furthermore, for $X \sim G(b, p)$ we obtain

$$E(X) = p/b \quad \text{and} \quad V(X) = p/b^2 \;. \tag{2.175}$$

The gamma distribution, which has been defined here, may be also derived as a waiting-time distribution.

2.4.3 Inverted Gamma Distribution

If a random variable X has the gamma distribution $X \sim G(b, p)$, then the random variable Z with $Z = 1/X$ has the *inverted gamma distribution* $Z \sim IG(b, p)$ with the density function

$$p(z|b, p) = \frac{b^p}{\Gamma(p)} \left(\frac{1}{z}\right)^{p+1} e^{-b/z} \quad \text{for} \quad b > 0 \,, \; p > 0 \,, \; 0 < z < \infty \quad (2.176)$$

and $p(z|b, p) = 0$ for the remaining values of Z. For $Z \sim IG(b, p)$ we obtain

$$E(Z) = b/(p - 1) \quad \text{for} \quad p > 1$$

and

$$V(Z) = b^2/[(p - 1)^2(p - 2)] \quad \text{for} \quad p > 2 \,. \tag{2.177}$$

2.4.4 Beta Distribution

Let the random variables Y and Z with $Y \sim G(b, \alpha)$ and $Z \sim G(b, \beta)$ be independent, then the random variable $X = Y/(Y + Z)$ has the *beta distribution* $B(\alpha, \beta)$ with the real-valued parameters α and β, thus $X \sim B(\alpha, \beta)$, defined by the density function

$$p(x|\alpha, \beta) = \frac{\Gamma(\alpha + \beta)}{\Gamma(\alpha)\Gamma(\beta)} x^{\alpha-1}(1 - x)^{\beta-1} \quad \text{for} \quad 0 < x < 1 \tag{2.178}$$

and $p(x|\alpha, \beta) = 0$ for the remaining values of x. The distribution function of the beta distribution is called the *incomplete beta function*. It may be computed by a series expansion.

2.4.5 χ^2-Distribution

Let the random variables $X_1, \ldots, X_n$ be independent and normally distributed according to $X_i \sim N(0, 1)$ with $i \in \{1, \ldots, n\}$, then the sum of squares $X = \sum_{i=1}^{n} X_i^2$ has the χ^2-distribution (chi-square distribution) $\chi^2(n)$ with the parameter n, thus $X \sim \chi^2(n)$. The density function is given by

$$p(x|n) = \frac{1}{2^{n/2}\Gamma(n/2)} x^{(n/2)-1} e^{-x/2} \quad \text{for} \quad 0 < x < \infty \tag{2.179}$$

and $p(x|n) = 0$ for the remaining values of X. The parameter n is also called *degree of freedom*. As a comparison with (2.172) shows, the χ^2-distribution follows as special case of the gamma distribution with $b = 1/2$ and $p = n/2$.

If an $n \times 1$ random vector $\mathbf{x}$ has the multivariate normal distribution $N(\mathbf{0}, \mathbf{\Sigma})$ defined in (2.195) by the vector $\mathbf{0}$ and the positive definite $n \times n$

matrix $\boldsymbol{\Sigma}$ as parameters, that is $\boldsymbol{x} \sim N(\boldsymbol{0}, \boldsymbol{\Sigma})$, then the quadratic form $\boldsymbol{x}'\boldsymbol{\Sigma}^{-1}\boldsymbol{x}$ has the χ^2-distribution with n as parameter

$$\boldsymbol{x}'\boldsymbol{\Sigma}^{-1}\boldsymbol{x} \sim \chi^2(n) \ . \tag{2.180}$$

To compute the distribution function $F(\chi^2; n)$ of the χ^2-distribution for the value χ^2, finite and infinite series exist depending on n. The lower α-percentage point $\chi^2_{\alpha;n}$ is corresponding to (2.171) defined by

$$F(\chi^2_{\alpha;n}; n) = \alpha \ . \tag{2.181}$$

The percentage points may be taken from tables or be iteratively computed.

2.4.6   *F*-Distribution

Let the random variables U and V be independently χ^2-distributed like $U \sim \chi^2(m)$ and $V \sim \chi^2(n)$, then the random variable $X = (U/m)/(V/n)$ has the *F-distribution* $F(m, n)$ with the parameters m and n, thus $X \sim F(m, n)$. The density function is given by

$$p(x|m, n) = \frac{\Gamma(\frac{m}{2} + \frac{n}{2})m^{\frac{m}{2}} n^{\frac{n}{2}} x^{\frac{m}{2}-1}}{\Gamma(\frac{m}{2})\Gamma(\frac{n}{2})(n + mx)^{\frac{m}{2}+\frac{n}{2}}} \quad \text{for} \quad 0 < x < \infty \tag{2.182}$$

and $p(x|m, n) = 0$ for the remaining values of X.

The distribution function $F(F_0; m, n)$ for the value F_0 may be computed by the infinite series of the incomplete beta function, the distribution function of the beta distribution. The lower α-percentage point $F_{\alpha;m,n}$ of the F-distribution is defined as in (2.171) by

$$F(F_{\alpha;m,n}; m, n) = \alpha \ . \tag{2.183}$$

The percentage points can be taken from tables or computed by approximate formulas whose results may be iteratively corrected.

2.4.7   *t*-Distribution

If two random variables Y and U with $Y \sim N(0, 1)$ and $U \sim \chi^2(k)$ are independent, the random variable X with

$$X = Y/\sqrt{U/k}$$

has the *t-distribution* $t(k)$ with the parameter k, thus $X \sim t(k)$, and with the density function

$$p(x|k) = \frac{\Gamma(\frac{k+1}{2})}{\sqrt{k\pi}\Gamma(\frac{k}{2})}\left(1 + \frac{x^2}{k}\right)^{-\frac{k+1}{2}} \quad \text{for} \quad -\infty < x < \infty \ . \tag{2.184}$$

Under the assumptions, which lead to the t-distribution, the F-distribution follows for X^2 because of (2.182), thus

$$X^2 \sim F(1,k) \quad \text{and} \quad X \sim t(k) \quad \text{with} \quad X = Y/\sqrt{U/k} \ . \tag{2.185}$$

Values of the distribution function of the t-distribution may therefore also be computed by using the distribution function of the F-distribution. The α-percentage point $F_{\alpha;1,k}$ of the random variable X^2 with the F-distribution $X^2 \sim F(1,k)$ follows from (2.183), if $p(x^2|m,n)$ denotes the density function for X^2,

$$P(X^2 < F_{\alpha;1,k}) = \int_0^{F_{\alpha;1,k}} p(x^2|m,n)dx^2 = \alpha \ .$$

Transforming the variable X^2 into X gives $P(\pm X < (F_{\alpha;1,k})^{1/2})$ and because of (2.185) the quantity $t_{\alpha;k}$ of the t-distribution which is equivalent to the α-percentage point $F_{\alpha;1,k}$ of the F-distribution

$$t_{\alpha;k} = (F_{\alpha;1,k})^{1/2} \tag{2.186}$$

with

$$P(-t_{\alpha;k} < X < t_{\alpha;k}) = \alpha \ , \tag{2.187}$$

since $X > -t_{\alpha;k}$ follows from $-X < t_{\alpha;k}$. However, one has to be aware that $t_{\alpha;k}$ is not the α-percentage point of the t-distribution, one rather gets because of the symmetry of the t-distribution

$$\int_{-\infty}^{t_{\alpha;k}} p(x|k)dx = 1 - (1-\alpha)/2 = (1+\alpha)/2 \ . \tag{2.188}$$

2.4.8 Exponential Distribution

A random variable X has the *exponential distribution* with the parameter μ, if its density function $p(x|\mu)$ is given by

$$p(x|\mu) = \frac{1}{\mu}e^{-x/\mu} \quad \text{for} \quad 0 \le x < \infty \quad \text{and} \quad \mu > 0 \tag{2.189}$$

and $p(x|\mu) = 0$ for the remaining values of X. The exponential distribution fulfills the two conditions (2.57).

If a random variable X has the exponential distribution, we obtain from Example 3 to (2.142) and Example 1 to (2.151)

$$E(X) = \mu \quad \text{and} \quad V(X) = \mu^2 \ . \tag{2.190}$$

The double exponential distribution

$$p(x|\mu) = \frac{1}{2\mu}e^{-|x|/\mu} \quad \text{for} \quad -\infty < x < \infty \quad \text{and} \quad \mu > 0 \tag{2.191}$$

is also called *Laplace distribution*.

2.4.9 Cauchy Distribution

A random variable X has the *Cauchy distribution* with the parameters θ and λ, if its density function $p(x|\theta, \lambda)$ is given by

$$p(x|\theta, \lambda) = (\pi\lambda)^{-1}\left[1 + \frac{1}{\lambda^2}(x - \theta)^2\right]^{-1}$$

$$\text{for} \quad -\infty < x < \infty \quad \text{and} \quad \lambda > 0 \, . \quad (2.192)$$

As is obvious, θ represents a translation and λ a scale parameter. The graph of the Cauchy distribution has a similar form as the normal distribution. For this reason, the Cauchy distribution may be used as the envelope of the normal distribution. This will be shown in Chapter 6.3.6.

The distribution function $F(x_c; \theta, \lambda)$ of the Cauchy distribution for the value x_c is computed according to (2.55) by

$$F(x_c; \theta, \lambda) = \frac{1}{\pi\lambda} \int_{-\infty}^{x_c} \left[1 + \frac{1}{\lambda^2}(x - \theta)^2\right]^{-1} dx$$

$$= \left[\frac{1}{\pi} \arctan\left(\frac{1}{\lambda}(x - \theta)\right)\right]_{-\infty}^{x_c}$$

or using $\arctan(-\infty) = -\pi/2$ by

$$F(x_c; \theta, \lambda) = \frac{1}{\pi} \arctan\left(\frac{1}{\lambda}(x_c - \theta)\right) + \frac{1}{2} \, . \tag{2.193}$$

Because of $\lambda > 0$ and

$$F(\infty; \theta, \lambda) = 1 \tag{2.194}$$

the conditions (2.57) are satisfied for the Cauchy distribution.

2.5 Multivariate Distributions

Multivariate distributions for continuous random variables, which will be needed later, are like the univariate distributions only shortly presented and some properties are mentioned. Derivations may be found for instance in BOX and TIAO (1973), JOHNSON and KOTZ (1972), KOCH (1990, 1999) and ZELLNER (1971).

2.5.1 Multivariate Normal Distribution

An $n \times 1$ random vector $\boldsymbol{x} = |X_1, \ldots, X_n|'$ is said to have the *multivariate normal distribution* $N(\boldsymbol{\mu}, \boldsymbol{\Sigma})$ with the $n \times 1$ vector $\boldsymbol{\mu}$ and the $n \times n$ positive definite matrix $\boldsymbol{\Sigma}$ as parameters, thus $\boldsymbol{x} \sim N(\boldsymbol{\mu}, \boldsymbol{\Sigma})$, if its density function $p(\boldsymbol{x}|\boldsymbol{\mu}, \boldsymbol{\Sigma})$ is given by

$$p(\boldsymbol{x}|\boldsymbol{\mu}, \boldsymbol{\Sigma}) = \frac{1}{(2\pi)^{n/2}(\det \boldsymbol{\Sigma})^{1/2}} e^{-\frac{1}{2}(\boldsymbol{x}-\boldsymbol{\mu})'\boldsymbol{\Sigma}^{-1}(\boldsymbol{x}-\boldsymbol{\mu})} \, . \tag{2.195}$$

It can be shown that the multivariate normal distribution fulfills the two conditions (2.74) and that it may be derived from the univariate normal distribution.

If the random vector $\boldsymbol{x}$ is normally distributed according to $\boldsymbol{x} \sim N(\boldsymbol{\mu}, \boldsymbol{\Sigma})$, then

$$E(\boldsymbol{x}) = \boldsymbol{\mu} \quad \text{and} \quad D(\boldsymbol{x}) = \boldsymbol{\Sigma} \; . \tag{2.196}$$

The parameters $\boldsymbol{\mu}$ and $\boldsymbol{\Sigma}$ of the normal distribution are therefore determined by the vector $E(\boldsymbol{x})$ of expected values and the covariance matrix $D(\boldsymbol{x})$ of the random vector $\boldsymbol{x}$.

Let the $n \times 1$ random vector $\boldsymbol{x}$ with the normal distribution $\boldsymbol{x} \sim N(\boldsymbol{\mu}, \boldsymbol{\Sigma})$ be decomposed with $\boldsymbol{x} = |\boldsymbol{x}_1', \boldsymbol{x}_2'|'$ into the $k \times 1$ and $(n - k) \times 1$ random vectors $\boldsymbol{x}_1$ and $\boldsymbol{x}_2$. With a corresponding partition of the parameters $\boldsymbol{\mu}$ and $\boldsymbol{\Sigma}$ into

$$\boldsymbol{\mu} = |\boldsymbol{\mu}_1', \boldsymbol{\mu}_2'|' \quad \text{and} \quad \boldsymbol{\Sigma} = \left| \begin{array}{cc} \boldsymbol{\Sigma}_{11} & \boldsymbol{\Sigma}_{12} \\ \boldsymbol{\Sigma}_{21} & \boldsymbol{\Sigma}_{22} \end{array} \right| \quad \text{with} \quad \boldsymbol{\Sigma}_{21} = \boldsymbol{\Sigma}_{12}'$$

the marginal distribution for $\boldsymbol{x}_1$ follows by

$$\boldsymbol{x}_1 \sim N(\boldsymbol{\mu}_1, \boldsymbol{\Sigma}_{11}) \tag{2.197}$$

and correspondingly for $\boldsymbol{x}_2$. The marginal distributions for $\boldsymbol{x}_1$ and $\boldsymbol{x}_2$ are therefore again normal distributions.

When decomposing the normally distributed random vector $\boldsymbol{x}$ into the random vectors $\boldsymbol{x}_1$ and $\boldsymbol{x}_2$ as for (2.197), the distribution for $\boldsymbol{x}_1$ under the condition that the second random vector takes on the values $\boldsymbol{x}_2$ is obtained by

$$\boldsymbol{x}_1 | \boldsymbol{x}_2 \sim N(\boldsymbol{\mu}_1 + \boldsymbol{\Sigma}_{12}\boldsymbol{\Sigma}_{22}^{-1}(\boldsymbol{x}_2 - \boldsymbol{\mu}_2), \; \boldsymbol{\Sigma}_{11} - \boldsymbol{\Sigma}_{12}\boldsymbol{\Sigma}_{22}^{-1}\boldsymbol{\Sigma}_{21}) \tag{2.198}$$

and the distribution for $\boldsymbol{x}_2$ under the condition of the values $\boldsymbol{x}_1$ by exchanging the two indices.

If random variables are independent, their covariances are equal to zero according to (2.153). The converse of this statement holds true for normally distributed random variables. If the random vector $\boldsymbol{x}$ with $\boldsymbol{x} \sim N(\boldsymbol{\mu}, \boldsymbol{\Sigma})$ is partitioned into the k random vectors $\boldsymbol{x}_i$ with $\boldsymbol{x} = |\boldsymbol{x}_1', \ldots, \boldsymbol{x}_k'|'$, the random vectors $\boldsymbol{x}_i$ are independent, if and only if for the corresponding partition of the covariance matrix $\boldsymbol{\Sigma} = (\boldsymbol{\Sigma}_{ij})$ the relation is valid

$$\boldsymbol{\Sigma}_{ij} = \mathbf{0} \quad \text{for} \quad i \neq j \quad \text{and} \quad i, j \in \{1, \ldots, k\} \; . \tag{2.199}$$

Example 1: Let the random variables X_i of the normally distributed $n \times 1$ random vector $\boldsymbol{x} = |X_1, \ldots, X_n|'$ be independent, then we get from (2.199)

$$\boldsymbol{x} \sim N(\boldsymbol{\mu}, \boldsymbol{\Sigma}) \quad \text{with} \quad \boldsymbol{\Sigma} = \mathrm{diag}(\sigma_1^2, \ldots, \sigma_n^2) \tag{2.200}$$

and from (2.197)

$$X_i \sim N(\mu_i, \sigma_i^2) \quad \text{for} \quad i \in \{1, \ldots, n\} \tag{2.201}$$

where μ_i denotes the expected value and σ_i^2 the variance of X_i. The density function $p(\boldsymbol{x}|\boldsymbol{\mu}, \boldsymbol{\Sigma})$ for $\boldsymbol{x}$ follows with (2.195) by

$$p(\boldsymbol{x}|\boldsymbol{\mu}, \boldsymbol{\Sigma}) = \frac{1}{(2\pi)^{n/2}(\prod_{i=1}^{n} \sigma_i^2)^{1/2}} \exp\Big(-\sum_{i=1}^{n} \frac{(x_i - \mu_i)^2}{2\sigma_i^2} \Big)$$

$$= \prod_{i=1}^{n} \Big(\frac{1}{\sqrt{2\pi}\sigma_i} e^{-\frac{(x_i-\mu_i)^2}{2\sigma_i^2}} \Big) .$$

Thus in agreement with (2.111), we obtain the joint distribution for the n independent random variables X_i by the product of their marginal distributions (2.201). $\triangle$

The $m \times 1$ random vector $\boldsymbol{z}$ which originates from the linear transformation $\boldsymbol{z} = \boldsymbol{A}\boldsymbol{x} + \boldsymbol{c}$, where $\boldsymbol{x}$ denotes an $n \times 1$ random vector with $\boldsymbol{x} \sim N(\boldsymbol{\mu}, \boldsymbol{\Sigma})$, $\boldsymbol{A}$ an $m \times n$ matrix of constants with $\text{rank}\boldsymbol{A} = m$ and $\boldsymbol{c}$ an $m \times 1$ vector of constants, has the normal distribution

$$\boldsymbol{z} \sim N(\boldsymbol{A}\boldsymbol{\mu} + \boldsymbol{c}, \ \boldsymbol{A}\boldsymbol{\Sigma}\boldsymbol{A}') . \tag{2.202}$$

Thus, normally distributed random vectors are again normally distributed after a linear transformation.

Example 2: Let the n independent random variables X_i be normally distributed with $X_i \sim N(\mu, \sigma^2)$, then the distribution of the random vector $\boldsymbol{x} = |X_1, \ldots, X_n|'$ is according to (2.200) and (2.201) given by

$$\boldsymbol{x} \sim N(\boldsymbol{\mu}, \boldsymbol{\Sigma}) \quad \text{with} \quad \boldsymbol{\mu} = |\mu, \ldots, \mu|' \quad \text{and} \quad \boldsymbol{\Sigma} = \text{diag}(\sigma^2, \ldots, \sigma^2) .$$

The mean $\bar{X} = \frac{1}{n}\sum_{i=1}^{n} X_i$ has therefore according to (2.202) with $\boldsymbol{A} = |1/n, 1/n, \ldots, 1/n|$ and $\boldsymbol{c} = \boldsymbol{0}$ the normal distribution

$$\bar{X} \sim N(\mu, \sigma^2/n) . \tag{2.203}$$

$\triangle$

2.5.2 Multivariate *t*-Distribution

Let the $k \times 1$ random vector $\boldsymbol{z} = |Z_1, \ldots, Z_k|'$ be normally distributed according to $\boldsymbol{z} \sim N(\boldsymbol{0}, \boldsymbol{N}^{-1})$ with the $k \times k$ positive definite matrix $\boldsymbol{N}$. Let furthermore the random variable H with values h have the distribution $H \sim \chi^2(\nu)$ with ν as parameter and let H and $\boldsymbol{z}$ be independent. The $k \times 1$ random vector $\boldsymbol{x}$ with values

$$x_i = z_i(h/\nu)^{-1/2} + \mu_i \quad \text{for} \quad i \in \{1, \ldots, k\}$$

then has the *multivariate t-distribution* with the $k \times 1$ vector $\boldsymbol{\mu} = (\mu_i)$, the matrix $\boldsymbol{N}^{-1}$ and ν as parameters, thus $\boldsymbol{x} \sim t(\boldsymbol{\mu}, \boldsymbol{N}^{-1}, \nu)$, whose density function $p(\boldsymbol{x}|\boldsymbol{\mu}, \boldsymbol{N}^{-1}, \nu)$ for $\boldsymbol{x}$ is given by

$$p(\boldsymbol{x}|\boldsymbol{\mu}, \boldsymbol{N}^{-1}, \nu) =$$
$$\frac{\nu^{\nu/2}\Gamma((k+\nu)/2)(\det \boldsymbol{N})^{1/2}}{\pi^{k/2}\Gamma(\nu/2)}(\nu + (\boldsymbol{x} - \boldsymbol{\mu})'\boldsymbol{N}(\boldsymbol{x} - \boldsymbol{\mu}))^{-(k+\nu)/2} \ .$$

$$(2.204)$$

The multivariate t-distribution is the multivariate generalization of the t-distribution, as becomes obvious from the following example.

Example: We obtain with $k = 1, \boldsymbol{x} = x, \boldsymbol{\mu} = \mu$ and $\boldsymbol{N} = f$ instead of (2.204)

$$p(x|\mu, 1/f, \nu) = \frac{\Gamma((\nu + 1)/2)}{\sqrt{\pi}\Gamma(\nu/2)} \left(\frac{f}{\nu}\right)^{1/2} \left(1 + \frac{f}{\nu}(x - \mu)^2\right)^{-(\nu+1)/2} \ . \quad (2.205)$$

This is the density function of a random variable X with the generalized t-distribution $t(\mu, 1/f, \nu)$, thus $X \sim t(\mu, 1/f, \nu)$. The standard form of this distribution leads to the density function (2.184) of the t-distribution and follows from the transformation of the variable x to z with

$$z = \sqrt{f}(x - \mu) \ . \tag{2.206}$$

The density function for the random variable Z follows with $dx/dz = 1/\sqrt{f}$ from, see for instance KOCH (1999, p.93),

$$p(z|\nu) = \frac{\Gamma((\nu + 1)/2)}{\sqrt{\nu\pi}\Gamma(\nu/2)}\left(1 + \frac{z^2}{\nu}\right)^{-(\nu+1)/2} \ . \tag{2.207}$$

The random variable Z has because of (2.184) the *t-distribution* $t(\nu)$ with the parameter ν, thus $Z \sim t(\nu)$. $\qquad\qquad\qquad\qquad\qquad\qquad\qquad\quad \triangle$

If the random vector $\boldsymbol{x}$ has the multivariate t-distribution $\boldsymbol{x} \sim t(\boldsymbol{\mu}, \boldsymbol{N}^{-1}, \nu)$, then

$$E(\boldsymbol{x}) = \boldsymbol{\mu} \quad \text{for} \quad \nu > 1$$

and

$$D(\boldsymbol{x}) = \nu(\nu - 2)^{-1}\boldsymbol{N}^{-1} \quad \text{for} \quad \nu > 2 \ . \tag{2.208}$$

If the $k \times 1$ random vector $\boldsymbol{x}$ with $\boldsymbol{x} \sim t(\boldsymbol{\mu}, \boldsymbol{N}^{-1}, \nu)$ is decomposed into the $(k - m) \times 1$ vector $\boldsymbol{x}_1$ and the $m \times 1$ vector $\boldsymbol{x}_2$ by $\boldsymbol{x} = |\boldsymbol{x}'_1, \boldsymbol{x}'_2|'$ and correspondingly $\boldsymbol{\mu} = |\boldsymbol{\mu}'_1, \boldsymbol{\mu}'_2|'$ as well as

$$\boldsymbol{N} = \begin{vmatrix} \boldsymbol{N}_{11} & \boldsymbol{N}_{12} \\ \boldsymbol{N}_{21} & \boldsymbol{N}_{22} \end{vmatrix} \quad \text{with} \quad \boldsymbol{N}^{-1} = \begin{vmatrix} \boldsymbol{I}_{11} & \boldsymbol{I}_{12} \\ \boldsymbol{I}_{21} & \boldsymbol{I}_{22} \end{vmatrix} ,$$

then the random vector $\boldsymbol{x}_2$ has also the multivariate t-distribution, that is, $\boldsymbol{x}_2$ has the marginal distribution

$$\boldsymbol{x}_2 \sim t(\boldsymbol{\mu}_2, \boldsymbol{I}_{22}, \nu) \tag{2.209}$$

with

$$\boldsymbol{I}_{22} = (\boldsymbol{N}_{22} - \boldsymbol{N}_{21}\boldsymbol{N}_{11}^{-1}\boldsymbol{N}_{12})^{-1} .$$

A corresponding marginal distribution is valid also for the random vector $\boldsymbol{x}_1$.

The $m \times 1$ random vector $\boldsymbol{y}$ which originates from the linear transformation $\boldsymbol{y} = \boldsymbol{A}\boldsymbol{x} + \boldsymbol{c}$, where $\boldsymbol{x}$ is a $k \times 1$ random vector with $\boldsymbol{x} \sim t(\boldsymbol{\mu}, \boldsymbol{N}^{-1}, \nu)$, $\boldsymbol{A}$ an $m \times k$ matrix of constants with rank$\boldsymbol{A} = m$ and $\boldsymbol{c}$ an $m \times 1$ vector of constants, has again a multivariate t-distribution

$$\boldsymbol{y} \sim t(\boldsymbol{A}\boldsymbol{\mu} + \boldsymbol{c}, \boldsymbol{A}\boldsymbol{N}^{-1}\boldsymbol{A}', \nu) . \tag{2.210}$$

Finally, there is a connection between the multivariate t-distribution and the F-distribution (2.182). If the $k \times 1$ random vector $\boldsymbol{x}$ is distributed according to $\boldsymbol{x} \sim t(\boldsymbol{\mu}, \boldsymbol{N}^{-1}, \nu)$, then the quadratic form $(\boldsymbol{x} - \boldsymbol{\mu})'\boldsymbol{N}(\boldsymbol{x} - \boldsymbol{\mu})/k$ has the F-distribution $F(k, \nu)$ with k and ν as parameters

$$(\boldsymbol{x} - \boldsymbol{\mu})'\boldsymbol{N}(\boldsymbol{x} - \boldsymbol{\mu})/k \sim F(k, \nu) . \tag{2.211}$$

2.5.3 Normal-Gamma Distribution

Let $\boldsymbol{x}$ be an $n \times 1$ random vector and τ be a random variable. Let the conditional density function $p(\boldsymbol{x}|\boldsymbol{\mu}, \tau^{-1}\boldsymbol{V})$ for $\boldsymbol{x}$ with values x_i under the condition that a value for τ is given be determined by the normal distribution $N(\boldsymbol{\mu}, \tau^{-1}\boldsymbol{V})$. Let τ have the gamma distribution $\tau \sim G(b, p)$ with the parameters b and p and the density function $p(\tau|b, p)$. The joint density function $p(\boldsymbol{x}, \tau|\boldsymbol{\mu}, \boldsymbol{V}, b, p)$ for $\boldsymbol{x}$ and τ then follows from (2.102) by

$$p(\boldsymbol{x}, \tau|\boldsymbol{\mu}, \boldsymbol{V}, b, p) = p(\boldsymbol{x}|\boldsymbol{\mu}, \tau^{-1}\boldsymbol{V})p(\tau|b, p) .$$

It is the density function of the so-called *normal-gamma distribution* $NG(\boldsymbol{\mu}, \boldsymbol{V}, b, p)$ with the parameters $\boldsymbol{\mu}, \boldsymbol{V}, b, p$, thus

$$\boldsymbol{x}, \tau \sim NG(\boldsymbol{\mu}, \boldsymbol{V}, b, p) .$$

The density function is obtained with (2.172), (2.195) and $(\det \tau^{-1}\boldsymbol{V})^{-1/2} = (\det \boldsymbol{V})^{-1/2}\tau^{n/2}$ by

$$p(\boldsymbol{x}, \tau|\boldsymbol{\mu}, \boldsymbol{V}, b, p) = (2\pi)^{-n/2}(\det \boldsymbol{V})^{-1/2}b^p(\Gamma(p))^{-1}$$
$$\tau^{n/2+p-1}\exp\left\{-\frac{\tau}{2}[2b + (\boldsymbol{x} - \boldsymbol{\mu})'\boldsymbol{V}^{-1}(\boldsymbol{x} - \boldsymbol{\mu})]\right\} \tag{2.212}$$

for $b > 0$, $p > 0$, $0 < \tau < \infty$, $-\infty < x_i < \infty$.

If $\boldsymbol{x}$ and τ have the normal-gamma distribution $\boldsymbol{x},\tau \sim NG(\boldsymbol{\mu},\boldsymbol{V},b,p)$, then $\boldsymbol{x}$ has as marginal distribution the multivariate t-distribution

$$\boldsymbol{x} \sim t(\boldsymbol{\mu},b\boldsymbol{V}/p,2p) \qquad (2.213)$$

and τ as marginal distribution the gamma distribution

$$\tau \sim G(b,p) \ . \qquad (2.214)$$

2.6   Prior Density Functions

To apply Bayes' theorem (2.122), the prior density function is needed by which the information is expressed that already exists for the unknown parameters. If no information on the unknown parameters is available, noninformative prior density functions are used. If information exists, it is important that the prior density function contains only the information which is available and beyond that no information. This is achieved by applying the principle of maximum entropy. If a prior density function can be treated analytically, it is helpful that the prior density and the posterior density function belong to the same class of distributions. Such a density function is called a conjugate prior.

2.6.1   Noninformative Priors

If nothing is known in advance about the unknown parameter X, it can take on values x between $-\infty$ and $+\infty$. Its *noninformative prior* $p(x|C)$ under the condition of the information C is then assumed to be

$$p(x|C) \propto \text{const} \quad \text{for} \quad -\infty < x < \infty \qquad (2.215)$$

where const denotes a constant. The density is an *improper density function*, since with $\int_{-\infty}^{\infty} p(x|C)dx \neq 1$ the condition (2.57) is not fulfilled so that the density function cannot be normalized. This is not a serious drawback, since the normalization of the posterior distribution can be achieved, if a likelihood function to be normalized is selected. For an $n \times 1$ vector $\boldsymbol{x}$ of parameters with values $\boldsymbol{x} = (x_i)$

$$p(\boldsymbol{x}|C) \propto \text{const} \quad \text{for} \quad -\infty < x_i < \infty \ , \ i \in \{1,\ldots,n\} \qquad (2.216)$$

is chosen corresponding to (2.215) as noninformative prior.

If an unknown parameter like the variance σ^2 from (2.151) can only take on values between 0 and ∞, we set

$$x = \ln \sigma^2 \qquad (2.217)$$

and again

$$p(x|C) \propto \text{const} \quad \text{for} \quad -\infty < x < \infty \ .$$

By the transformation of x to σ^2 with $dx/d\sigma^2 = 1/\sigma^2$ from (2.217), see for instance KOCH (1999, p.93), the noninformative prior for the variance σ^2 follows by

$$p(\sigma^2|C) \propto 1/\sigma^2 \quad \text{for} \quad 0 < \sigma^2 < \infty . \tag{2.218}$$

Very often it is more convenient to introduce the *weight* or *precision parameter* τ instead of σ^2 with

$$\tau = 1/\sigma^2 . \tag{2.219}$$

By transforming σ^2 to τ with $d\sigma^2/d\tau = -1/\tau^2$ the noninformative prior density function for τ follows instead of (2.218) by

$$p(\tau|C) \propto 1/\tau \quad \text{for} \quad 0 < \tau < \infty . \tag{2.220}$$

The prior density function (2.218) for σ^2 is invariant with respect to the transformation (2.219), since independent from choosing the prior density function (2.218) or (2.220) identical probabilities are obtained by the posterior density functions. To show this, the probability that σ^2 lies in the interval $d\sigma^2$ is computed from (2.56) by the posterior density function $p(\sigma^2|\boldsymbol{y}, C)$, which follows with (2.218) and the likelihood function $p(\boldsymbol{y}|\sigma^2, C)$ from Bayes' theorem (2.122),

$$p(\sigma^2|\boldsymbol{y}, C)d\sigma^2 \propto \frac{1}{\sigma^2}p(\boldsymbol{y}|\sigma^2, C)d\sigma^2 .$$

If the parameter τ from (2.219) is used, we obtain with the identical likelihood function

$$p(\tau|\boldsymbol{y}, C)d\tau \propto \frac{1}{\tau}p(\boldsymbol{y}|\sigma^2, C)d\tau .$$

Because of $d\sigma^2/d\tau = -1/\tau^2$ and $d\sigma^2/\sigma^2 \propto d\tau/\tau$ finally

$$p(\sigma^2|\boldsymbol{y}, C)d\sigma^2 \propto p(\tau|\boldsymbol{y}, C)d\tau \tag{2.221}$$

follows so that independent of the parameters σ^2 or τ the posterior density functions lead to identical probabilities.

Based on the invariance of a transformation JEFFREYS (1961, p.179) derived a general formula for obtaining noninformative priors which include the density functions (2.215), (2.216) and (2.218) as special cases, see for instance KOCH (1990, p.11).

2.6.2 Maximum Entropy Priors

Entropy is a measure of uncertainty. The principle of *maximum entropy* is applied to derive prior density functions which contain the prior information on the unknown parameters, but beyond this they are as uncertain as possible. By performing an experiment or a measurement to obtain information about an unknown phenomenon, uncertainty is removed which was existing before the experiment or the measurement took place. The uncertainty which is eliminated by the experiment corresponds to the information which was gained by the experiment.

The uncertainty or the information $I(A)$ of a random event A should be equal to zero, if $P(A) = 1$ holds. Furthermore, we require $I(A_1) > I(A_2)$ for $P(A_1) < P(A_2)$ so that the smaller the probability the larger the uncertainty becomes. Finally, the uncertainty should add for independent events. Then it may be shown, see for instance KOCH (1990, p.16), that

$$I(A) = -c \ln P(A) \tag{2.222}$$

where c denotes a constant. If the expected value of the uncertainty is formed for a discrete random variable with the density function $p(x_i|C)$ for $i \in \{1 \ldots, n\}$ given the information C, then the discrete entropy H_n follows with $c = 1$ by

$$H_n = -\sum_{i=1}^{n} p(x_i|C) \ln p(x_i|C) \tag{2.223}$$

and correspondingly the continuous entropy for a continuous random variable with values x in the interval $a \leq x \leq b$ and the density function $p(x|C)$ by

$$H = -\int_a^b p(x|C) \ln p(x|C) dx \; . \tag{2.224}$$

Since the prior information is in general incomplete, the prior density function should be except for the prior information as uncertain as possible. The prior density function is therefore derived such that under the constraint of the given prior information, like the known expected value and the variance of a random variable, the entropy becomes maximal.

It can be shown that a random variable which is defined in the interval $[a, b]$ and whose density function maximizes the entropy has the uniform distribution (2.59). A random variable X with given expected value $E(X) = \mu$ and variance $V(X) = \sigma^2$ which is defined in the interval $(-\infty, \infty)$ and whose density function maximizes the entropy has the normal distribution (2.166). A random variable X with known expected value $E(X) = \mu$ which is defined in the interval $[0, \infty)$ and whose density function maximizes the entropy has the exponential distribution (2.189). A random variable with given expected value and variance which is defined in the interval $[0, \infty)$ and whose density function maximizes the entropy has the truncated normal distribution, see for instance KOCH (1990, p.17).

2.6.3   Conjugate Priors

A density function is called a *conjugate prior*, if it leads after being multiplied by the likelihood function to a posterior density function which belongs to the same family of distributions as the prior density function. This property is important, if one starts with an analytically tractable distribution, for instance with the normal distribution, and obtains as posterior density function again an analytically tractable density function. Without entering into the derivation of conjugate priors, see for instance BERNARDO and SMITH (1994, p.265), DEGROOT (1970, p.159), RAIFFA and SCHLAIFER (1961, p.43) or ROBERT (1994, p.97), it will be shown in the following that for the linear model which will be treated in Chapter 4 the normal distribution and the normal-gamma distribution lead to conjugate priors, if the likelihood function is determined by the normal distribution. Without mentioning it, the density function of the beta distribution was obtained as conjugate prior in Example 2 to (2.134). The prior density and the posterior density function in (2.135) result from a beta distribution, if the likelihood function is defined by a binomial distribution.

Let $\boldsymbol{\beta}$ be the $u \times 1$ random vector of unknown parameters, $\boldsymbol{X}$ the $n \times u$ matrix of given coefficients with $\text{rank}\boldsymbol{X} = u$, $\boldsymbol{y}$ the $n \times 1$ random vector of observations with the vector $\boldsymbol{X}\boldsymbol{\beta} = E(\boldsymbol{y}|\boldsymbol{\beta})$ of expected values and the covariance matrix $D(\boldsymbol{y}|\sigma^2) = \sigma^2 \boldsymbol{P}^{-1}$ where σ^2 designates the variance factor, which frequently is an unknown random variable, and $\boldsymbol{P}$ the positive definite known matrix of weights of the observations from (2.159). Let the observations be normally distributed so that with (2.196) the distribution follows

$$\boldsymbol{y}|\boldsymbol{\beta}, \sigma^2 \sim N(\boldsymbol{X}\boldsymbol{\beta}, \sigma^2 \boldsymbol{P}^{-1}) \tag{2.225}$$

whose density function gives the likelihood function.

As mentioned in connection with (2.138) and (2.139), it is now necessary to indicate by $E(\boldsymbol{y}|\boldsymbol{\beta})$ the condition that the expected value of $\boldsymbol{y}$ is computed for given values of $\boldsymbol{\beta}$, since $\boldsymbol{\beta}$ is a random vector. Correspondingly, $D(\boldsymbol{y}|\sigma^2)$ and $\boldsymbol{y}|\boldsymbol{\beta}, \sigma^2$ in (2.225) have to be interpreted such that $\boldsymbol{\beta}$ and σ^2 mean given values.

First it will be assumed that the variance factor σ^2 is known. The density function of the normal distribution

$$\boldsymbol{\beta} \sim N(\boldsymbol{\mu}, \sigma^2 \boldsymbol{\Sigma}) \tag{2.226}$$

is chosen as prior for the unknown parameters $\boldsymbol{\beta}$. It is a conjugate prior, since the posterior density function for $\boldsymbol{\beta}$ follows again from the normal distribution

$$\boldsymbol{\beta}|\boldsymbol{y} \sim N\big(\boldsymbol{\mu}_0, \sigma^2 (\boldsymbol{X}'\boldsymbol{P}\boldsymbol{X} + \boldsymbol{\Sigma}^{-1})^{-1}\big) \tag{2.227}$$

with

$$\boldsymbol{\mu}_0 = (\boldsymbol{X}'\boldsymbol{P}\boldsymbol{X} + \boldsymbol{\Sigma}^{-1})^{-1}(\boldsymbol{X}'\boldsymbol{P}\boldsymbol{y} + \boldsymbol{\Sigma}^{-1}\boldsymbol{\mu}) \, .$$

The reason is that according to Bayes' theorem (2.122) the posterior density function $p(\boldsymbol{\beta}|\boldsymbol{y}, C)$ for $\boldsymbol{\beta}$ results with (2.195) and (2.225) in

$$p(\boldsymbol{\beta}|\boldsymbol{y}, C) \propto$$
$$\exp\left\{-\frac{1}{2\sigma^2}[(\boldsymbol{\beta} - \boldsymbol{\mu})'\boldsymbol{\Sigma}^{-1}(\boldsymbol{\beta} - \boldsymbol{\mu}) + (\boldsymbol{y} - \boldsymbol{X\beta})'\boldsymbol{P}(\boldsymbol{y} - \boldsymbol{X\beta})]\right\}.$$

We obtain for the expression in brackets of the exponent

$$\boldsymbol{y}'\boldsymbol{Py} + \boldsymbol{\mu}'\boldsymbol{\Sigma}^{-1}\boldsymbol{\mu} - 2\boldsymbol{\beta}'(\boldsymbol{X}'\boldsymbol{Py} + \boldsymbol{\Sigma}^{-1}\boldsymbol{\mu}) + \boldsymbol{\beta}'(\boldsymbol{X}'\boldsymbol{PX} + \boldsymbol{\Sigma}^{-1})\boldsymbol{\beta}$$
$$= \boldsymbol{y}'\boldsymbol{Py} + \boldsymbol{\mu}'\boldsymbol{\Sigma}^{-1}\boldsymbol{\mu} - \boldsymbol{\mu}_0'(\boldsymbol{X}'\boldsymbol{PX} + \boldsymbol{\Sigma}^{-1})\boldsymbol{\mu}_0$$
$$+ (\boldsymbol{\beta} - \boldsymbol{\mu}_0)'(\boldsymbol{X}'\boldsymbol{PX} + \boldsymbol{\Sigma}^{-1})(\boldsymbol{\beta} - \boldsymbol{\mu}_0). \tag{2.228}$$

If the term dependent on $\boldsymbol{\beta}$ is substituted only, since constants do not have to be considered, we find

$$p(\boldsymbol{\beta}|\boldsymbol{y}, C) \propto \exp\left\{-\frac{1}{2\sigma^2}[(\boldsymbol{\beta} - \boldsymbol{\mu}_0)'(\boldsymbol{X}'\boldsymbol{PX} + \boldsymbol{\Sigma}^{-1})(\boldsymbol{\beta} - \boldsymbol{\mu}_0)]\right\}$$

and with (2.195) the normal distribution (2.227).

Example: Let an unknown quantity s, for instance, a distance or an angle be measured n times so that the observation vector $\boldsymbol{y} = |y_1, y_2, \ldots, y_n|'$ is obtained. Let the measurements be independent and normally distributed like $y_i|s \sim N(s, \sigma^2)$ where the variance σ^2 is known. Thus

$$\begin{aligned}
s &= E(y_1|s) \quad \text{with} \quad V(y_1) = \sigma^2 \\
s &= E(y_2|s) \quad \text{with} \quad V(y_2) = \sigma^2 \\
& \cdots\cdots\cdots\cdots\cdots\cdots\cdots\cdots\cdots\cdots\cdots \\
s &= E(y_n|s) \quad \text{with} \quad V(y_n) = \sigma^2,
\end{aligned} \tag{2.229}$$

and we obtain in (2.225) $\boldsymbol{X} = |1, 1, \ldots, 1|', \boldsymbol{\beta} = s$ and $\boldsymbol{P} = \boldsymbol{I}$ because of (2.199).

Let the prior density function for the quantity s be normally distributed with given expected value $E(s) = \mu_s$ and given variance $V(s) = \sigma^2\sigma_s^2$. Then we get in (2.226) $\boldsymbol{\mu} = \mu_s$ and $\boldsymbol{\Sigma} = \sigma_s^2$ because of (2.196). From (2.227) with $\boldsymbol{\mu}_0 = \mu_{0s}$

$$\mu_{0s} = \frac{\sum_{i=1}^{n} y_i + \frac{\mu_s}{\sigma_s^2}}{n + \frac{1}{\sigma_s^2}} = \frac{\left(\frac{1}{n}\right)^{-1}\frac{1}{n}\sum_{i=1}^{n} y_i + (\sigma_s^2)^{-1}\mu_s}{\left(\frac{1}{n}\right)^{-1} + (\sigma_s^2)^{-1}} \tag{2.230}$$

follows and with

$$\sigma_{0s}^2 = \frac{1}{\left(\frac{1}{n}\right)^{-1} + (\sigma_s^2)^{-1}} \tag{2.231}$$

as posterior distribution for s the normal distribution

$$s|\boldsymbol{y} \sim N(\mu_{0s}, \sigma^2\sigma_{0s}^2). \tag{2.232}$$

The quantity μ_{0s} is because of (2.167) the expected value of s. It is computed from (2.230) as weighted mean of the prior information μ_s for s and of the mean $(1/n)\sum_{i=1}^{n} y_i$ of the observations, because the variance of μ_s is $\sigma^2\sigma_s^2$ and the variance of the mean σ^2/n according to (2.203). The reciprocal values of these variances give according to (2.160) with $c = \sigma^2$ the weights in (2.230). Δ

Let the variance factor σ^2 now be a random variable and unknown. To obtain the conjugate prior for the unknown parameters $\boldsymbol{\beta}$ and σ^2 we introduce with $\tau = 1/\sigma^2$ according to (2.219) instead of σ^2 the unknown weight parameter τ. The likelihood function then follows with (2.195) and $(\det \tau^{-1}\boldsymbol{P}^{-1})^{-1/2} = (\det \boldsymbol{P})^{1/2}\tau^{n/2}$ from (2.225) by

$$p(\boldsymbol{y}|\boldsymbol{\beta},\tau,C) = (2\pi)^{-n/2}(\det \boldsymbol{P})^{1/2}\tau^{n/2}\exp[-\frac{\tau}{2}(\boldsymbol{y}-\boldsymbol{X}\boldsymbol{\beta})'\boldsymbol{P}(\boldsymbol{y}-\boldsymbol{X}\boldsymbol{\beta})]. \quad (2.233)$$

As prior for $\boldsymbol{\beta}$ and τ the density function (2.212) of the normal-gamma distribution

$$\boldsymbol{\beta},\tau \sim NG(\boldsymbol{\mu},\boldsymbol{V},b,p) \quad (2.234)$$

is chosen. It is a conjugate prior, since the posterior density function for $\boldsymbol{\beta}$ and τ is again obtained from the normal-gamma distribution

$$\boldsymbol{\beta},\tau|\boldsymbol{y} \sim NG(\boldsymbol{\mu}_0,\boldsymbol{V}_0,b_0,p_0) \quad (2.235)$$

with

$$\begin{aligned}
\boldsymbol{\mu}_0 &= (\boldsymbol{X}'\boldsymbol{P}\boldsymbol{X} + \boldsymbol{V}^{-1})^{-1}(\boldsymbol{X}'\boldsymbol{P}\boldsymbol{y} + \boldsymbol{V}^{-1}\boldsymbol{\mu})\\
\boldsymbol{V}_0 &= (\boldsymbol{X}'\boldsymbol{P}\boldsymbol{X} + \boldsymbol{V}^{-1})^{-1}\\
b_0 &= [2b + (\boldsymbol{\mu} - \boldsymbol{\mu}_0)'\boldsymbol{V}^{-1}(\boldsymbol{\mu} - \boldsymbol{\mu}_0) + (\boldsymbol{y} - \boldsymbol{X}\boldsymbol{\mu}_0)'\boldsymbol{P}(\boldsymbol{y} - \boldsymbol{X}\boldsymbol{\mu}_0)]/2\\
p_o &= (n + 2p)/2 \, .
\end{aligned} \quad (2.236)$$

To show this results, the posterior density function $p(\boldsymbol{\beta},\tau|\boldsymbol{y},C)$ for $\boldsymbol{\beta}$ and τ is derived with (2.122), (2.212) and (2.233) by

$$p(\boldsymbol{\beta},\tau|\boldsymbol{y},C) \propto \tau^{u/2+p-1}\exp\left\{-\frac{\tau}{2}[2b + (\boldsymbol{\beta} - \boldsymbol{\mu})'\boldsymbol{V}^{-1}(\boldsymbol{\beta} - \boldsymbol{\mu})]\right\}$$

$$\tau^{n/2}\exp[-\frac{\tau}{2}(\boldsymbol{y} - \boldsymbol{X}\boldsymbol{\beta})'\boldsymbol{P}(\boldsymbol{y} - \boldsymbol{X}\boldsymbol{\beta})]$$

$$\propto \tau^{n/2+p+u/2-1}$$

$$\exp\left\{-\frac{\tau}{2}[2b + (\boldsymbol{\beta} - \boldsymbol{\mu})'\boldsymbol{V}^{-1}(\boldsymbol{\beta} - \boldsymbol{\mu}) + (\boldsymbol{y} - \boldsymbol{X}\boldsymbol{\beta})'\boldsymbol{P}(\boldsymbol{y} - \boldsymbol{X}\boldsymbol{\beta})]\right\} \, .$$

The expression in brackets of the exponent follows with

$$2b + y'Py + \mu'V^{-1}\mu - 2\beta'(X'Py + V^{-1}\mu) + \beta'(X'PX + V^{-1})\beta$$
$$= 2b + y'Py + \mu'V^{-1}\mu - \mu_0'(X'PX + V^{-1})\mu_0$$
$$+ (\beta - \mu_0)'(X'PX + V^{-1})(\beta - \mu_0)$$
$$= 2b + y'Py + \mu'V^{-1}\mu - 2\mu_0'(X'Py + V^{-1}\mu)$$
$$+ \mu_0'(X'PX + V^{-1})\mu_0 + (\beta - \mu_0)'(X'PX + V^{-1})(\beta - \mu_0)$$
$$= 2b + (\mu - \mu_0)'V^{-1}(\mu - \mu_0) + (y - X\mu_0)'P(y - X\mu_0)$$
$$+ (\beta - \mu_0)'(X'PX + V^{-1})(\beta - \mu_0) \ . \tag{2.237}$$

Substituting this result gives by a comparison with (2.212) the normal-gamma distribution (2.235).

3 Parameter Estimation, Confidence Regions and Hypothesis Testing

As already indicated in Chapter 2.2.8, the knowledge of the posterior density function for the unknown parameters from Bayes' theorem allows to estimate the unknown parameters, to establish regions which contain the unknown parameters with given probability and to test hypotheses for the parameters. These methods will be presented in the following for continuous random vectors. The integrals which appear to compute the probabilities, marginal distributions and expected values have to be replaced for discrete random vectors by summations, in order to obtain the probabilities according to (2.69), the marginal distributions from (2.85) and the expected values according to (2.140).

3.1 Bayes Rule

The task to estimate parameters or to test hypotheses can be simply formulated as a problem of decision theory. This is obvious for the test of hypotheses where one has to decide between a null hypothesis and an alternative hypothesis. When estimating parameters one has to decide on the estimates themselves.

To solve a problem, different actions are possible, and one has to decide for one. The decision needs to be judged, because one should know whether the decision was appropriate. This depends on the state of the system for which the decision has to be made. Let the system be represented by the continuous random vector $\boldsymbol{x}$ of the unknown parameters. Let data with information on the system be available. They are collected in the random vector $\boldsymbol{y}$. To make a decision, a *decision rule* $\delta(\boldsymbol{y})$ is established. It determines which action is started depending on the available data $\boldsymbol{y}$.

The loss which occurs with the action triggered by $\delta(\boldsymbol{y})$ is used as a criterion to judge the decision. Depending on $\boldsymbol{x}$ and $\delta(\boldsymbol{y})$ the *loss function* $L(\boldsymbol{x}, \delta(\boldsymbol{y}))$ is therefore introduced. The expected loss will be considered which is obtained by forming the expected value of the loss function. Let $p(\boldsymbol{x}|\boldsymbol{y}, C)$ be the posterior density function for the parameter vector $\boldsymbol{x}$ which follows from Bayes' theorem (2.122) with the prior density function $p(\boldsymbol{x}|C)$ for $\boldsymbol{x}$ and with the likelihood function $p(\boldsymbol{y}|\boldsymbol{x}, C)$ where C denotes the information on the system. The *posterior expected loss* which is computed by the posterior density function $p(\boldsymbol{x}|\boldsymbol{y}, C)$ is obtained with (2.144) by

$$E[L(\boldsymbol{x}, \delta(\boldsymbol{y}))] = \int_{\mathcal{X}} L(\boldsymbol{x}, \delta(\boldsymbol{y}))p(\boldsymbol{x}|\boldsymbol{y}, C)d\boldsymbol{x} \tag{3.1}$$

where the parameter space $\mathcal{X}$ introduced in connection with Bayes' theorem (2.122) denotes the domain for integrating the values $\boldsymbol{x}$ of the unknown parameters. A very reasonable and plausible choice for the decision rule $\delta(\boldsymbol{y})$ is such that the posterior expected loss (3.1) attains a minimum. This is called *Bayes rule*.

Bayes rule may be also established like this. In traditional statistics not the unknown parameters $\boldsymbol{x}$ but the data $\boldsymbol{y}$ are random variables, as already mentioned in Chapter 2.2.8. The expected loss is therefore computed by the likelihood function $p(\boldsymbol{y}|\boldsymbol{x}, C)$ and is called the *risk function* $R(\boldsymbol{x}, \delta)$

$$R(\boldsymbol{x}, \delta) = \int_{\mathcal{Y}} L(\boldsymbol{x}, \delta(\boldsymbol{y}))p(\boldsymbol{y}|\boldsymbol{x}, C)d\boldsymbol{y} \tag{3.2}$$

where $\mathcal{Y}$ denotes the domain for integrating the random vector $\boldsymbol{y}$ of observations.

The risk function depends on the unknown parameters $\boldsymbol{x}$. To obtain the expected loss, one therefore has to average also over the values of $\boldsymbol{x}$. By the prior density function $p(\boldsymbol{x}|C)$ for the unknown parameters $\boldsymbol{x}$ we compute the so-called *Bayes risk* by

$$r(\delta) = \int_{\mathcal{X}} R(\boldsymbol{x}, \delta)p(\boldsymbol{x}|C)d\boldsymbol{x} \ . \tag{3.3}$$

Bayes rule minimizes Bayes risk because of

$$r(\delta) = \int_{\mathcal{Y}} E[L(\boldsymbol{x}, \delta(\boldsymbol{y}))]p(\boldsymbol{y}|C)d\boldsymbol{y} \ . \tag{3.4}$$

Thus, $r(\delta)$ is minimized, if $E[L(\boldsymbol{x}, \delta(\boldsymbol{y}))]$ from (3.1) attains a minimum. The function $p(\boldsymbol{y}|C)$ is according to (2.128) the marginal density function for $\boldsymbol{y}$. The relation (3.4) follows from (3.3) with (2.121), (3.1) and (3.2)

$$\begin{aligned}
r(\delta) &= \int_{\mathcal{X}} \int_{\mathcal{Y}} L(\boldsymbol{x}, \delta(\boldsymbol{y}))p(\boldsymbol{y}|\boldsymbol{x}, C)p(\boldsymbol{x}|C)d\boldsymbol{y}d\boldsymbol{x} \\
&= \int_{\mathcal{Y}} \big[\int_{\mathcal{X}} L(\boldsymbol{x}, \delta(\boldsymbol{y}))p(\boldsymbol{x}|\boldsymbol{y}, C)d\boldsymbol{x}\big]p(\boldsymbol{y}|C)d\boldsymbol{y} \\
&= \int_{\mathcal{Y}} E[L(\boldsymbol{x}, \delta(\boldsymbol{y}))]p(\boldsymbol{y}|C)d\boldsymbol{y} \ .
\end{aligned}$$

In the following we will assume that decisions have to be made in a system which is represented by the complete vector $\boldsymbol{x}$ of unknown parameters. However, if with $\boldsymbol{x} = |\boldsymbol{x}_1', \boldsymbol{x}_2'|'$ and with the parameter spaces $\mathcal{X}_1 \subset \mathcal{X}$ and

$\mathcal{X}_2 \subset \mathcal{X}$ the vector $\boldsymbol{x}$ is partitioned into the vectors $\boldsymbol{x}_1 \in \mathcal{X}_1$ and $\boldsymbol{x}_2 \in \mathcal{X}_2$ and if one is interested only in the vector $\boldsymbol{x}_1$ of unknown parameters, the posterior marginal density function $p(\boldsymbol{x}_1|\boldsymbol{y}, C)$ for $\boldsymbol{x}_1$ from (2.91) needs to be computed with

$$p(\boldsymbol{x}_1|\boldsymbol{y}, C) = \int_{\mathcal{X}_2} p(\boldsymbol{x}_1, \boldsymbol{x}_2|\boldsymbol{y}, C)d\boldsymbol{x}_2 \ . \tag{3.5}$$

It is used in (3.1) instead of the posterior density function $p(\boldsymbol{x}|\boldsymbol{y}, C) = p(\boldsymbol{x}_1, \boldsymbol{x}_2|\boldsymbol{y}, C)$.

3.2 Point Estimation

Values for the vector $\boldsymbol{x}$ of unknown parameters shall be estimated. This is called *point estimation* in contrast to the estimation of confidence regions for the unknown parameters to be dealt with in Chapter 3.3. Bayes rule is applied. Possible decisions are represented by possible estimates $\hat{\boldsymbol{x}}$ of the unknown parameters $\boldsymbol{x}$ which are obtained by the observations $\boldsymbol{y}$, thus $\delta(\boldsymbol{y}) = \hat{\boldsymbol{x}}$. The true state of the system is characterized by the true values $\boldsymbol{x}$ of the unknown parameters. The loss is a function of the estimates $\hat{\boldsymbol{x}}$ and of the true values $\boldsymbol{x}$ of the parameters. The loss function $L(\boldsymbol{x}, \hat{\boldsymbol{x}})$ must express how good or how bad the estimate $\hat{\boldsymbol{x}}$ is. It therefore has to increase for bad estimates. Three different loss functions are introduced in the following.

3.2.1 Quadratic Loss Function

A simple loss function results from the sum of squares $(\boldsymbol{x} - \hat{\boldsymbol{x}})'(\boldsymbol{x} - \hat{\boldsymbol{x}})$ of the errors $\boldsymbol{x} - \hat{\boldsymbol{x}}$ of the estimates $\hat{\boldsymbol{x}}$ of the unknown parameters $\boldsymbol{x}$. This sum of squares is generalized by means of the covariance matrix $D(\boldsymbol{x}) = \boldsymbol{\Sigma}$ defined in (2.157) for the random vector $\boldsymbol{x}$. It shall be positive definite. Its inverse $\boldsymbol{\Sigma}^{-1}$ is according to (2.159) proportional to the weight matrix of $\boldsymbol{x}$. Thus, the quadratic loss function

$$L(\boldsymbol{x}, \hat{\boldsymbol{x}}) = (\boldsymbol{x} - \hat{\boldsymbol{x}})'\boldsymbol{\Sigma}^{-1}(\boldsymbol{x} - \hat{\boldsymbol{x}}) \tag{3.6}$$

is chosen where the squares of the errors $\boldsymbol{x} - \hat{\boldsymbol{x}}$ are weighted by $\boldsymbol{\Sigma}^{-1}$. This loss function leads to the well known method of least squares, as will be shown in Chapter 4.2.2.

To determine the posterior expected loss (3.1) of the quadratic loss function (3.6), the expected value of (3.6) has to be computed by the posterior density function $p(\boldsymbol{x}|\boldsymbol{y}, C)$ for $\boldsymbol{x}$. We obtain

$$E[L(\boldsymbol{x}, \hat{\boldsymbol{x}})] = \int_{\mathcal{X}} (\boldsymbol{x} - \hat{\boldsymbol{x}})'\boldsymbol{\Sigma}^{-1}(\boldsymbol{x} - \hat{\boldsymbol{x}})p(\boldsymbol{x}|\boldsymbol{y}, C)d\boldsymbol{x} \tag{3.7}$$

where $\mathcal{X}$ denotes again the parameter space over which the values of $\boldsymbol{x}$ have to be integrated. With (2.146) and the following identity we obtain the expected value of the loss function by

$$\begin{aligned}
E[L(\boldsymbol{x}, \hat{\boldsymbol{x}})] &= E[(\boldsymbol{x} - \hat{\boldsymbol{x}})'\boldsymbol{\Sigma}^{-1}(\boldsymbol{x} - \hat{\boldsymbol{x}})] \\
&= E\{[\boldsymbol{x} - E(\boldsymbol{x}) - (\hat{\boldsymbol{x}} - E(\boldsymbol{x}))]'\boldsymbol{\Sigma}^{-1}[\boldsymbol{x} - E(\boldsymbol{x}) - (\hat{\boldsymbol{x}} - E(\boldsymbol{x}))]\} \\
&= E[(\boldsymbol{x} - E(\boldsymbol{x}))'\boldsymbol{\Sigma}^{-1}(\boldsymbol{x} - E(\boldsymbol{x}))] + (\hat{\boldsymbol{x}} - E(\boldsymbol{x}))'\boldsymbol{\Sigma}^{-1}(\hat{\boldsymbol{x}} - E(\boldsymbol{x})) \quad (3.8)
\end{aligned}$$

because of

$$2E[(\boldsymbol{x} - E(\boldsymbol{x}))']\boldsymbol{\Sigma}^{-1}(\hat{\boldsymbol{x}} - E(\boldsymbol{x})) = 0 \quad \text{with} \quad E[\boldsymbol{x} - E(\boldsymbol{x})] = \boldsymbol{0} \ .$$

The first term on the right-hand side of (3.8) does not depend on the estimate $\hat{\boldsymbol{x}}$, while the second one attains a minimum for

$$\hat{\boldsymbol{x}}_B = E(\boldsymbol{x}|\boldsymbol{y}) \ , \tag{3.9}$$

since $\boldsymbol{\Sigma}^{-1}$ is positive definite. It is stressed by the notation $E(\boldsymbol{x}|\boldsymbol{y})$, as already mentioned in connection with (2.225), that the expected value is computed according to (3.7) with the posterior density function for $\boldsymbol{x}$ whose values for $\boldsymbol{y}$ are given.

Thus, Bayes rule leads to the estimate $\hat{\boldsymbol{x}}_B$ which is called *Bayes estimate*, if the quadratic loss function (3.6) is applied. By forming the expected value as in (3.7) the Bayes estimate $\hat{\boldsymbol{x}}_B$ follows with (3.9) by

$$\hat{\boldsymbol{x}}_B = \int_X \boldsymbol{x} p(\boldsymbol{x}|\boldsymbol{y}, C) d\boldsymbol{x} \ . \tag{3.10}$$

As was already pointed out in Chapter 2.2.8, the vector $\boldsymbol{x}$ of unknown parameters is a random vector. Its Bayes estimate $\hat{\boldsymbol{x}}_B$ from (3.9) or (3.10) is a fixed quantity. The same holds true for the median $\hat{\boldsymbol{x}}_{med}$ of $\boldsymbol{x}$ in (3.21) and the MAP estimate $\hat{\boldsymbol{x}}_M$ of $\boldsymbol{x}$ in (3.30). In traditional statistics it is the opposite, the vector of unknown parameters is a fixed quantity, while its estimate is a random variable.

To express the accuracy of the Bayes estimate $\hat{\boldsymbol{x}}_B$, the covariance matrix $D(\boldsymbol{x}|\boldsymbol{y})$ for the vector $\boldsymbol{x}$ of unknown parameters is introduced with (2.150) and (2.157) by the posterior density function $p(\boldsymbol{x}|\boldsymbol{y}, C)$. The covariance matrix expresses because of (3.9) the dispersion of $\boldsymbol{x}$ about the Bayes estimate $\hat{\boldsymbol{x}}_B$. It is therefore representative for the accuracy of the estimate

$$\begin{aligned}
D(\boldsymbol{x}|\boldsymbol{y}) &= E[(\boldsymbol{x} - E(\boldsymbol{x}|\boldsymbol{y}))(\boldsymbol{x} - E(\boldsymbol{x}|\boldsymbol{y}))'] \\
&= \int_X (\boldsymbol{x} - \hat{\boldsymbol{x}}_B)(\boldsymbol{x} - \hat{\boldsymbol{x}}_B)' p(\boldsymbol{x}|\boldsymbol{y}, C) d\boldsymbol{x} \ . \tag{3.11}
\end{aligned}$$

The posterior expected loss (3.1) follows for the Bayes estimate $\hat{\boldsymbol{x}}_B$ from (3.8) with (3.9) and (3.11) by

$$\begin{aligned}
E[L(\boldsymbol{x}, \hat{\boldsymbol{x}}_B)] &= E\{\text{tr}[\boldsymbol{\Sigma}^{-1}(\boldsymbol{x} - \hat{\boldsymbol{x}}_B)(\boldsymbol{x} - \hat{\boldsymbol{x}}_B)']\} \\
&= \text{tr}[\boldsymbol{\Sigma}^{-1}D(\boldsymbol{x}|\boldsymbol{y})] \ . \tag{3.12}
\end{aligned}$$

Example: In the example to (2.227) the posterior distribution for the unknown quantity s, which is measured n times and for which prior information is available, was derived as the normal distribution (2.232)

$$s|\boldsymbol{y} \sim N(\mu_{0s}, \sigma^2 \sigma_{0s}^2)$$

with μ_{0s} and σ_{0s}^2 from (2.230) and (2.231). The Bayes estimate $\hat{s}_B$ of the quantity s therefore follows because of (2.167) from (3.9) by

$$\hat{s}_B = \mu_{0s} \tag{3.13}$$

and the variance $V(s|\boldsymbol{y})$ of s from (3.11) by

$$V(s|\boldsymbol{y}) = \sigma^2 \sigma_{0s}^2 \ . \tag{3.14}$$

$\boldsymbol{\Delta}$

The Bayes estimate distributes large errors of individual observations because of the weighted sum of squares of the loss function (3.6). This effect is disadvantageous, if outliers are suspected in the observations and need to be detected. This will be dealt with in Chapter 4.2.5.

3.2.2 Loss Function of the Absolute Errors

To diminish the effects of large errors $\boldsymbol{x} - \hat{\boldsymbol{x}}$, the absolute value of errors is introduced instead of the squares of errors as loss function. With $\boldsymbol{x} = (x_i)$, $\hat{\boldsymbol{x}} = (\hat{x}_i)$ and $i \in \{1, \ldots, u\}$ we obtain the loss function

$$L(x_i, \hat{x}_i) = |x_i - \hat{x}_i| \ . \tag{3.15}$$

Bayes rule requires that the posterior expected loss (3.1)

$$E[L(x_i, \hat{x}_i)] = \int_{\mathcal{X}} |x_i - \hat{x}_i| p(\boldsymbol{x}|\boldsymbol{y}, C) d\boldsymbol{x} \tag{3.16}$$

attains a minimum.

Let the parameter space $\mathcal{X}$ over which one has to integrate be defined by the intervals

$$x_{li} < x_i < x_{ri} \quad \text{for} \quad i \in \{1, \ldots, u\} \ . \tag{3.17}$$

We then obtain with (2.74) instead of (3.16), since $|x_i - \hat{x}_i|$ is positive,

$$
\begin{aligned}
E[L(x_i, \hat{x}_i)] &= \int_{x_{lu}}^{\hat{x}_u} \ldots \int_{x_{l1}}^{\hat{x}_1} (\hat{x}_i - x_i) p(\boldsymbol{x}|\boldsymbol{y}, C) dx_1 \ldots dx_u \\
&\quad + \int_{\hat{x}_u}^{x_{ru}} \ldots \int_{\hat{x}_1}^{x_{r1}} (x_i - \hat{x}_i) p(\boldsymbol{x}|\boldsymbol{y}, C) dx_1 \ldots dx_u \\
&= \hat{x}_i F(\hat{\boldsymbol{x}}) - \int_{x_{lu}}^{\hat{x}_u} \ldots \int_{x_{l1}}^{\hat{x}_1} x_i p(\boldsymbol{x}|\boldsymbol{y}, C) dx_1 \ldots dx_u \\
&\quad + \int_{\hat{x}_u}^{x_{ru}} \ldots \int_{\hat{x}_1}^{x_{r1}} x_i p(\boldsymbol{x}|\boldsymbol{y}, C) dx_1 \ldots dx_u - \hat{x}_i(1 - F(\hat{\boldsymbol{x}}))
\end{aligned} \tag{3.18}
$$

where $F(\hat{\boldsymbol{x}})$ with

$$F(\hat{\boldsymbol{x}}) = \int_{x_{lu}}^{\hat{x}_u} \ldots \int_{x_{l1}}^{\hat{x}_1} p(\boldsymbol{x}|\boldsymbol{y}, C)dx_1 \ldots dx_u \tag{3.19}$$

denotes the distribution function defined by (2.73) for the values $\hat{\boldsymbol{x}}$. To find the minimum of (3.18), we differentiate with respect to $\hat{x}_i$ and set the derivatives equal to zero. Since the derivatives with respect to the limits $\hat{x}_i$ of the integrals in (3.18) and (3.19) cancel, we obtain

$$\partial E[L(x_i, \hat{x}_i)]/\partial \hat{x}_i = F(\hat{\boldsymbol{x}}) - 1 + F(\hat{\boldsymbol{x}}) = 0 \tag{3.20}$$

or

$$F(\hat{\boldsymbol{x}}_{\mathrm{med}}) = 0.5 \; . \tag{3.21}$$

Thus, if the absolute error (3.15) is introduced as loss function, Bayes rule gives the estimate $\hat{\boldsymbol{x}}_{\mathrm{med}}$ which is also called the *median* of the posterior density function $p(\boldsymbol{x}|\boldsymbol{y}, C)$. The median is determined by the value for which the distribution function (3.19) is equal to 0.5. The median minimizes (3.16), since the second derivatives $\partial^2 E[L(x_i, \hat{x}_i)]/\partial \hat{x}_i^2$ are positive, as becomes obvious from (3.20). The application of the loss function (3.15) for the search of outliers in the observations is dealt with in Chapter 4.2.5.

Example: In the example to (2.227) the posterior distribution of an unknown quantity s, which is determined by n measurements $\boldsymbol{y} = |y_1, y_2, \ldots, y_n|'$ and by prior information, was obtained by the normal distribution

$$s|\boldsymbol{y} \sim N(\mu_{0s}, \sigma^2 \sigma_{0s}^2)$$

with μ_{0s} and σ_{0s}^2 from (2.230) and (2.231). By the transformation (2.169) of the random variable s into the random variable y with

$$y = (s - \mu_{0s})/(\sigma \sigma_{0s}) \tag{3.22}$$

the standard normal distribution follows for y

$$y \sim N(0, 1) \; .$$

Because of the symmetry of the standard normal distribution we obtain with (2.170)

$$F(0; 0, 1) = 0.5 \; .$$

Thus, the median $\hat{y}_{\mathrm{med}}$ of y follows with

$$\hat{y}_{\mathrm{med}} = 0$$

and from (3.22) the median $\hat{s}_{\mathrm{med}}$ of s with

$$\hat{s}_{\mathrm{med}} = \mu_{0s} \; . \tag{3.23}$$

This estimate is for the example identical with the Bayes estimate (3.13). The reason is the symmetry of the normal distribution for $s|\boldsymbol{y}$.

Let the quantity s now be a discrete random variable with the n values $y_1, y_2, \ldots, y_n$ of the measurements. Let the discrete densities be identical with

$$p(y_i|C) = \frac{1}{n} \quad \text{for} \quad i \in \{1, \ldots, n\} \tag{3.24}$$

because of (2.47). If the data are ordered according to increasing magnitude

$$y_1 \leq y_2 \leq \cdots \leq y_n \;,$$

the median $\hat{s}_{\mathrm{med}}$ of the quantity s follows with (2.50) by

$$\hat{s}_{\mathrm{med}} = y_{(n+1)/2} \;, \tag{3.25}$$

if n is odd, and

$$y_{n/2} \leq \hat{s}_{\mathrm{med}} \leq y_{n/2+1} \;, \tag{3.26}$$

if n is even. For $y_{n/2} < y_{n/2+1}$ the estimate is nonunique, and one generally chooses

$$\hat{s}_{\mathrm{med}} = \frac{1}{2}(y_{n/2} + y_{n/2+1}) \;, \tag{3.27}$$

if n is even.

As will be shown in Chapter 4.2.5, the results (3.25) and (3.26) are also obtained by a direct application of the loss function (3.15). $\triangle$

3.2.3   Zero-One Loss

Zero-one loss means costs or no costs. It is suited for the estimation but also for hypothesis tests, as will be shown in Chapter 3.4. It leads to the loss function

$$L(x_i, \hat{x}_i) = \begin{cases} 0 & \text{for} \quad |x_i - \hat{x}_i| < b \\ a & \text{for} \quad |x_i - \hat{x}_i| \geq b \end{cases} \tag{3.28}$$

where a and b with $b > 0$ mean constants and where $a = 1$ may be set without restricting the generality. Bayes rule requires that the posterior expected loss (3.1) attains a minimum. By introducing the intervals (3.17) in order to define the parameter space for the integration, we obtain with (2.74)

$$\begin{aligned}
E[L(x_i, \hat{x}_i)] &= \int_{x_{lu}}^{\hat{x}_u - b} \cdots \int_{x_{l1}}^{\hat{x}_1 - b} p(\boldsymbol{x}|\boldsymbol{y}, C) dx_1 \ldots dx_u \\
&\quad + \int_{\hat{x}_u + b}^{x_{ru}} \cdots \int_{\hat{x}_1 + b}^{x_{r1}} p(\boldsymbol{x}|\boldsymbol{y}, C) dx_1 \ldots dx_u \\
&= 1 - \int_{\hat{x}_u - b}^{\hat{x}_u + b} \cdots \int_{\hat{x}_1 - b}^{\hat{x}_1 + b} p(\boldsymbol{x}|\boldsymbol{y}, C) dx_1 \ldots dx_u \;.
\end{aligned}$$

The posterior expected loss $E[L(x_i, \hat{x}_i)]$ is minimized, if

$$\int_{\hat{x}_u-b}^{\hat{x}_u+b} \dots \int_{\hat{x}_1-b}^{\hat{x}_1+b} p(\boldsymbol{x}|\boldsymbol{y}, C)dx_1 \dots dx_u \tag{3.29}$$

attains a maximum.

The special case $b \to 0$ only will be discussed. Thus, the estimate $\hat{\boldsymbol{x}}_M$ of $\boldsymbol{x}$ follows from (3.29) as value which maximizes the posterior density function $p(\boldsymbol{x}|\boldsymbol{y}, C)$

$$\hat{\boldsymbol{x}}_M = \arg\max_{\boldsymbol{x}} p(\boldsymbol{x}|\boldsymbol{y}, C) \ . \tag{3.30}$$

We call $\hat{\boldsymbol{x}}_M$ the *MAP estimate*, i.e. the *maximum a posteriori estimate*.

Example: In the example to (2.227) the normal distribution (2.232) was derived as posterior distribution for the unknown quantity s

$$s|\boldsymbol{y} \sim N(\mu_{0s}, \sigma^2 \sigma_{0s}^2)$$

whose density function is given according to (2.166) by

$$p(s|\boldsymbol{y}) = \frac{1}{\sqrt{2\pi}\sigma\sigma_{0s}} e^{-\frac{(s-\mu_{0s})^2}{2\sigma^2\sigma_{0s}^2}} \ . \tag{3.31}$$

The MAP estimate $\hat{s}_M$ of the quantity s follows according to (3.30) with the value μ_{0s} for which the density function (3.31) becomes maximal. This result may be also obtained by differentiating the density function with respect to s and by setting the derivative equal to zero. This gives

$$\frac{1}{\sqrt{2\pi}\sigma\sigma_{0s}} \exp\left[-\frac{(s-\mu_{0s})^2}{2\sigma^2\sigma_{0s}^2}\right]\left(-\frac{s-\mu_{0s}}{\sigma^2\sigma_{0s}^2}\right) = 0$$

and again the solution

$$\hat{s}_M = \mu_{0s} \ . \tag{3.32}$$

The MAP estimate for this example agrees with the Bayes estimate (3.13) and the median (3.23). The reason is the symmetric normal distribution for $s|\boldsymbol{y}$, as already mentioned in connection with (3.23). $\triangle$

If the prior density function $p(\boldsymbol{x}|C)$, with which the posterior density function $p(\boldsymbol{x}|\boldsymbol{y}, C)$ follows from Bayes' theorem (2.122), is a constant, as mentioned with (2.216), the *maximum-likelihood estimate* $\hat{\boldsymbol{x}}_{ML}$ of the vector $\boldsymbol{x}$ of unknown parameters

$$\hat{\boldsymbol{x}}_{ML} = \arg\max_{\boldsymbol{x}} p(\boldsymbol{y}|\boldsymbol{x}, C) \tag{3.33}$$

is obtained instead of the MAP estimate. It is determined by the value for which the likelihood function $p(\boldsymbol{y}|\boldsymbol{x}, C)$ in Bayes' theorem (2.122) becomes maximal. The maximum-likelihood estimate is often applied in traditional statistics.

3.3 Estimation of Confidence Regions

In contrast to the point estimation of Chapter 3.2, for which estimates of
the values of the vector x of unknown parameters are sought, a region shall
now be determined in which the parameter vector x is situated with a given
probability. This problem can be immediately solved by Bayesian statistics,
since the parameter vector x is a random vector with a probability density
function, whereas in the traditional statistics the estimation of confidence
regions is generally derived via hypothesis testing.

3.3.1 Confidence Regions

By means of the posterior density function $p(x|y, C)$ for the parameter vector
x from Bayes' theorem (2.122) the probability is determined according to
(2.71)

$$P(x \in \mathcal{X}_u|y, C) = \int_{\mathcal{X}_u} p(x|y, C)dx \tag{3.34}$$

that the vector x belongs to the subspace $\mathcal{X}_u$ of the parameter space $\mathcal{X}$ with
$\mathcal{X}_u \subset \mathcal{X}$. One is often interested to find the subspace where most of the
probability, for instance 95%, is concentrated. Given a probability there are
obviously many possibilities to establish such a subspace. A region of values
for x within the subspace, however, should be more probable than a region of
equal size outside the subspace. It will be therefore required that the density
of each point within the subspace is equal to or greater than the density of
a point outside the subspace. The region of highest posterior density, also
called H.P.D. region, is thus obtained.

Let $p(x|y, C)$ be the posterior density function for the vector x of un-
known parameters, the subspace $\mathcal{X}_B$ with $\mathcal{X}_B \subset \mathcal{X}$ is then called a $1 - \alpha$
H.P.D. region, Bayesian confidence region or shortly *confidence region*, if

$$P(x \in \mathcal{X}_B|y, C) = \int_{\mathcal{X}_B} p(x|y, C)dx = 1 - \alpha$$

and

$$p(x_1|y, C) \geq p(x_2|y, C) \quad \text{for} \quad x_1 \in \mathcal{X}_B \,, \; x_2 \notin \mathcal{X}_B \,. \tag{3.35}$$

If the vector x contains only the random variable X as component, the *confidence interval* for X is defined by (3.35). The value for α is generally chosen
to be $\alpha = 0.05$, but also $\alpha = 0.1$ or $\alpha = 0.01$ are selected.

If the posterior density function $p(x|y, C)$ is constant in the area of the
boundary of the confidence region (3.35), the boundary cannot be determined
uniquely. But the confidence region has the property that its hypervolume is
minimal in comparison to any region with probability mass $1 - \alpha$. To show

this, let such a region be denoted by $\mathcal{X}_B'$. We then find with (3.35)

$$\int_{\mathcal{X}_B} p(\boldsymbol{x}|\boldsymbol{y}, C)d\boldsymbol{x} = \int_{\mathcal{X}_B'} p(\boldsymbol{x}|\boldsymbol{y}, C)d\boldsymbol{x} = 1 - \alpha \ .$$

If the integration over the intersection $\mathcal{X}_B \cap \mathcal{X}_B'$ is eliminated, we obtain with the complements $\bar{\mathcal{X}}_B = \mathcal{X} \setminus \mathcal{X}_B$ and $\bar{\mathcal{X}}_B' = \mathcal{X} \setminus \mathcal{X}_B'$

$$\int_{\mathcal{X}_B \cap \bar{\mathcal{X}}_B'} p(\boldsymbol{x}|\boldsymbol{y}, C)d\boldsymbol{x} = \int_{\mathcal{X}_B' \cap \bar{\mathcal{X}}_B} p(\boldsymbol{x}|\boldsymbol{y}, C)d\boldsymbol{x} \ .$$

The confidence region $\mathcal{X}_B$ fulfills $p(\boldsymbol{x}_1|\boldsymbol{y}, C) \geq p(\boldsymbol{x}_2|\boldsymbol{y}, C)$ for $\boldsymbol{x}_1 \in \mathcal{X}_B \cap \bar{\mathcal{X}}_B'$ and $\boldsymbol{x}_2 \in \mathcal{X}_B' \cap \bar{\mathcal{X}}_B$ because of (3.35). Thus,

$$\mathrm{hypervolume}_{\mathcal{X}_B \cap \bar{\mathcal{X}}_B'} \leq \mathrm{hypervolume}_{\mathcal{X}_B' \cap \bar{\mathcal{X}}_B}$$

follows. If the hypervolume of $\mathcal{X}_B \cap \mathcal{X}_B'$ is added to both sides, we finally get

$$\mathrm{hypervolume}_{\mathcal{X}_B} \leq \mathrm{hypervolume}_{\mathcal{X}_B'} \ .$$

Example: Let the posterior density function for the unknown $u \times 1$ parameter vector $\boldsymbol{x}$ be determined by the normal distribution $N(\boldsymbol{\mu}, \boldsymbol{\Sigma})$ with the density function $p(\boldsymbol{x}|\boldsymbol{\mu}, \boldsymbol{\Sigma})$ from (2.195). A hypersurface of equal density is determined by the relation

$$(\boldsymbol{x} - \boldsymbol{\mu})'\boldsymbol{\Sigma}^{-1}(\boldsymbol{x} - \boldsymbol{\mu}) = \mathrm{const}$$

which follows from the exponent of (2.195). It has the shape of a hyperell-ipsoid with the center at $\boldsymbol{\mu}$, see for instance KOCH (1999, p.298). A *confidence hyperellipsoid* is therefore obtained. The density function $p(\boldsymbol{x}|\boldsymbol{\mu}, \boldsymbol{\Sigma})$ is a monotonically decreasing function of the quadratic form $(\boldsymbol{x} - \boldsymbol{\mu})'\boldsymbol{\Sigma}^{-1}(\boldsymbol{x} - \boldsymbol{\mu})$. It has according to (2.180) and (2.202) the χ^2-distribution $\chi^2(u)$ with u as parameter. The $1 - \alpha$ confidence hyperellipsoid for the parameter vector $\boldsymbol{x}$ is therefore determined according to (3.35) by

$$P\big((\boldsymbol{x} - \boldsymbol{\mu})'\boldsymbol{\Sigma}^{-1}(\boldsymbol{x} - \boldsymbol{\mu}) < \chi_{1-\alpha;u}^2\big) = 1 - \alpha$$

and its shape by

$$(\boldsymbol{x} - \boldsymbol{\mu})'\boldsymbol{\Sigma}^{-1}(\boldsymbol{x} - \boldsymbol{\mu}) = \chi_{1-\alpha;u}^2 \tag{3.36}$$

where $\chi_{1-\alpha;u}^2$ denotes the upper α-percentage point of the χ^2-distribution defined in (2.181). The orientation of the axes of the confidence hyperellipsoid is obtained by the matrix $\boldsymbol{C}$ of the eigenvectors of $\boldsymbol{\Sigma}$ with

$$\boldsymbol{C}'\boldsymbol{C} = \boldsymbol{I} \quad \text{and therefore} \quad \boldsymbol{C}' = \boldsymbol{C}^{-1}, \ \boldsymbol{C} = (\boldsymbol{C}')^{-1} \ . \tag{3.37}$$

The semi-axes c_i of the confidence hyperellipsoid follow with the matrix $\boldsymbol{\Lambda}$ of eigenvalues of $\boldsymbol{\Sigma}$

$$\boldsymbol{C'\Sigma C} = \boldsymbol{\Lambda} \quad \text{and} \quad \boldsymbol{\Lambda} = \text{diag}(\lambda_1, \ldots, \lambda_u) \tag{3.38}$$

by

$$c_i = (\lambda_i \chi^2_{1-\alpha;u})^{1/2} \quad \text{for} \quad i \in \{1, \ldots, u\} \,. \tag{3.39}$$

Δ

3.3.2 Boundary of a Confidence Region

One has to ascertain for one method of testing hypotheses, whether a point $\boldsymbol{x}_0$ lies within the confidence region $\mathcal{X}_B$ defined by (3.35). This is the case, if the inequality

$$p(\boldsymbol{x}_0|\boldsymbol{y}, C) > p(\boldsymbol{x}_B|\boldsymbol{y}, C) \tag{3.40}$$

is fulfilled where $\boldsymbol{x}_B$ denotes a point at the boundary of the confidence region $\mathcal{X}_B$. If its density p_B is introduced by

$$p_B = p(\boldsymbol{x}_B|\boldsymbol{y}, C) \,, \tag{3.41}$$

a point $\boldsymbol{x}_0$ lies within the confidence region $\mathcal{X}_B$, if

$$p(\boldsymbol{x}_0|\boldsymbol{y}, C) > p_B \,. \tag{3.42}$$

Example: In the example to (3.35) a $1 - \alpha$ confidence hyperellipsoid was determined by (3.36). The density p_B of a point at the boundary of the confidence hyperellipsoid follows with the density function (2.195) of the normal distribution by

$$p_B = \frac{1}{(2\pi)^{u/2}(\det \boldsymbol{\Sigma})^{1/2}} \exp\left(-\frac{1}{2}\chi^2_{1-\alpha;u}\right) \,. \tag{3.43}$$

If this result is substituted on the right-hand side of (3.42) and if on the left-hand side the density function of the normal distribution is introduced, one obtains instead of (3.42)

$$(\boldsymbol{x}_0 - \boldsymbol{\mu})'\boldsymbol{\Sigma}^{-1}(\boldsymbol{x}_0 - \boldsymbol{\mu}) < \chi^2_{1-\alpha;u} \,. \tag{3.44}$$

If this inequality is fulfilled for a point $\boldsymbol{x}_0$, it is situated within the confidence hyperellipsoid. Δ

3.4 Hypothesis Testing

Propositions concerning the unknown parameters may be formulated as hypotheses, and methods for deciding whether to accept or to reject the hypotheses are called *hypothesis tests*.

3.4.1   Different Hypotheses

Let $\mathcal{X}_0 \subset \mathcal{X}$ and $\mathcal{X}_1 \subset \mathcal{X}$ be subspaces of the parameter space $\mathcal{X}$ and let $\mathcal{X}_0$ and $\mathcal{X}_1$ be disjoint, i.e. $\mathcal{X}_0 \cap \mathcal{X}_1 = \emptyset$. The assumption that the vector $\boldsymbol{x}$ of unknown parameters is element of the subspace $\mathcal{X}_0$ is called the *null hypothesis* H_0 and the assumption that $\boldsymbol{x}$ is element of $\mathcal{X}_1$ the *alternative* hypothesis H_1. The null hypothesis H_0 is tested against the alternative hypothesis H_1, i.e.

$$H_0 : \boldsymbol{x} \in \mathcal{X}_0 \quad \text{versus} \quad H_1 : \boldsymbol{x} \in \mathcal{X}_1 \; . \tag{3.45}$$

The null hypothesis and the alternative hpothesis are mutually exclusive because of $\mathcal{X}_0 \cap \mathcal{X}_1 = \emptyset$.

Often $\mathcal{X}_1$ is the complement of $\mathcal{X}_0$, thus

$$\mathcal{X}_1 = \mathcal{X} \setminus \mathcal{X}_0 \; . \tag{3.46}$$

The hypothesis is then exhaustive which means either H_0 or H_1 is true.

It is assumed that the subspaces $\mathcal{X}_0$ and $\mathcal{X}_1$ contain more than one vector, respectively. We therefore call (3.45) a *composite hypothesis* contrary to the *simple hypothesis*

$$H_0 : \boldsymbol{x} = \boldsymbol{x}_0 \quad \text{versus} \quad H_1 : \boldsymbol{x} = \boldsymbol{x}_1 \tag{3.47}$$

where the subspace $\mathcal{X}_0$ only contains the given vector $\boldsymbol{x}_0$ and the subspace $\mathcal{X}_1$ only the given vector $\boldsymbol{x}_1$.

If there is only the vector $\boldsymbol{x}_0$ in $\mathcal{X}_0$ and if $\mathcal{X}_1$ according to (3.46) is the complement of $\mathcal{X}_0$, the *point null hypothesis* follows

$$H_0 : \boldsymbol{x} = \boldsymbol{x}_0 \quad \text{versus} \quad H_1 : \boldsymbol{x} \neq \boldsymbol{x}_0 \; . \tag{3.48}$$

If instead of the parameter vector $\boldsymbol{x}$ itself a linear transformation $\boldsymbol{Hx}$ of $\boldsymbol{x}$ shall be tested where $\boldsymbol{H}$ denotes a given matrix of constants, we formulate instead of (3.45)

$$H_0 : \boldsymbol{Hx} \in \mathcal{X}_{H1} \quad \text{versus} \quad H_1 : \boldsymbol{Hx} \in \mathcal{X}_{H2} \tag{3.49}$$

where $\mathcal{X}_{H1}$ and $\mathcal{X}_{H2}$ denote subspaces of the parameter space $\mathcal{X}_H$ of the transformed parameters $\boldsymbol{Hx}$. The corresponding point null hypothesis follows with the given vector $\boldsymbol{w}$ by

$$H_0 : \boldsymbol{Hx} = \boldsymbol{w} \quad \text{versus} \quad H_1 : \boldsymbol{Hx} \neq \boldsymbol{w} \; . \tag{3.50}$$

This hypothesisis is often tested in the linear model, as will be shown in the Chapters 4.2.1, 4.2.6, 4.3.1 and 4.3.2.

The point null hypothesis (3.48) or (3.50) is not always realistic, since the information to be tested might be better described by a small region around the given point $\boldsymbol{x}_0$ or $\boldsymbol{w}$ rather than by the identity with $\boldsymbol{x}_0$ or $\boldsymbol{w}$. Thus, the composite hypothesis (3.45) or (3.49) should have been formulated. As a consequence of a nonrealistic point null hypothesis the hypothesis test of traditional statistics reacts too sensitive, i.e. the null hypothesis is rejected, although additional information not entering the test speaks against it. By means of Bayesian statitistics less sensitive tests can be derived (KOCH 1990, p.88; RIESMEIER 1984).

3.4.2 Test of Hypotheses

Let the two mutually exclusive and exhaustive hypotheses (3.45) satisfying (3.46) be tested. By means of Bayes rule explained in Chapter 3.1 we will decide whether to accept the null hypothesis H_0 or the alternative hypothesis H_1. The system for which the decision needs to be made is characterized by the two states $\boldsymbol{x} \in \mathcal{X}_0$ or $\boldsymbol{x} \in \mathcal{X}_1$ which trigger the two actions accept H_0 or accept H_1. Four values of the loss function need to be defined, and it is reasonable to work with the zero-one loss (3.28), which introduces no costs for the correct decision. We obtain

$$L(\boldsymbol{x} \in \mathcal{X}_i, H_i) = 0 \quad \text{for} \quad i \in \{0,1\}$$
$$L(\boldsymbol{x} \in \mathcal{X}_i, H_j) \neq 0 \quad \text{for} \quad i,j \in \{0,1\}, \ i \neq j \ . \tag{3.51}$$

Thus, for the correct decision to accept H_0, if $\boldsymbol{x} \in \mathcal{X}_0$ is valid, and H_1, if $\boldsymbol{x} \in \mathcal{X}_1$, the loss function is equal to zero, and for the wrong decision it is unequal to zero.

The expected value $E[L(H_0)]$ of the loss function for accepting H_0 is computed according to (3.1) with the discrete posterior density functions $p(H_0|\boldsymbol{y}, C)$ and $p(H_1|\boldsymbol{y}, C)$ for H_0 and H_1 given the data $\boldsymbol{y}$ and the information C about the system by

$$E[L(H_0)] = p(H_0|\boldsymbol{y}, C)L(\boldsymbol{x} \in \mathcal{X}_0, H_0) + p(H_1|\boldsymbol{y}, C)L(\boldsymbol{x} \in \mathcal{X}_1, H_0)$$
$$= p(H_1|\boldsymbol{y}, C)L(\boldsymbol{x} \in \mathcal{X}_1, H_0) \tag{3.52}$$

because of (3.51). Correspondingly, the posterior expected loss for accepting H_1 follows with

$$E[L(H_1)] = p(H_0|\boldsymbol{y}, C)L(\boldsymbol{x} \in \mathcal{X}_0, H_1) + p(H_1|\boldsymbol{y}, C)L(\boldsymbol{x} \in \mathcal{X}_1, H_1)$$
$$= p(H_0|\boldsymbol{y}, C)L(\boldsymbol{x} \in \mathcal{X}_0, H_1) \ . \tag{3.53}$$

Bayes rule requires to make the decision which minimizes this posterior loss. Thus, the null hypothesis H_0 is accepted, if

$$E[L(H_0)] < E[L(H_1)]$$

holds true. This means after substituting (3.52) and (3.53), if

$$\frac{p(H_0|\boldsymbol{y}, C)L(\boldsymbol{x} \in \mathcal{X}_0, H_1)}{p(H_1|\boldsymbol{y}, C)L(\boldsymbol{x} \in \mathcal{X}_1, H_0)} > 1 \ , \ \text{accept} \quad H_0 \ . \tag{3.54}$$

Otherwise accept H_1.

This decision is also valid, if not the hypothesis (3.45) in connection with (3.46) is tested but the general hypothesis (3.45). To the two states $x \in \mathcal{X}_0$ or $x \in \mathcal{X}_1$ of the system, for which the decision is needed, the state $x \in \overline{\mathcal{X}_0 \cup \mathcal{X}_1} = \mathcal{X} \setminus (\mathcal{X}_0 \cup \mathcal{X}_1)$ has then to be added which is determined by the complement of $\mathcal{X}_0 \cup \mathcal{X}_1$. In addition to the two actions accept H_0 or accept H_1, the action $\bar{H}$ needs to be considered that neither H_0 nor H_1 are accepted. But this action is not investigated so that only the two values have to be added to the loss function (3.51)

$$L(x \in \overline{\mathcal{X}_0 \cup \mathcal{X}_1}, H_i) = a \quad \text{for} \quad i \in \{0,1\} \tag{3.55}$$

which are assumed to be identical and have the constant magnitude a. The posterior expected loss for accepting H_0 or H_1 is then computed with the posterior density function for H_0 and H_1 and with $p(\bar{H}|y,C)$ for $\bar{H}$ instead of (3.52) by

$$
\begin{aligned}
E[L(H_0)] &= p(H_0|y,C)L(x \in \mathcal{X}_0, H_0) + p(H_1|y,C)L(x \in \mathcal{X}_1, H_0) \\
&\quad + p(\bar{H}|y,C)L(x \in \overline{\mathcal{X}_0 \cup \mathcal{X}_1}, H_0) \\
&= p(H_1|y,C)L(x \in \mathcal{X}_1, H_0) + ap(\bar{H}|y,C)
\end{aligned}
\tag{3.56}
$$

and instead of (3.53) by

$$
\begin{aligned}
E[L(H_1)] &= p(H_0|y,C)L(x \in \mathcal{X}_0, H_1) + p(H_1|y,C)L(x \in \mathcal{X}_1, H_1) \\
&\quad + p(\bar{H}|y,C)L(x \in \overline{\mathcal{X}_0 \cup \mathcal{X}_1}, H_1) \\
&= p(H_0|y,C)L(x \in \mathcal{X}_0, H_1) + ap(\bar{H}|y,C) \ .
\end{aligned}
\tag{3.57}
$$

Bayes rule requires to accept the null hypothesis H_0, if

$$E[L(H_0)] < E[L(H_1)]$$

which leads to the decision (3.54).

If wrong decisions obtain identical losses which will be assumed for the following, Bayes rule requires according to (3.54), if

$$\frac{p(H_0|y,C)}{p(H_1|y,C)} > 1 \ , \ \text{accept} \quad H_0 \ . \tag{3.58}$$

Otherwise accept H_1. The ratio V

$$V = \frac{p(H_0|y,C)}{p(H_1|y,C)} \tag{3.59}$$

in (3.58) is called the *posterior odds ratio* for H_0 to H_1.

The posterior density functions $p(H_0|y,C)$ and $p(H_1|y,C)$ in (3.52) and (3.53) are discrete density functions. They express according to (2.46) probabilities. If (3.46) holds, either H_0 or H_1 is true and one obtains $p(H_0|y,C) +$

$p(H_1|\boldsymbol{y},C) = 1$ from (2.22). The probability that H_0 is true or that H_1 is true is then computed with (3.59) by

$$P(H_0|\boldsymbol{y},C) = \frac{V}{1+V} \quad \text{and} \quad P(H_1|\boldsymbol{y},C) = \frac{1}{1+V} \ . \tag{3.60}$$

The hypotheses are formulated for the unknown parameters $\boldsymbol{x}$. The discrete posterior density functions in (3.58) for H_0 and H_1 may therefore be determined by the posterior density function for the parameters $\boldsymbol{x}$. Since $p(H_0|\boldsymbol{y},C)$ and $p(H_1|\boldsymbol{y},C)$ express probabilities according to (2.46), we obtain with (2.71)

$$p(H_i|\boldsymbol{y},C) = \int_{\mathcal{X}_i} p(\boldsymbol{x}|\boldsymbol{y},C)d\boldsymbol{x} \quad \text{for} \quad i \in \{0,1\} \ . \tag{3.61}$$

For the test of the composite hypothesis (3.45)

$$H_0 : \boldsymbol{x} \in \mathcal{X}_0 \quad \text{versus} \quad H_1 : \boldsymbol{x} \in \mathcal{X}_1$$

Bayes rule therefore leads according to (3.58) to the decision, if the posterior odds ratio

$$\frac{\int_{\mathcal{X}_0} p(\boldsymbol{x}|\boldsymbol{y},C)d\boldsymbol{x}}{\int_{\mathcal{X}_1} p(\boldsymbol{x}|\boldsymbol{y},C)d\boldsymbol{x}} > 1 \ , \quad \text{accept} \quad H_0 \ . \tag{3.62}$$

Otherwise accept H_1.

To test the point null hypothesis (3.48) by means of (3.62), the subspace $\mathcal{X}_0$ has to shrink to the point $\boldsymbol{x}_0$. This, however, gives $p(H_0|\boldsymbol{y},C) = 0$ from (3.61), since the posterior density function is assumed to be continuous, and furthermore $p(H_1|\boldsymbol{y},C) \neq 0$. The point null hypothesis (3.48) therefore cannot be tested unless a special prior density function is introduced for the hypotheses, as explained in the following Chapter 3.4.3.

In order to test the simple hypothesis (3.47), we let the subspaces $\mathcal{X}_0$ and $\mathcal{X}_1$ shrink to the points $\boldsymbol{x}_0$ and $\boldsymbol{x}_1$ where the posterior density function $p(\boldsymbol{x}|\boldsymbol{y},C)$ may be considered as constant and obtain instead of (3.62)

$$\frac{\lim_{\Delta\mathcal{X}_0 \to 0} \int_{\Delta\mathcal{X}_0} p(\boldsymbol{x}|\boldsymbol{y},C)d\boldsymbol{x}/\Delta\mathcal{X}_0}{\lim_{\Delta\mathcal{X}_1 \to 0} \int_{\Delta\mathcal{X}_1} p(\boldsymbol{x}|\boldsymbol{y},C)d\boldsymbol{x}/\Delta\mathcal{X}_1} = \frac{p(\boldsymbol{x}_0|\boldsymbol{y},C)}{p(\boldsymbol{x}_1|\boldsymbol{y},C)} \tag{3.63}$$

where $\Delta\mathcal{X}_0$ and $\Delta\mathcal{X}_1$, respectively, with $\Delta\mathcal{X}_0 = \Delta\mathcal{X}_1$ represent small spaces around $\boldsymbol{x}_0$ and $\boldsymbol{x}_1$. Thus, we find for the test of the simple hypothesis (3.47)

$$H_0 : \boldsymbol{x} = \boldsymbol{x}_0 \quad \text{versus} \quad H_1 : \boldsymbol{x} = \boldsymbol{x}_1$$

according to (3.62), if

$$\frac{p(\boldsymbol{x}_0|\boldsymbol{y},C)}{p(\boldsymbol{x}_1|\boldsymbol{y},C)} > 1 \ , \quad \text{accept} \quad H_0 \ . \tag{3.64}$$

Otherwise accept H_1.

Example: In the example to (2.227) the normal distribution (2.232)

$$s|\boldsymbol{y} \sim N(\mu_{0s}, \sigma^2\sigma_{0s}^2)$$

was obtained as posterior distribution for the unknown quantity s. The simple hypothesis

$$H_0 : s = s_0 \quad \text{versus} \quad H_1 : s = s_1$$

shall be tested. With the density function (2.166) of the normal distribution we obtain according to (3.64), if

$$\frac{\exp\left(-\frac{1}{2\sigma^2\sigma_{0s}^2}(s_0 - \mu_{0s})^2\right)}{\exp\left(-\frac{1}{2\sigma^2\sigma_{0s}^2}(s_1 - \mu_{0s})^2\right)} > 1 \ , \ \text{accept} \quad H_0 \ .$$

Thus, if $|s_0 - \mu_{0s}| < |s_1 - \mu_{0s}|$ holds true, then H_0 is accepted, otherwise H_1.

$$\triangle$$

3.4.3   Special Priors for Hypotheses

The tests of hypotheses treated in the preceding chapter are based according to (3.61) on the computation of the posterior density functions for the hypotheses by the posterior density functions for the parameters. Special prior density functions shall now be associated with the hypotheses. If the posterior density function $p(H_0|\boldsymbol{y}, C)$ in (3.58) for the null hypothesis is expressed by Bayes' theorem (2.122), we obtain

$$p(H_0|\boldsymbol{y}, C) = \frac{p(H_0|C)p(\boldsymbol{y}|H_0, C)}{p(\boldsymbol{y}|C)} \tag{3.65}$$

where $p(H_0|C)$ denotes the prior density function for the hypothesis. With a corresponding expression for the posterior density function $p(H_1|\boldsymbol{y}, C)$ the posterior odds ratio V follows from (3.59) with

$$V = \frac{p(H_0|\boldsymbol{y}, C)}{p(H_1|\boldsymbol{y}, C)} = \frac{p(H_0|C)p(\boldsymbol{y}|H_0, C)}{p(H_1|C)p(\boldsymbol{y}|H_1, C)} \ . \tag{3.66}$$

The ratio B of this relation

$$B = \frac{p(\boldsymbol{y}|H_0, C)}{p(\boldsymbol{y}|H_1, C)} \tag{3.67}$$

is called *Bayes factor*. It expresses the change of the ratio $p(H_0|C)/p(H_1|C)$ of the prior density functions for the hypotheses by the data $\boldsymbol{y}$, since with (3.66) we have

$$B = \frac{p(H_0|\boldsymbol{y}, C)/p(H_1|\boldsymbol{y}, C)}{p(H_0|C)/p(H_1|C)} \ . \tag{3.68}$$

The hypotheses are formulated for the unknown parameters $\boldsymbol{x}$. Special prior density functions are therefore associated with the hypotheses by introducing a special prior density function $p(\boldsymbol{x}|C)$ for the unknown parameters $\boldsymbol{x}$ with

$$p(\boldsymbol{x}|C) = \begin{cases} p(H_0|C)p_0(\boldsymbol{x}|C) & \text{for} \quad \boldsymbol{x} \in \mathcal{X}_0 \\ p(H_1|C)p_1(\boldsymbol{x}|C) & \text{for} \quad \boldsymbol{x} \in \mathcal{X}_1 \,. \end{cases} \tag{3.69}$$

The density functions $p_0(\boldsymbol{x}|C)$ and $p_1(\boldsymbol{x}|C)$ are defined in the subspaces $\mathcal{X}_0$ and $\mathcal{X}_1$ of the hypotheses H_0 and H_1. They fulfill (2.74) and describe the manner, how the density functions are distributed over the subspaces. If $\mathcal{X}_1$ forms with (3.46) the complement of $\mathcal{X}_0$, we get

$$p(\boldsymbol{x}|C) = \begin{cases} p(H_0|C)p_0(\boldsymbol{x}|C) & \text{for} \quad \boldsymbol{x} \in \mathcal{X}_0 \\ (1 - p(H_0|C))p_1(\boldsymbol{x}|C) & \text{for} \quad \boldsymbol{x} \in \mathcal{X} \setminus \mathcal{X}_0 \,, \end{cases} \tag{3.70}$$

since the condition $\int_{\mathcal{X}} p(\boldsymbol{x}|C)d\boldsymbol{x} = 1$ from (2.74) has also to be fulfilled.

We express in the posterior odds ratio (3.66) the posterior density function for the hypotheses as in (3.61) by the posterior density functions of the unknown parameters $\boldsymbol{x}$ with the prior density function (3.69) and the likelihood function. We then obtain for testing the composite hypothesis (3.45)

$$H_0 : \boldsymbol{x} \in \mathcal{X}_0 \quad \text{versus} \quad H_1 : \boldsymbol{x} \in \mathcal{X}_1 \,,$$

if the posterior odds ratio

$$\frac{p(H_0|C) \int_{\mathcal{X}_0} p_0(\boldsymbol{x}|C)p(\boldsymbol{y}|\boldsymbol{x}, C)d\boldsymbol{x}}{p(H_1|C) \int_{\mathcal{X}_1} p_1(\boldsymbol{x}|C)p(\boldsymbol{y}|\boldsymbol{x}, C)d\boldsymbol{x}} > 1 \,, \text{ accept } H_0 \,. \tag{3.71}$$

Otherwise accept H_1. The Bayes factor B follows for this test with (3.66) and (3.68) from (3.71) by

$$B = \frac{\int_{\mathcal{X}_0} p_0(\boldsymbol{x}|C)p(\boldsymbol{y}|\boldsymbol{x}, C)d\boldsymbol{x}}{\int_{\mathcal{X}_1} p_1(\boldsymbol{x}|C)p(\boldsymbol{y}|\boldsymbol{x}, C)d\boldsymbol{x}} \,. \tag{3.72}$$

To test the simple hypothesis (3.47)

$$H_0 : \boldsymbol{x} = \boldsymbol{x}_0 \quad \text{versus} \quad H_1 : \boldsymbol{x} = \boldsymbol{x}_1 \,,$$

we let again as in (3.63) the two subspaces $\mathcal{X}_0$ and $\mathcal{X}_1$ shrink to the points $\boldsymbol{x}_0$ and $\boldsymbol{x}_1$. Because of $\int_{\mathcal{X}_0} p_0(\boldsymbol{x}|C)d\boldsymbol{x} = 1$ and $\int_{\mathcal{X}_1} p_1(\boldsymbol{x}|C)d\boldsymbol{x} = 1$ from (2.74) we obtain instead of (3.71), if

$$\frac{p(H_0|C)p(\boldsymbol{y}|\boldsymbol{x}_0, C)}{p(H_1|C)p(\boldsymbol{y}|\boldsymbol{x}_1, C)} > 1 \,, \text{ accept } H_0 \,. \tag{3.73}$$

Otherwise accept H_1. For testing the point null hypothesis (3.48)

$$H_0 : \boldsymbol{x} = \boldsymbol{x}_0 \quad \text{versus} \quad H_1 : \boldsymbol{x} \neq \boldsymbol{x}_0$$

we finally get with (3.70) instead of (3.71), if the posterior odds ratio

$$\frac{p(H_0|C)p(\boldsymbol{y}|\boldsymbol{x}_0,C)}{(1-p(H_0|C))\int_{\{\boldsymbol{x}\neq\boldsymbol{x}_0\}} p_1(\boldsymbol{x}|C)p(\boldsymbol{y}|\boldsymbol{x},C)d\boldsymbol{x}} > 1 \text{ , accept } H_0 \text{ . } \quad (3.74)$$

Otherwise accept H_1. The two Bayes factors for the tests (3.73) and (3.74) follow again corresponding to (3.72).

If a point null hypothesis is tested with (3.74), results may follow which do not agree with the ones of traditional statistics. This happens, if the prior density function $p_1(\boldsymbol{x}|C)$ is spread out considerably because of a large variance resulting from an uncertain prior information. Then, the likelihood function averaged by the integral in (3.74) over the space of the alternative hypothesis becomes smaller than the likelihood function $p(\boldsymbol{y}|\boldsymbol{x}_0,C)$ for the null hypothesis. Thus, the null hypothesis is accepted although a test of traditional statistics may reject it. This discrepancy was first detected by LINDLEY (1957), and it is therefore called *Lindley's paradox*, see also BERGER (1985, p.156).

Example: Let the point null hypothesis

$$H_0 : \boldsymbol{\beta} = \boldsymbol{\beta}_0 \quad \text{versus} \quad H_1 : \boldsymbol{\beta} \neq \boldsymbol{\beta}_0 \text{ ,} \qquad (3.75)$$

where $\boldsymbol{\beta}$ denotes a $u \times 1$ vector of unknown parameters and $\boldsymbol{\beta}_0$ a $u \times 1$ vector of given values, be tested by the posterior odds ratio (3.74). Let the likelihood function be determined by the normal distribution (2.225) where $\boldsymbol{y}$ denotes the $n \times 1$ vector of observations in a linear model. Let the prior density function for $\boldsymbol{\beta}$ be determined by the normal distribution (2.226). The likelihood function in (3.74) then follows with the density function (2.195) of the normal distribution by

$$p(\boldsymbol{y}|\boldsymbol{\beta}_0,C) = \frac{1}{(2\pi)^{n/2}(\det \sigma^2 \boldsymbol{P}^{-1})^{1/2}}$$
$$\exp\left\{-\frac{1}{2\sigma^2}(\boldsymbol{y}-\boldsymbol{X}\boldsymbol{\beta}_0)'\boldsymbol{P}(\boldsymbol{y}-\boldsymbol{X}\boldsymbol{\beta}_0)\right\} \text{ . } \quad (3.76)$$

The exponent is transformed as in (2.228) and one obtains

$$(\boldsymbol{y}-\boldsymbol{X}\boldsymbol{\beta}_0)'\boldsymbol{P}(\boldsymbol{y}-\boldsymbol{X}\boldsymbol{\beta}_0) = \boldsymbol{y}'\boldsymbol{P}\boldsymbol{y} - \bar{\boldsymbol{\mu}}'\boldsymbol{X}'\boldsymbol{P}\boldsymbol{X}\bar{\boldsymbol{\mu}}$$
$$+ (\boldsymbol{\beta}_0 - \bar{\boldsymbol{\mu}})'\boldsymbol{X}'\boldsymbol{P}\boldsymbol{X}(\boldsymbol{\beta}_0 - \bar{\boldsymbol{\mu}}) \quad (3.77)$$

with

$$\bar{\boldsymbol{\mu}} = (\boldsymbol{X}'\boldsymbol{P}\boldsymbol{X})^{-1}\boldsymbol{X}'\boldsymbol{P}\boldsymbol{y} \text{ . }$$

The prior density function from (2.226) is determined by

$$p_1(\boldsymbol{\beta}|C) = \frac{1}{(2\pi)^{u/2}(\det \sigma^2 \boldsymbol{\Sigma})^{1/2}} \exp\left\{-\frac{1}{2\sigma^2}(\boldsymbol{\beta}-\boldsymbol{\mu})'\boldsymbol{\Sigma}^{-1}(\boldsymbol{\beta}-\boldsymbol{\mu})\right\} \text{ . }$$

The posterior density function for $\boldsymbol{\beta}$ then follows with

$$p_1(\boldsymbol{\beta}|C)p(\boldsymbol{y}|\boldsymbol{\beta},C) = \frac{1}{(2\pi)^{(n+u)/2}(\det \sigma^2\boldsymbol{\Sigma} \det \sigma^2\boldsymbol{P}^{-1})^{1/2}}$$

$$\exp\left\{-\frac{1}{2\sigma^2}[(\boldsymbol{\beta}-\boldsymbol{\mu})'\boldsymbol{\Sigma}^{-1}(\boldsymbol{\beta}-\boldsymbol{\mu})+(\boldsymbol{y}-\boldsymbol{X}\boldsymbol{\beta})'\boldsymbol{P}(\boldsymbol{y}-\boldsymbol{X}\boldsymbol{\beta})]\right\} . \quad (3.78)$$

The exponent is transformed as in (2.228)

$$\begin{aligned}
(\boldsymbol{\beta}-\boldsymbol{\mu})'\boldsymbol{\Sigma}^{-1}&(\boldsymbol{\beta}-\boldsymbol{\mu})+(\boldsymbol{y}-\boldsymbol{X}\boldsymbol{\beta})'\boldsymbol{P}(\boldsymbol{y}-\boldsymbol{X}\boldsymbol{\beta})\\
&= \boldsymbol{y}'\boldsymbol{P}\boldsymbol{y}+\boldsymbol{\mu}'\boldsymbol{\Sigma}^{-1}\boldsymbol{\mu}-\boldsymbol{\mu}_0'(\boldsymbol{X}'\boldsymbol{P}\boldsymbol{X}+\boldsymbol{\Sigma}^{-1})\boldsymbol{\mu}_0\\
&\quad +(\boldsymbol{\beta}-\boldsymbol{\mu}_0)'(\boldsymbol{X}'\boldsymbol{P}\boldsymbol{X}+\boldsymbol{\Sigma}^{-1})(\boldsymbol{\beta}-\boldsymbol{\mu}_0)
\end{aligned} \quad (3.79)$$

with

$$\boldsymbol{\mu}_0 = (\boldsymbol{X}'\boldsymbol{P}\boldsymbol{X}+\boldsymbol{\Sigma}^{-1})^{-1}(\boldsymbol{X}'\boldsymbol{P}\boldsymbol{y}+\boldsymbol{\Sigma}^{-1}\boldsymbol{\mu}) .$$

The posterior density function for $\boldsymbol{\beta}$ has to be integrated to find the posterior odds ratio (3.74). The point $\boldsymbol{\beta}_0$ to be excluded is of no concern, since the posterior density function is continuous. After substituting (3.79) in (3.78) we find

$$\frac{1}{(2\pi)^{u/2}[\det \sigma^2(\boldsymbol{X}'\boldsymbol{P}\boldsymbol{X}+\boldsymbol{\Sigma}^{-1})^{-1}]^{1/2}}$$

$$\int_{-\infty}^{\infty}\cdots\int_{-\infty}^{\infty}\exp\left\{-\frac{1}{2\sigma^2}(\boldsymbol{\beta}-\boldsymbol{\mu}_0)'(\boldsymbol{X}'\boldsymbol{P}\boldsymbol{X}+\boldsymbol{\Sigma}^{-1})(\boldsymbol{\beta}-\boldsymbol{\mu}_0)\right\}d\boldsymbol{\beta} = 1$$
$$(3.80)$$

where the constants are chosen according to (2.195) such that the integration gives the value one. Then instead of (3.74), the decision follows with (3.76) to (3.80), if the posterior odds ratio

$$\frac{p(H_0|C)[\det \boldsymbol{\Sigma}]^{1/2}}{(1-p(H_0|C))[\det(\boldsymbol{X}'\boldsymbol{P}\boldsymbol{X}+\boldsymbol{\Sigma}^{-1})^{-1}]^{1/2}}$$

$$\exp\left\{-\frac{1}{2\sigma^2}[(\boldsymbol{\beta}_0-\bar{\boldsymbol{\mu}})'\boldsymbol{X}'\boldsymbol{P}\boldsymbol{X}(\boldsymbol{\beta}_0-\bar{\boldsymbol{\mu}})-\bar{\boldsymbol{\mu}}'\boldsymbol{X}'\boldsymbol{P}\boldsymbol{X}\bar{\boldsymbol{\mu}}\right.$$

$$\left.-\boldsymbol{\mu}'\boldsymbol{\Sigma}^{-1}\boldsymbol{\mu}+\boldsymbol{\mu}_0'(\boldsymbol{X}'\boldsymbol{P}\boldsymbol{X}+\boldsymbol{\Sigma}^{-1})\boldsymbol{\mu}_0]\right\} > 1 , \text{ accept } H_0 . \quad (3.81)$$

With large variances for the prior information $\det \boldsymbol{\Sigma}$ becomes large so that the posterior odds ratio (3.81) becomes greater than one. This was already pointed out in connection with (3.74).

To apply (3.81), the point null hypothesis

$$H_0 : s = s_0 \quad \text{versus} \quad H_1 : s \neq s_0$$

shall be tested for the example to (2.227) in addition to the hypothesis of the example to (3.64). With $\boldsymbol{X} = |1,1,\ldots,1|', \boldsymbol{\beta} = s, \boldsymbol{P} = \boldsymbol{I}, \boldsymbol{\mu} = \mu_s, \boldsymbol{\Sigma} =$

$\sigma_s^2, \boldsymbol{\beta}_0 = s_0$ we obtain

$$\bar{\boldsymbol{\mu}} = \bar{s} = \frac{1}{n}\sum_{i=1}^{n} y_i$$

as well as $\boldsymbol{\mu}_0 = \mu_{0s}$ and σ_{0s}^2 from (2.230) and (2.231). The posterior odds ratio (3.81) then follows with

$$\frac{p(H_0|C)}{1 - p(H_0|C)}\left(\frac{\sigma_s^2 + 1/n}{1/n}\right)^{1/2}$$
$$\exp\left\{-\frac{1}{2}\left[\frac{(s_0 - \bar{s})^2}{\sigma^2/n} - \frac{\bar{s}^2}{\sigma^2/n} - \frac{\mu_s^2}{\sigma^2\sigma_s^2} + \frac{\mu_{0s}^2}{\sigma^2\sigma_{0s}^2}\right]\right\}.$$

Furthermore we get

$$\frac{\mu_{0s}^2}{\sigma^2\sigma_{0s}^2} = \frac{1}{\sigma^2\sigma_s^2 + \sigma^2/n}\left(n\sigma_s^2\bar{s}^2 + 2\bar{s}\mu_s + \frac{\mu_s^2}{n\sigma_s^2}\right)$$

and therefore the decision, if

$$\frac{p(H_0|C)}{1 - p(H_0|C)}\left(\frac{\sigma_s^2 + 1/n}{1/n}\right)^{1/2}$$
$$\exp\left\{-\frac{1}{2}\left[\frac{(s_0 - \bar{s})^2}{\sigma^2/n} - \frac{(\mu_s - \bar{s})^2}{\sigma^2\sigma_s^2 + \sigma^2/n}\right]\right\} > 1\,,$$

accept H_0.

It is obvious like in (3.81) that for large values of σ_s^2 the decision is reached in favor of the null hypothesis. Δ

3.4.4   Test of the Point Null Hypothesis by Confidence Regions

Lindley's paradox of the preceding chapter indicates a test of a point null hypothesis by Bayesian statistics whose results need not agree with the ones of traditional statistics. In general, different decisions have to be expected, since the hypotheses are differently treated in Bayesian statistics and traditional statistics. In traditional statistics the null hypothesis is as long maintained as knowledge from the data speaks against it. In Bayesian statistics, however, the null and the alternative hypothesis are treated equally. If according to (3.58) the posterior density function of the null hypothesis becomes larger than the one of the alternative hypothesis, the null hypothesis is accepted, or if it is smaller, the alternative hypothesis.

However, if one proceeds when testing hypotheses as in traditional statistics, one gets by Bayesian statistics the test procedures of traditional statistics. For instance, the point null hypothesis (3.48) in the linear model of traditional statistics for the vector of unknown parameters or the more general point null hypothesis (3.50) for a vector of linear functions of the parameters are accepted, if the point of the null hypothesis lies within the $1 - \alpha$ confidence region for the parameter vector or for the vector of linear functions of the parameters, see for instance KOCH (1999, p.301).

The problem to decide whether a point $\boldsymbol{x}_0$ lies within a confidence region was already solved by the inequality (3.42)

$$p(\boldsymbol{x}_0|\boldsymbol{y}, C) > p_B$$

where p_B denotes the posterior density of a point at the boundary of the confidence region for the parameter vector $\boldsymbol{x}$. If the inequality is fulfilled, the point lies within the confidence region. Thus, for the test of the point null hypothesis (3.48)

$$H_0 : \boldsymbol{x} = \boldsymbol{x}_0 \quad \text{versus} \quad H_1 : \boldsymbol{x} \neq \boldsymbol{x}_0$$

the density p_B of a point $\boldsymbol{x}_B$ at the boundary of the $1-\alpha$ confidence region $\mathcal{X}_B$ for the parameter vector $\boldsymbol{x}$ is needed according to (3.41). The test procedure of traditional statistics is obtained with (3.42), i.e. if

$$p(\boldsymbol{x}_0|\boldsymbol{y}, C) > p_B \ , \ \text{accept} \quad H_0 \tag{3.82}$$

with a significance level of α. Otherwise H_0 is rejected. Correspondingly, the density of a point at the boundary of the confidence region for a linear transformation of the parameters has to be introduced for the test of the general point null hypothesis (3.50). As was already shown with (3.44) and as will be demonstrated with (4.26), (4.98), (4.145) and (4.175), the density p_B can be replaced in the linear model by an upper percentage point. Thus, the test procedures of traditional statistics are readily obtained.

The test procedure (3.82) of traditional statistics for the point null hypothesis (3.48) may be also derived as a test of a hypothesis of Bayesian statistics which is different from (3.48). The test of the simple hypothesis

$$H_0 : \boldsymbol{x} = \boldsymbol{x}_0 \quad \text{versus} \quad H_1 : \boldsymbol{x} = \boldsymbol{x}_B \ , \tag{3.83}$$

where $\boldsymbol{x}_B$ again denotes a point at the boundary of the confidence region $\mathcal{X}_B$ for the parameter vector $\boldsymbol{x}$, leads with (3.64) to the decision, if

$$\frac{p(\boldsymbol{x}_0|\boldsymbol{y}, C)}{p_B} > 1 \ , \ \text{accept} \quad H_0 \ .$$

Otherwise accept H_1. This decision is identical with (3.82).

4 Linear Model

Measurements are taken and data are collected to gain information about unknown parameters. In order to estimate the values of the unknown parameters by the methods explained in Chapter 3.2, the functional relations between the unknown parameters and the observations need to be defined and the statistical properties of the observations have to be specified. These definitions determine the model of the data analysis.

Frequently linear relations are given between the unknown parameters and the observations which lead to a linear model. Nonlinear relations may in general be transferred, as will be shown in Chapter 4.1, by a linearization into a linear model. The linear model is therefore often applied and will be explained in detail in the following.

4.1 Definition and Likelihood Function

Let $\boldsymbol{X}$ be an $n \times u$ matrix of given coefficients with full column rank, i.e. rank$\boldsymbol{X} = u$, $\boldsymbol{\beta}$ a $u \times 1$ random vector of unknown parameters, $\boldsymbol{y}$ an $n \times 1$ random vector of observations, $D(\boldsymbol{y}|\sigma^2) = \sigma^2 \boldsymbol{P}^{-1}$ the $n \times n$ covariance matrix of $\boldsymbol{y}$, σ^2 the unknown random variable which is called *variance factor* or *variance of unit weight* and $\boldsymbol{P}$ the known positive definite weight matrix of the observations. Then

$$\boldsymbol{X}\boldsymbol{\beta} = E(\boldsymbol{y}|\boldsymbol{\beta}) \quad \text{with} \quad D(\boldsymbol{y}|\sigma^2) = \sigma^2 \boldsymbol{P}^{-1} \tag{4.1}$$

is called a *linear model*.

Out of reasons which were explained in Chapter 2.4.1 the observations $\boldsymbol{y}$ are assumed as normally distributed so that with (2.196) and (4.1)

$$\boldsymbol{y}|\boldsymbol{\beta}, \sigma^2 \sim N(\boldsymbol{X}\boldsymbol{\beta}, \sigma^2 \boldsymbol{P}^{-1}) \tag{4.2}$$

follows. The expected values of the observations, i.e. the mean values or the "true" values of the observations, are expressed in (4.1) as linear combinations $\boldsymbol{X}\boldsymbol{\beta}$ of the unknown parameters $\boldsymbol{\beta}$ subject to the condition that the values for $\boldsymbol{\beta}$ are given. This is indicated, as already mentioned in connection with (2.225), by the notation $E(\boldsymbol{y}|\boldsymbol{\beta})$. In the same way $\boldsymbol{\beta}$ and σ^2 in $D(\boldsymbol{y}|\sigma^2)$ and in $\boldsymbol{y}|\boldsymbol{\beta}, \sigma^2$ denote given values.

Because of rank$\boldsymbol{X} = u$ we have $n \geq u$. But one should strive for keeping the number n of observations larger than the number u of unknown parameters in order to diminish the influence of the variances and covariances of the observations $\boldsymbol{y}$ on the estimates of the unknown parameters $\boldsymbol{\beta}$. If, however, prior information is introduced for the unknown parameters $\boldsymbol{\beta}$ as in Chapters 4.2.6, 4.2.7, 4.3.2 and 4.4.2, $n < u$ may be valid.

For $n > u$ the system $\boldsymbol{X\beta} = \boldsymbol{y}$ of equations is in general not consistent. By adding the $n \times 1$ random vector $\boldsymbol{e}$ of *errors* of the observations the consistent system

$$\boldsymbol{X\beta} = \boldsymbol{y} + \boldsymbol{e} \quad \text{with} \quad E(\boldsymbol{e}|\boldsymbol{\beta}) = \boldsymbol{0}$$
$$\text{and} \quad D(\boldsymbol{e}|\boldsymbol{\beta},\sigma^2) = D(\boldsymbol{y}|\sigma^2) = \sigma^2 \boldsymbol{P}^{-1} \quad (4.3)$$

is obtained, since $E(\boldsymbol{e}|\boldsymbol{\beta}) = \boldsymbol{0}$ follows with $E(\boldsymbol{y}|\boldsymbol{\beta}) = \boldsymbol{X\beta}$ from (4.1) and $D(\boldsymbol{e}|\boldsymbol{\beta},\sigma^2) = \sigma^2 \boldsymbol{P}^{-1}$ with $\boldsymbol{e} = -\boldsymbol{y} + \boldsymbol{X\beta}$ from (2.158). An alternative formulation of the model (4.1) is therefore obtained with (4.3). The equations $\boldsymbol{X\beta} = E(\boldsymbol{y}|\boldsymbol{\beta}) = \boldsymbol{y} + \boldsymbol{e}$ are called *observation equations*.

If the positive definite covariance matrix of $\boldsymbol{y}$ is denoted by $\boldsymbol{\Sigma}$, $D(\boldsymbol{y}|\sigma^2) = \boldsymbol{\Sigma} = \sigma^2 \boldsymbol{P}^{-1}$ or $\boldsymbol{P} = \sigma^2 \boldsymbol{\Sigma}^{-1}$ follows from (4.1). Thus, $c = \sigma^2$ has been set in the definition (2.159) of the weight matrix $\boldsymbol{P}$ to obtain the covariance matrix $D(\boldsymbol{y}|\sigma^2)$ in (4.1). With $\boldsymbol{P} = \boldsymbol{I}$ follows $D(\boldsymbol{y}|\sigma^2) = \sigma^2 \boldsymbol{I}$. The variances of the observations then result from the variance factor σ^2.

Example 1: Let the unknown quantity s again be measured n times like in the example to (2.227) so that the observation vector $\boldsymbol{y} = |y_1, y_2, \ldots, y_n|'$ is obtained. Let the observations be independent and have different weights p_i for $i \in \{1, \ldots, n\}$ defined in (2.160). The following observation equations with the errors e_i and the variances $V(y_i|\sigma^2)$ are then obtained

$$s = E(y_1|s) = y_1 + e_1 \quad \text{with} \quad V(y_1|\sigma^2) = \sigma^2/p_1$$
$$s = E(y_2|s) = y_2 + e_2 \quad \text{with} \quad V(y_2|\sigma^2) = \sigma^2/p_2$$
$$\ldots\ldots\ldots\ldots\ldots\ldots\ldots\ldots\ldots\ldots\ldots\ldots\ldots\ldots\ldots\ldots\ldots$$
$$s = E(y_n|s) = y_n + e_n \quad \text{with} \quad V(y_n|\sigma^2) = \sigma^2/p_n \ .$$

By setting $\boldsymbol{X} = |1, \ldots, 1|'$, $\boldsymbol{e} = |e_1, \ldots, e_n|'$ and $\boldsymbol{P}^{-1} = \mathrm{diag}(1/p_1, \ldots, 1/p_n)$ the linear model (4.1) follows or its alternative formulation (4.3). Δ

Since the observations $\boldsymbol{y}$ are normally distributed according to (4.2), the likelihood function $p(\boldsymbol{y}|\boldsymbol{\beta},\sigma^2)$ follows with (2.195) by

$$p(\boldsymbol{y}|\boldsymbol{\beta},\sigma^2) = \frac{1}{(2\pi)^{n/2}(\det \sigma^2 \boldsymbol{P}^{-1})^{1/2}} e^{-\frac{1}{2\sigma^2}(\boldsymbol{y} - \boldsymbol{X\beta})'\boldsymbol{P}(\boldsymbol{y} - \boldsymbol{X\beta})}$$

or

$$p(\boldsymbol{y}|\boldsymbol{\beta},\sigma^2) = \frac{1}{(2\pi\sigma^2)^{n/2}(\det \boldsymbol{P})^{-1/2}} e^{-\frac{1}{2\sigma^2}(\boldsymbol{y} - \boldsymbol{X\beta})'\boldsymbol{P}(\boldsymbol{y} - \boldsymbol{X\beta})} \ . \quad (4.4)$$

In general, nonlinear relations will exist between the unknown parameters and the observations so that we get corresponding to (4.3)

$$h_1(\beta_1, \ldots, \beta_u) = y_1^* + e_1$$
$$h_2(\beta_1, \ldots, \beta_u) = y_2^* + e_2$$
$$\ldots\ldots\ldots\ldots\ldots\ldots\ldots\ldots\ldots \quad (4.5)$$
$$h_n(\beta_1, \ldots, \beta_u) = y_n^* + e_n$$

where $h_i(\beta_1,\ldots,\beta_u)$ with $i \in \{1,\ldots,n\}$ are real-valued differentiable functions of the unknown parameters $\beta_1,\ldots,\beta_u$ and y_i^* the observations with their errors e_i. Let approximate values β_{j0} with $\beta_j = \beta_{j0} + \Delta\beta_j$ and $j \in \{1,\ldots,u\}$ be given for the parameters β_j so that the corrections $\Delta\beta_j$ are unknown and have to be estimated. Then, we may linearize by the Taylor expansion which is cut off at the linear term. We find with $\boldsymbol{\beta}_0 = (\beta_{j0})$

$$
\begin{aligned}
h_i(\beta_1,\ldots,\beta_u) &= h_i(\beta_{10}+\Delta\beta_1,\ldots,\beta_{u0}+\Delta\beta_u)\\
&= h_i(\beta_{10},\ldots,\beta_{u0}) + \frac{\partial h_i}{\partial\beta_1}\Big|_{\boldsymbol{\beta}_0}\Delta\beta_1 + \ldots + \frac{\partial h_i}{\partial\beta_u}\Big|_{\boldsymbol{\beta}_0}\Delta\beta_u \; .
\end{aligned}
\tag{4.6}
$$

By substituting

$$
\begin{aligned}
\boldsymbol{y} &= |y_1^* - h_1(\beta_{10},\ldots,\beta_{u0}),\ldots,y_n^* - h_n(\beta_{10},\ldots,\beta_{u0})|'\\
\boldsymbol{\beta} &= |\Delta\beta_1,\ldots,\Delta\beta_u|'
\end{aligned}
\tag{4.7}
$$

and

$$
\boldsymbol{X} = \left|
\begin{array}{ccc}
\dfrac{\partial h_1}{\partial\beta_1}\Big|_{\boldsymbol{\beta}_0} & \cdots & \dfrac{\partial h_1}{\partial\beta_u}\Big|_{\boldsymbol{\beta}_0}\\
\cdots\cdots\cdots\cdots\cdots\cdots\cdots\cdots\\
\dfrac{\partial h_n}{\partial\beta_1}\Big|_{\boldsymbol{\beta}_0} & \cdots & \dfrac{\partial h_n}{\partial\beta_u}\Big|_{\boldsymbol{\beta}_0}
\end{array}
\right|
\tag{4.8}
$$

the linear model (4.1) or (4.3) is obtained instead of (4.5). The expansion (4.6) is only valid, if the corrections $\Delta\beta_j$ are small quantitites. If these prerequisites are not fulfilled, one has to estimate iteratively by introducing the approximate value β_{j0} plus the estimate of $\Delta\beta_j$ as approximate value β_{j0} for the next estimate. At each step of the iterations the observations according to (4.7) and the derivatives from (4.8) have to be recomputed.

Example 2: The coordinates x_i, y_i of points in a plane with the approximate coordinates x_{i0}, y_{i0} which are collected in the vector $\boldsymbol{\beta}_0$ shall be determined by measuring the distances between the points. Let s_{ij} be the planar distance between the points i and j and e_{ij} its error, we then obtain by the theorem of Pythagoras instead of (4.5)

$$
\left((x_i - x_j)^2 + (y_i - y_j)^2\right)^{1/2} = s_{ij} + e_{ij} \; .
\tag{4.9}
$$

The coefficients of the matrix $\boldsymbol{X}$ follow with (4.8) by

$$
\frac{\partial s_{ij}}{\partial x_i}\Big|_{\boldsymbol{\beta}_0} = \frac{x_{i0} - x_{j0}}{s_{ij0}} \quad , \quad \frac{\partial s_{ij}}{\partial y_i}\Big|_{\boldsymbol{\beta}_0} = \frac{y_{i0} - y_{j0}}{s_{ij0}}
$$

$$
\frac{\partial s_{ij}}{\partial x_j}\Big|_{\boldsymbol{\beta}_0} = -\frac{x_{i0} - x_{j0}}{s_{ij0}} \quad , \quad \frac{\partial s_{ij}}{\partial y_j}\Big|_{\boldsymbol{\beta}_0} = -\frac{y_{i0} - y_{j0}}{s_{ij0}}
\tag{4.10}
$$

with

$$
s_{ij0} = \left((x_{i0} - x_{j0})^2 + (y_{i0} - y_{j0})^2\right)^{1/2} \; .
$$

$$\triangle$$

Only one observation appears in each of the nonlinear relations (4.5) between the observations and the unknown parameters. However, if we want to fit a line through the measured coordinates of points in a plane or to fit a curve or a surface through points in the three-dimensional space, more than one observations shows up in the relations for determining the parameters of the fit. The same holds true, if the parameters of transformations between points have to be computed whose coordinates are measured in two different coordinate systems. By introducing additional unknown parameters these relations may be transformed to the equations (4.5). This will be shown for a general case after presenting Example 3.

Example 3: Let a circle be fitted through the points P_i in a plane with the measured coordinates x_i and y_i and $i \in \{1,\ldots,m\}$. The coordinates a and b of the midpoint of the circle and the radius r are the unknown parameters. The nonlinear relations are obtained, if e_{x_i} denotes the error of x_i and e_{y_i} the error of y_i

$$(x_i + e_{x_i} - a)^2 + (y_i + e_{y_i} - b)^2 = r^2 \ .$$

In contrast to (4.5) not only the observation x_i appears, but also the observation y_i. As will be shown in the following, this relation may be transferred to the form (4.5) by introducing an additional unknown parameter. $\triangle$

Let $\boldsymbol{y}_1, \boldsymbol{y}_2, \ldots, \boldsymbol{y}_k$ be independent vectors of observations with $\boldsymbol{y}_i = (y_{ij})$, $i \in \{1,\ldots,m\}, j \in \{1,\ldots,k\}$ and the covariance matrices $D(\boldsymbol{y}_i|\sigma^2) = \sigma^2 \boldsymbol{P}_i^{-1}$ like in (4.1) and (4.3) or dependent vectors with a joint covariance matrix. Instead of (4.5) we have the nonlinear relations

$$
\begin{aligned}
f_1(y_{11} + e_{11}, y_{12} + e_{12}, \ldots, y_{1k} + e_{1k}, \beta_1, \ldots, \beta_u) &= 0 \\
f_2(y_{21} + e_{21}, y_{22} + e_{22}, \ldots, y_{2k} + e_{2k}, \beta_1, \ldots, \beta_u) &= 0 \\
\cdots\cdots\cdots\cdots\cdots\cdots\cdots\cdots\cdots\cdots\cdots\cdots\cdots\cdots\cdots\cdots & \\
f_m(y_{m1} + e_{m1}, y_{m2} + e_{m2}, \ldots, y_{mk} + e_{mk}, \beta_1, \ldots, \beta_u) &= 0
\end{aligned}
$$

where e_{ij} denotes the error of y_{ij}. These relations are solved for $y_{i1} + e_{i1}$ with $i \in \{1,\ldots,m\}$ so that the relations $g_i(\ldots)$ are obtained which shall be differentiable and where $y_{ij} + e_{ij}$ with $i \in \{1,\ldots,m\}$, $j \in \{2,\ldots,k\}$ are regarded as unknown parameters which are denoted by $\bar{y}_{ij}$. To determine each additional unknown parameter $\bar{y}_{ij}$, the additional observation equation $\bar{y}_{ij} = y_{ij} + e_{ij}$ is introduced. Thus, we obtain

$$
\begin{aligned}
g_1(\bar{y}_{12}, \ldots, \bar{y}_{1k}, \beta_1, \ldots, \beta_u) &= y_{11} + e_{11} \\
\bar{y}_{12} = y_{12} + e_{12}, \ldots, \bar{y}_{1k} &= y_{1k} + e_{1k} \\
g_2(\bar{y}_{22}, \ldots, \bar{y}_{2k}, \beta_1, \ldots, \beta_u) &= y_{21} + e_{21} \\
\bar{y}_{22} = y_{22} + e_{22}, \ldots, \bar{y}_{2k} &= y_{2k} + e_{2k} \\
\cdots\cdots\cdots\cdots\cdots\cdots\cdots\cdots\cdots\cdots\cdots\cdots\cdots & \\
g_m(\bar{y}_{m2}, \ldots, \bar{y}_{mk}, \beta_1, \ldots, \beta_u) &= y_{m1} + e_{m1} \\
\bar{y}_{m2} = y_{m2} + e_{m2}, \ldots, \bar{y}_{mk} &= y_{mk} + e_{mk} \ .
\end{aligned}
$$

These relations correspond to (4.5), since only one observation appears in each equation between the observations and the unknown parameters. They are linearized by means of approximate values for the unknown parameters, as is shown with (4.6) to (4.8). The approximate values cause in the observation equations $\bar{y}_{ij} = y_{ij} + e_{ij}$ with $i \in \{1, \ldots, m\}$ and $j \in \{2, \ldots, k\}$, which are already linear, an increase of the accuracy of the numerical computation. An application can be found in KOCH et al. (2000) where the parameters of the tranformation between points are estimated whose coordinates are determined in two three-dimensional coordinate systems.

Real-valued parameters $\boldsymbol{\beta}$ are assumed in the linear model (4.1). When analyzing data of the Global Positioning System (GPS), however, baseline coordinates appear as unknown parameters which are real-valued and phase ambiguities which are integers. For estimating the unknown parameters and the confidence regions for the baseline coordinates in this special linear model see for instance BETTI et al. (1993) and GUNDLICH and KOCH (2002).

4.2 Linear Model with Known Variance Factor

The variance factor σ^2 in (4.1) is first assumed to be known so that only the parameter vector $\boldsymbol{\beta}$ is unknown. In Chapter 4.3 then follows the linear model with the unknown variance factor. As prior density function for $\boldsymbol{\beta}$ a noninformative prior is assumed for the following, in Chapters 4.2.6 and 4.2.7 an informative prior.

4.2.1 Noninformative Priors

With the noninformative prior density function (2.216), which is determined by a constant, the posterior density function $p(\boldsymbol{\beta}|\boldsymbol{y})$ for the vector $\boldsymbol{\beta}$ of unknown parameters results with Bayes' theorem (2.122) immediately from the likelihood function (4.4) by

$$p(\boldsymbol{\beta}|\boldsymbol{y}) \propto e^{-\frac{1}{2\sigma^2}(\boldsymbol{y} - \boldsymbol{X}\boldsymbol{\beta})'\boldsymbol{P}(\boldsymbol{y} - \boldsymbol{X}\boldsymbol{\beta})} \tag{4.11}$$

where terms which do not depend on $\boldsymbol{\beta}$ are not considered, since they are constant. The statement C refering to the background information, which enters Bayes' theorem (2.122) as a condition, is omitted for the sake of simplifying the notation. This happens for all applications of the linear model.

The exponent in (4.11) is transformed like (2.228) by

$$\begin{aligned}
(\boldsymbol{y} - \boldsymbol{X}\boldsymbol{\beta})'\boldsymbol{P}(\boldsymbol{y} - \boldsymbol{X}\boldsymbol{\beta}) &= \boldsymbol{y}'\boldsymbol{P}\boldsymbol{y} - 2\boldsymbol{\beta}'\boldsymbol{X}'\boldsymbol{P}\boldsymbol{y} + \boldsymbol{\beta}'\boldsymbol{X}'\boldsymbol{P}\boldsymbol{X}\boldsymbol{\beta} \\
&= \boldsymbol{y}'\boldsymbol{P}\boldsymbol{y} - \boldsymbol{\mu}_0'\boldsymbol{X}'\boldsymbol{P}\boldsymbol{X}\boldsymbol{\mu}_0 + (\boldsymbol{\beta} - \boldsymbol{\mu}_0)'\boldsymbol{X}'\boldsymbol{P}\boldsymbol{X}(\boldsymbol{\beta} - \boldsymbol{\mu}_0)
\end{aligned} \tag{4.12}$$

with

$$\boldsymbol{\mu}_0 = (\boldsymbol{X}'\boldsymbol{P}\boldsymbol{X})^{-1}\boldsymbol{X}'\boldsymbol{P}\boldsymbol{y} \, . \tag{4.13}$$

After substituting (4.12) in (4.11) a comparison of the term depending on $\boldsymbol{\beta}$ with (2.195) reveals that the posterior distribution for the vector $\boldsymbol{\beta}$ of unknown parameters is the normal distribution

$$\boldsymbol{\beta}|\boldsymbol{y} \sim N(\boldsymbol{\mu}_0, \sigma^2(\boldsymbol{X}'\boldsymbol{P}\boldsymbol{X})^{-1}) \, . \tag{4.14}$$

The Bayes estimate $\hat{\boldsymbol{\beta}}_B$ of the unknown parameters $\boldsymbol{\beta}$ therefore follows from (3.9) with (2.196) by

$$\hat{\boldsymbol{\beta}}_B = (\boldsymbol{X}'\boldsymbol{P}\boldsymbol{X})^{-1}\boldsymbol{X}'\boldsymbol{P}\boldsymbol{y} \tag{4.15}$$

and the associated covariance matrix $D(\boldsymbol{\beta}|\boldsymbol{y})$ from (3.11) by

$$D(\boldsymbol{\beta}|\boldsymbol{y}) = \sigma^2(\boldsymbol{X}'\boldsymbol{P}\boldsymbol{X})^{-1} \, . \tag{4.16}$$

Because of rank$\boldsymbol{X} = u$ also rank$(\boldsymbol{X}'\boldsymbol{P}\boldsymbol{X}) = u$ holds true so that $(\boldsymbol{X}'\boldsymbol{P}\boldsymbol{X})^{-1}$ exits and $\hat{\boldsymbol{\beta}}_B$ and $D(\boldsymbol{\beta}|\boldsymbol{y})$ are uniquely determined. The linear equations $\boldsymbol{X}'\boldsymbol{P}\boldsymbol{X}\hat{\boldsymbol{\beta}}_B = \boldsymbol{X}'\boldsymbol{P}\boldsymbol{y}$ for $\hat{\boldsymbol{\beta}}_B$ are called *normal equations* and $\boldsymbol{X}'\boldsymbol{P}\boldsymbol{X}$ the *matrix of normal equations*.

Since the posterior density function (4.11) results immediately from the likelihood function, the MAP estimate $\hat{\boldsymbol{\beta}}_M$ of the vector $\boldsymbol{\beta}$ of unknown parameters is because of (3.30) identical with the maximum likelihood estimate (3.33). The MAP estimate $\hat{\boldsymbol{\beta}}_M$ is found at the maximum of the posterior density function for $\boldsymbol{\beta}$, i.e. at the point where the quadratic form $S(\boldsymbol{\beta})$ in the exponent of (4.11) has a minimum, thus

$$S(\boldsymbol{\beta}) = (\boldsymbol{y} - \boldsymbol{X}\boldsymbol{\beta})'\boldsymbol{P}(\boldsymbol{y} - \boldsymbol{X}\boldsymbol{\beta})/\sigma^2 \to \min \, . \tag{4.17}$$

With (4.12) and

$$\partial S(\boldsymbol{\beta})/\partial\boldsymbol{\beta} = (-2\boldsymbol{X}'\boldsymbol{P}\boldsymbol{y} + 2\boldsymbol{X}'\boldsymbol{P}\boldsymbol{X}\boldsymbol{\beta})/\sigma^2 = \boldsymbol{0} \tag{4.18}$$

the MAP estimate follows by

$$\hat{\boldsymbol{\beta}}_M = (\boldsymbol{X}'\boldsymbol{P}\boldsymbol{X})^{-1}\boldsymbol{X}'\boldsymbol{P}\boldsymbol{y} \tag{4.19}$$

which is identical with the Bayes estimate $\hat{\boldsymbol{\beta}}_B$ from (4.15). The reason is the symmetry of the density function of the posterior distribution (4.14) for $\boldsymbol{\beta}$, which is a normal distribution. The MAP estimate $\hat{\boldsymbol{\beta}}_M$ is also obtained by the posterior density function from (4.14) which attains at $\boldsymbol{\mu}_0$ the maximum, as can bee seen with (2.195).

For the numerical computation of the estimates and for the check of the computations see for instance Koch (1999, p.165).

Example 1: For the Example 1 to (4.3) we obtain with $\hat{\boldsymbol{\beta}}_B = \hat{\boldsymbol{\beta}}_M = \hat{s}$ from (4.15) and (4.19) and with $\boldsymbol{X}'\boldsymbol{P} = |p_1, \ldots, p_n|$ the estimate $\hat{s}$ as weighted arithmetic mean by

$$\hat{s} = \frac{1}{\sum_{i=1}^{n} p_i} \sum_{i=1}^{n} p_i y_i \tag{4.20}$$

with the variance $V(s|\boldsymbol{y})$ from (4.16)

$$V(s|\boldsymbol{y}) = \sigma^2 / \sum_{i=1}^{n} p_i \; . \qquad\qquad (4.21)$$

Δ

Example 2: To compute the estimates $\hat{\boldsymbol{\beta}}_B$ or $\hat{\boldsymbol{\beta}}_M$ of the unknown parameters $\boldsymbol{\beta}$, the matrix $\boldsymbol{X'PX}$ of normal equations and the absolute term $\boldsymbol{X'Py}$ need to be built up by means of the coefficient matrix $\boldsymbol{X}$. If the weight matrix $\boldsymbol{P}$ is a diagonal matrix, for instance, for independent observations because of (2.153) and (2.159), it might be computationally advantageous to avoid setting up the coefficient matrix $\boldsymbol{X}$ and to compute directly the contribution of each observation equation to the normal equations.

Let the weight matrix be diagonal with $\boldsymbol{P} = \text{diag}(p_1, \ldots, p_n)$ and let the coefficient matrix $\boldsymbol{X}$ be represented by its rows $\boldsymbol{x}'_i$ with $\boldsymbol{X} = |\boldsymbol{x}_1, \ldots, \boldsymbol{x}_n|'$ and $i \in \{1, \ldots, n\}$. With $\boldsymbol{y} = (y_i)$ and $\boldsymbol{e} = (e_i)$ from (4.1) and (4.3) the observation equation for y_i then follows with

$$\boldsymbol{x}'_i\boldsymbol{\beta} = E(y_i|\boldsymbol{\beta}) = y_i + e_i \quad \text{and} \quad V(y_i) = \sigma^2/p_i \; .$$

The matrix $\boldsymbol{X'PX}$ of normal equations and the vector $\boldsymbol{X'Py}$ of absolute terms result with $\boldsymbol{X'P} = |p_1\boldsymbol{x}_1, \ldots, p_n\boldsymbol{x}_n|$ from

$$\boldsymbol{X'PX} = |p_1\boldsymbol{x}_1\boldsymbol{x}'_1 + \ldots + p_n\boldsymbol{x}_n\boldsymbol{x}'_n|$$
$$\boldsymbol{X'Py} = |p_1 y_1\boldsymbol{x}_1 + \ldots + p_n y_n\boldsymbol{x}_n| \; .$$

Thus, the contribution of each observation equation is added to form the normal equations. Δ

Example 3: Let a periodic function, which is determined at points t_n by the independent measurements $y(t_n)$ with identical variances, be approximated by a sum of harmonic oscillations of different amplitudes and frequencies. The observation equation where $e(t_n)$ denotes the error is then obtained from (4.3) by, see for instance KOCH and SCHMIDT (1994, p.8),

$$\frac{A_0}{2} + \sum_{k=1}^{K}(A_k \cos k\omega_0 t_n + B_k \sin k\omega_0 t_n) = y(t_n) + e(t_n)$$

with

$$V(y(t_n)) = \sigma^2$$

and

$$t_n = -\frac{\pi}{\omega_0} + \frac{2\pi}{\omega_0}\frac{n}{N} \quad \text{for} \quad n \in \{0, 1, \ldots, N-1\} \; .$$

The coefficients A_0, A_k and B_k for $k \in \{1, \ldots, K\}$ are the unknown parameters and ω_0 the given fundamental frequency. With N observations $2K+1$ unknown parameters have to be estimated so that

$$N \geq 2K + 1$$

must hold true.

The coefficient matrix $\boldsymbol{X}$ follows from (4.3) or (4.8) by

$$
\boldsymbol{X} = \begin{vmatrix}
1/2 & \cos\omega_0 t_0 & \sin\omega_0 t_0 & \ldots & \cos K\omega_0 t_0 & \sin K\omega_0 t_0 \\
1/2 & \cos\omega_0 t_1 & \sin\omega_0 t_1 & \ldots & \cos K\omega_0 t_1 & \sin K\omega_0 t_1 \\
\multicolumn{6}{c}{\dotfill} \\
1/2 & \cos\omega_0 t_{N-1} & \sin\omega_0 t_{N-1} & \ldots & \cos K\omega_0 t_{N-1} & \sin K\omega_0 t_{N-1}
\end{vmatrix}
$$

and with it the matrix $\boldsymbol{X'X}$ of normal equations because of $\boldsymbol{P} = \boldsymbol{I}$. The sine and cosine functions are mutually orthogonal with

$$
\sum_{n=0}^{N-1} \cos k\omega_0 t_n \cos m\omega_0 t_n = \begin{cases} 0 & \text{for} \quad k \neq m \\ N/2 & \text{for} \quad k = m \quad \text{and} \quad k > 0 \end{cases}
$$

and the corresponding equation for the sine function and with

$$
\sum_{n=0}^{N-1} \sin k\omega_0 t_n \cos m\omega_0 t_n = 0 \ .
$$

Furthermore,

$$
\sum_{n=0}^{N-1} \cos k\omega_0 t_n = 0
$$

is valid and the corresponding equation for the sine function. The matrix of normal equations therefore results as the diagonal matrix

$$
\boldsymbol{X'X} = (N/2)\mathrm{diag}(1/2, 1, 1, \ldots, 1) \ .
$$

With its inverse and the vector of absolute terms the estimates $\hat{A}_k$ and $\hat{B}_k$ of the unknown parameters A_k and B_k follow from (4.15) or (4.19) by

$$
\hat{A}_k = \frac{2}{N} \sum_{n=0}^{N-1} y(t_n) \cos k\omega_0 t_n \quad \text{for} \quad k \in \{0, 1, \ldots, K\}
$$

$$
\hat{B}_k = \frac{2}{N} \sum_{n=0}^{N-1} y(t_n) \sin k\omega_0 t_n \quad \text{for} \quad k \in \{1, 2, \ldots, K\} \ .
$$

If the periodic function is not determined at discrete points, but if it is given continuously, the Fourier series follows from these results. $\quad\quad\quad\quad\Delta$

The confidence region for the vector $\boldsymbol{\beta}$ of unknown parameters is obtained with the normal distribution (4.14) as posterior distribution. As in the example to (3.35) a hypersurface of equal density follows by the relation, which results from the exponent of the density function (2.195) of the normal distribution,

$$
(\boldsymbol{\beta} - \boldsymbol{\mu}_0)' \boldsymbol{X'PX} (\boldsymbol{\beta} - \boldsymbol{\mu}_0)/\sigma^2 = \text{const} \ .
$$

It has the shape of a hyperellipsoid, and the $1 - \alpha$ confidence hyperellipsoid for $\boldsymbol{\beta}$ is given according to (3.36) by

$$(\boldsymbol{\beta} - \boldsymbol{\mu}_0)' \boldsymbol{X}' \boldsymbol{P} \boldsymbol{X} (\boldsymbol{\beta} - \boldsymbol{\mu}_0)/\sigma^2 = \chi^2_{1-\alpha;u} \tag{4.22}$$

where $\chi^2_{1-\alpha;u}$ denotes the upper α-percentage point (2.181) of the χ^2-distribution with u as parameter. The axes of the confidence hyperellipsoid and their orientations are found as in (3.38) and (3.39).

With the normal distribution (4.14) as posterior distribution for $\boldsymbol{\beta}$ hypotheses may be tested, too, as explained in Chapter 3.4.

Example 4: The point null hypothesis (3.50)

$$H_0 : \boldsymbol{H}\boldsymbol{\beta} = \boldsymbol{w} \quad \text{versus} \quad H_1 : \boldsymbol{H}\boldsymbol{\beta} \neq \boldsymbol{w} \tag{4.23}$$

where $\boldsymbol{H}$ denotes an $r \times u$ matrix with $\text{rank}\,\boldsymbol{H} = r$ and $r < u$ shall be tested according to (3.82) by the confidence hyperellipsoid for $\boldsymbol{H}\boldsymbol{\beta}$. The posterior distribution for $\boldsymbol{H}\boldsymbol{\beta}$ follows from the posterior distribution (4.14) for $\boldsymbol{\beta}$ with (2.202) by

$$\boldsymbol{H}\boldsymbol{\beta}|\boldsymbol{y} \sim N(\boldsymbol{H}\boldsymbol{\mu}_0, \sigma^2 \boldsymbol{H}(\boldsymbol{X}'\boldsymbol{P}\boldsymbol{X})^{-1}\boldsymbol{H}') \; . \tag{4.24}$$

The $1 - \alpha$ confidence hyperellipsoid for $\boldsymbol{H}\boldsymbol{\beta}$ is obtained corresponding to (4.22) by

$$(\boldsymbol{H}\boldsymbol{\beta} - \boldsymbol{H}\boldsymbol{\mu}_0)'(\boldsymbol{H}(\boldsymbol{X}'\boldsymbol{P}\boldsymbol{X})^{-1}\boldsymbol{H}')^{-1}(\boldsymbol{H}\boldsymbol{\beta} - \boldsymbol{H}\boldsymbol{\mu}_0)/\sigma^2 = \chi^2_{1-\alpha;r} \; . \tag{4.25}$$

If $\boldsymbol{H} = |\boldsymbol{0}, \boldsymbol{I}, \boldsymbol{0}|$, for instance, is substituted, the confidence hyperellipsoid for a subset of unknown parameters in $\boldsymbol{\beta}$ follows.

The point null hypothesis (4.23) is accepted according to (3.44) and (3.82), if

$$(\boldsymbol{H}\boldsymbol{\mu}_0 - \boldsymbol{w})'(\boldsymbol{H}(\boldsymbol{X}'\boldsymbol{P}\boldsymbol{X})^{-1}\boldsymbol{H}')^{-1}(\boldsymbol{H}\boldsymbol{\mu}_0 - \boldsymbol{w})/\sigma^2 < \chi^2_{1-\alpha;r} \; . \tag{4.26}$$
$$\Delta$$

4.2.2 Method of Least Squares

The observations $\boldsymbol{y}$ contain information about the unknown parameters $\boldsymbol{\beta}$, the expected values $E(\boldsymbol{y}|\boldsymbol{\beta})$ of the observations therefore depend on $\boldsymbol{\beta}$. Let the deviations of the data $\boldsymbol{y}$ from their estimated expected values $s[E(\boldsymbol{y}|\boldsymbol{\beta})]$ determine the loss of the estimation. The quadratic loss function (3.6) is therefore chosen which is caused by the difference $\boldsymbol{y} - s[E(\boldsymbol{y}|\boldsymbol{\beta})]$ and weighted by the inverse of the covariance matrix $D(\boldsymbol{y}|\boldsymbol{\beta}) = \boldsymbol{\Sigma}$ of the observations $\boldsymbol{y}$. The unknown parameters $\boldsymbol{\beta}$ are estimated such that the loss function attains a minimum, thus

$$(\boldsymbol{y} - s[E(\boldsymbol{y}|\boldsymbol{\beta})])'\boldsymbol{\Sigma}^{-1}(\boldsymbol{y} - s[E(\boldsymbol{y}|\boldsymbol{\beta})]) \to \min \; . \tag{4.27}$$

The estimation of the unknown parameters $\boldsymbol{\beta}$ according to (4.27) is called *method of least squares* or *least squares adjustment*, see for instance GRAFAREND and SCHAFFRIN (1993) and WOLF (1968, 1975, 1979). It is frequently applied in traditional statistics. For the linear model (4.1) we obtain instead of (4.27) the quadratic form $S(\boldsymbol{\beta})$ to be minimized

$$S(\boldsymbol{\beta}) = (\boldsymbol{y} - \boldsymbol{X}\boldsymbol{\beta})'\boldsymbol{P}(\boldsymbol{y} - \boldsymbol{X}\boldsymbol{\beta})/\sigma^2 \to \min . \tag{4.28}$$

If $\boldsymbol{y}$ now denotes given data, i.e. values of the random vector $\boldsymbol{y}$, then (4.28) is identical with (4.17). The estimate $\hat{\boldsymbol{\beta}}$ of the unknown parameters $\boldsymbol{\beta}$ by the method of least squares therefore follows with

$$\hat{\boldsymbol{\beta}} = (\boldsymbol{X}'\boldsymbol{P}\boldsymbol{X})^{-1}\boldsymbol{X}'\boldsymbol{P}\boldsymbol{y} \tag{4.29}$$

and agrees with the MAP estimate $\hat{\boldsymbol{\beta}}_M$ from (4.19) and also with the Bayes estimate $\hat{\boldsymbol{\beta}}_B$ from (4.15).

The estimate of the unknown parameters $\boldsymbol{\beta}$ by the method of least squares in traditional statistics proceeds with corresponding considerations so that the result is identical with (4.29). The interpretation of the estimate, however, is different. The estimates $\hat{\boldsymbol{\beta}}_B, \hat{\boldsymbol{\beta}}_M$ and $\hat{\boldsymbol{\beta}}$ of the random parameters $\boldsymbol{\beta}$ of Bayesian statistics are fixed quantities, which was already mentioned in connection with (3.10), since the estimates are determined by given fixed values for the observations $\boldsymbol{y}$. In traditional statistics, however, the vector $\boldsymbol{\beta}$ of the linear model, which is called Gauss-Markov model in traditional statistics, is a vector of fixed unknown parameters and its estimate $\hat{\boldsymbol{\beta}}$, which is identical with (4.29), is a random vector obtained by a linear transformation of the random vector $\boldsymbol{y}$ of observations. Also the variance factor σ^2 is like $\boldsymbol{\beta}$ a fixed and in general an unknown parameter.

An *unbiased estimation* is a property which is often required in traditional statistics. It means that the expected value of the estimate has to be equal to the quantity to be estimated. The estimate $\hat{\boldsymbol{\beta}}$ of traditional statistics fulfills this requirement because of (4.29), since we obtain with (2.146) and $E(\boldsymbol{y}) = \boldsymbol{X}\boldsymbol{\beta}$ from (4.1)

$$E(\hat{\boldsymbol{\beta}}) = (\boldsymbol{X}'\boldsymbol{P}\boldsymbol{X})^{-1}\boldsymbol{X}'\boldsymbol{P}E(\boldsymbol{y}) = \boldsymbol{\beta} . \tag{4.30}$$

The covariance matrix $D(\hat{\boldsymbol{\beta}})$ of the estimate $\hat{\boldsymbol{\beta}}$ of traditional statistics follows with (2.158) and $D(\boldsymbol{y}) = \sigma^2\boldsymbol{P}^{-1}$ from (4.1) by

$$\begin{aligned}
D(\hat{\boldsymbol{\beta}}) &= \sigma^2(\boldsymbol{X}'\boldsymbol{P}\boldsymbol{X})^{-1}\boldsymbol{X}'\boldsymbol{P}\boldsymbol{P}^{-1}\boldsymbol{P}\boldsymbol{X}(\boldsymbol{X}'\boldsymbol{P}\boldsymbol{X})^{-1} \\
&= \sigma^2(\boldsymbol{X}'\boldsymbol{P}\boldsymbol{X})^{-1} .
\end{aligned} \tag{4.31}$$

This matrix is identical with $D(\boldsymbol{\beta}|\boldsymbol{y})$ from (4.16).

4.2.3   Estimation of the Variance Factor in Traditional Statistics

The Bayes estimate of the variance factor σ^2 will be dealt with in Chapter 4.3.1 and 4.3.2, since the variance factor σ^2 is assumed as known for the Chapter 4.2. Nevertheless, the maximum likelihood estimate $\bar{\sigma}^2$ of σ^2 of traditional statistics shall be derived for a comparison with the results of Chapter 4.3.1 and 4.3.2. Hence, the estimate $\bar{\sigma}^2$ is according to (3.33) determined such that the likelihood function (4.4) attains a maximum. The likelihood function is therefore differentiated with respect to σ^2 and the derivative is set equal to zero. To simplify the derivation, not $p(\boldsymbol{y}|\boldsymbol{\beta},\sigma^2)$ but $\ln p(\boldsymbol{y}|\boldsymbol{\beta},\sigma^2)$ is differentiated. It is admissible, since the likelihood function like the density function of the normal distribution is positive and with $\partial \ln p(\boldsymbol{y}|\boldsymbol{\beta},\sigma^2)/\partial\sigma^2 = 0$ follows $[1/p(\boldsymbol{y}|\boldsymbol{\beta},\sigma^2)]\partial p(\boldsymbol{y}|\boldsymbol{\beta},\sigma^2)/\partial\sigma^2 = 0$ and therefore $\partial p(\boldsymbol{y}|\boldsymbol{\beta},\sigma^2)/\partial\sigma^2 = 0$. One gets from (4.4)

$$\ln p(\boldsymbol{y}|\boldsymbol{\beta},\sigma^2) = -\frac{n}{2}\ln(2\pi) - \frac{n}{2}\ln\sigma^2 + \frac{1}{2}\ln\det\boldsymbol{P}$$
$$- \frac{1}{2\sigma^2}(\boldsymbol{y}-\boldsymbol{X}\boldsymbol{\beta})'\boldsymbol{P}(\boldsymbol{y}-\boldsymbol{X}\boldsymbol{\beta}) .$$

With

$$\frac{\partial \ln p(\boldsymbol{y}|\boldsymbol{\beta},\sigma^2)}{\partial\sigma^2} = -\frac{n}{2\sigma^2} + \frac{1}{2(\sigma^2)^2}(\boldsymbol{y}-\boldsymbol{X}\boldsymbol{\beta})'\boldsymbol{P}(\boldsymbol{y}-\boldsymbol{X}\boldsymbol{\beta}) = 0$$

the estimate $\bar{\sigma}^2$ of σ^2 follows with

$$\bar{\sigma}^2 = \frac{1}{n}(\boldsymbol{y}-\boldsymbol{X}\hat{\boldsymbol{\beta}})'\boldsymbol{P}(\boldsymbol{y}-\boldsymbol{X}\hat{\boldsymbol{\beta}}) \tag{4.32}$$

where the vector $\boldsymbol{\beta}$ of unknown parameters is replaced by the vector $\hat{\boldsymbol{\beta}}$ of estimates from (4.29).

If the estimates $\hat{\boldsymbol{\beta}}$ are substituted in the observation equations $\boldsymbol{X}\boldsymbol{\beta} = \boldsymbol{y}+\boldsymbol{e}$ from (4.3), the vector $\hat{\boldsymbol{e}}$ of *residuals* is obtained instead of the vector $\boldsymbol{e}$ of errors

$$\hat{\boldsymbol{e}} = \boldsymbol{X}\hat{\boldsymbol{\beta}} - \boldsymbol{y} . \tag{4.33}$$

Thus, the variance factor is estimated by the weighted sum of squares of the residuals Ω

$$\Omega = \hat{\boldsymbol{e}}'\boldsymbol{P}\hat{\boldsymbol{e}} \tag{4.34}$$

which is minimal because of (4.28). This can be easily shown, see for instance KOCH (1999, p.158).

The requirement of unbiasedness of an estimate, which was discussed in connection with (4.30), is not fulfilled for the estimate $\bar{\sigma}^2$ of σ^2. To show this, we substitute $\hat{\boldsymbol{\beta}}$ from (4.29) in (4.33) and obtain

$$\hat{\boldsymbol{e}} = -(\boldsymbol{I} - \boldsymbol{X}(\boldsymbol{X}'\boldsymbol{P}\boldsymbol{X})^{-1}\boldsymbol{X}'\boldsymbol{P})\boldsymbol{y} \tag{4.35}$$

and for Ω from (4.34)

$$\Omega = \boldsymbol{y}'(\boldsymbol{P} - \boldsymbol{P}\boldsymbol{X}(\boldsymbol{X}'\boldsymbol{P}\boldsymbol{X})^{-1}\boldsymbol{X}'\boldsymbol{P})\boldsymbol{y} . \tag{4.36}$$

The expected value $E(\Omega)$ follows from (2.165) and (4.1) with

$$
\begin{aligned}
E(\Omega) &= \sigma^2 \mathrm{tr}(\boldsymbol{I} - \boldsymbol{P}\boldsymbol{X}(\boldsymbol{X}'\boldsymbol{P}\boldsymbol{X})^{-1}\boldsymbol{X}') \\
&\quad + \boldsymbol{\beta}'\boldsymbol{X}'(\boldsymbol{P} - \boldsymbol{P}\boldsymbol{X}(\boldsymbol{X}'\boldsymbol{P}\boldsymbol{X})^{-1}\boldsymbol{X}'\boldsymbol{P})\boldsymbol{X}\boldsymbol{\beta} \\
&= \sigma^2(n - \mathrm{tr}[(\boldsymbol{X}'\boldsymbol{P}\boldsymbol{X})^{-1}\boldsymbol{X}'\boldsymbol{P}\boldsymbol{X}]) \\
&= \sigma^2(n - u) \,.
\end{aligned}
\tag{4.37}
$$

We therefore obtain from (4.32)

$$
E(\bar{\sigma}^2) = \frac{1}{n}E(\Omega) = \frac{\sigma^2}{n}(n - u) \neq \sigma^2 \,.
\tag{4.38}
$$

However, the estimate $\hat{\sigma}^2$ of σ^2 with

$$
\hat{\sigma}^2 = \frac{1}{n - u}\Omega
\tag{4.39}
$$

is unbiasesd, as can be readily seen by forming the expected value $E(\hat{\sigma}^2)$.

Example: For the Example 1 to (4.3) we get from (4.20) and (4.33) $\hat{\boldsymbol{e}} = |\hat{s} - y_1, \hat{s} - y_2, \ldots, \hat{s} - y_n|'$ and $\hat{\boldsymbol{e}}'\boldsymbol{P} = |p_1(\hat{s} - y_1), p_2(\hat{s} - y_2), \ldots, p_n(\hat{s} - y_n)|$. The unbiased estimate $\hat{\sigma}^2$ of the variance factor σ^2 is therefore obtained with

$$
\hat{\sigma}^2 = \frac{1}{n - 1}\sum_{i=1}^{n} p_i(\hat{s} - y_i)^2 \,.
\tag{4.40}
$$

$$\Delta$$

4.2.4 Linear Model with Constraints in Traditional Statistics

The parameters of a *linear model with constraints* in traditional statistics are estimated in a simple manner by the method of least squares. This model is defined by

$$
\boldsymbol{X}\boldsymbol{\beta} = E(\boldsymbol{y}) \quad \text{with} \quad \boldsymbol{H}\boldsymbol{\beta} = \boldsymbol{w} \quad \text{and} \quad D(\boldsymbol{y}) = \sigma^2\boldsymbol{P}^{-1} \,.
\tag{4.41}
$$

The $r \times u$ matrix $\boldsymbol{H}$ with $\mathrm{rank}\,\boldsymbol{H} = r$ and $r < u$ contains known coefficients and $\boldsymbol{w}$ is a known $r \times 1$ vector. In the interpretation of traditional statistics $\boldsymbol{\beta}$ denotes a vector of fixed unknown parameters and $\boldsymbol{y}$ a random vector. To estimate the unknown parameters $\boldsymbol{\beta}$ by the method of least squares, (4.28) has to be minimized subject to the constraints $\boldsymbol{H}\boldsymbol{\beta} = \boldsymbol{w}$. The Lagrange function $w(\boldsymbol{\beta})$ is therefore set up

$$
w(\boldsymbol{\beta}) = (\boldsymbol{y} - \boldsymbol{X}\boldsymbol{\beta})'\boldsymbol{P}(\boldsymbol{y} - \boldsymbol{X}\boldsymbol{\beta})/\sigma^2 + 2\boldsymbol{k}'(\boldsymbol{H}\boldsymbol{\beta} - \boldsymbol{w})/\sigma^2
$$

where the $r \times 1$ vector $2\boldsymbol{k}/\sigma^2$ contains Lagrange multipliers. The derivative of $w(\boldsymbol{\beta})$ with respect to $\boldsymbol{\beta}$ is set equal to zero. One obtains with (4.18)

$$
\partial w(\boldsymbol{\beta})/\partial \boldsymbol{\beta} = (-2\boldsymbol{X}'\boldsymbol{P}\boldsymbol{y} + 2\boldsymbol{X}'\boldsymbol{P}\boldsymbol{X}\boldsymbol{\beta} + 2\boldsymbol{H}'\boldsymbol{k})/\sigma^2 = \boldsymbol{0} \,.
$$

Thus, together with the constraints the estimate $\tilde{\beta}$ of β follows from

$$\left|\begin{array}{cc} X'PX & H' \\ H & 0 \end{array}\right| \left|\begin{array}{c} \tilde{\beta} \\ k \end{array}\right| = \left|\begin{array}{c} X'Py \\ w \end{array}\right| . \tag{4.42}$$

The values for $\tilde{\beta}$ and k are uniquely determined, since we have with rank$H = r$

$$\det\left|\begin{array}{cc} X'PX & H' \\ H & 0 \end{array}\right| = \det(X'PX)\det(-H(X'PX)^{-1}H') \neq 0 .$$

If $\tilde{\beta}$ is eliminated from (4.42), we obtain

$$-H(X'PX)^{-1}H'k = w - H(X'PX)^{-1}X'Py . \tag{4.43}$$

This result is substituted in (4.42), the estimate $\tilde{\beta}$ then follows with

$$\tilde{\beta} = (X'PX)^{-1}[X'Py + H'(H(X'PX)^{-1}H')^{-1}$$
$$(w - H(X'PX)^{-1}X'Py)] \tag{4.44}$$

and from (2.158) the covariance matrix $D(\tilde{\beta})$ of $\tilde{\beta}$ with

$$D(\tilde{\beta}) = \sigma^2[(X'PX)^{-1} - (X'PX)^{-1}H'(H(X'PX)^{-1}H')^{-1}$$
$$H(X'PX)^{-1}] . \tag{4.45}$$

For inverting the matrix on the left-hand side of (4.42) the matrix identity is needed, see for instance KOCH (1999, p.33),

$$\left|\begin{array}{cc} A & B \\ C & D \end{array}\right|^{-1} =$$
$$\left|\begin{array}{cc} A^{-1} + A^{-1}B(D - CA^{-1}B)^{-1}CA^{-1} & -A^{-1}B(D - CA^{-1}B)^{-1} \\ -(D - CA^{-1}B)^{-1}CA^{-1} & (D - CA^{-1}B)^{-1} \end{array}\right| , \tag{4.46}$$

where A and D denote regular matrices. In the following, two identities will be used in addition, see KOCH (1999, p.34),

$$(A - BD^{-1}C)^{-1} = A^{-1} + A^{-1}B(D - CA^{-1}B)^{-1}CA^{-1} \tag{4.47}$$

and

$$D^{-1}C(A - BD^{-1}C)^{-1} = (D - CA^{-1}B)^{-1}CA^{-1} . \tag{4.48}$$

Substituting $N = X'PX$ in (4.42) gives with (4.46)

$$\left|\begin{array}{cc} N & H' \\ H & 0 \end{array}\right|^{-1} =$$
$$\left|\begin{array}{cc} N^{-1} - N^{-1}H'(HN^{-1}H')^{-1}HN^{-1} & N^{-1}H'(HN^{-1}H')^{-1} \\ (HN^{-1}H')^{-1}HN^{-1} & -(HN^{-1}H')^{-1} \end{array}\right| . \tag{4.49}$$

The matrix which is needed for computing $D(\tilde{\beta})$ from (4.45) stands at the position of $X'PX$ in the inverse matrix. The estimate $\tilde{\beta}$ and its covariance matrix $D(\tilde{\beta})$ may therefore be obtained from (4.42).

To estimate the variance factor σ^2 of the model (4.41) by the maximum likelihood method, the Lagrange function following from the likelihood function (4.4) and the constraints $H\beta = w$ is differentiated with respect to σ^2 and the derivative is set equal to zero. Corresponding to (4.32) the estimate $\bar{\bar{\sigma}}^2$ of σ^2 is obtained by

$$\bar{\bar{\sigma}}^2 = \frac{1}{n}(y - X\tilde{\beta})'P(y - X\tilde{\beta}) . \tag{4.50}$$

By introducing as in (4.33) the vector $\tilde{e}$ of residuals

$$\tilde{e} = X\tilde{\beta} - y \tag{4.51}$$

it becomes obvious that the estimate $\bar{\bar{\sigma}}^2$ follows from the weighted sum of squares Ω_H of the residuals which is minimal like (4.34)

$$\Omega_H = \tilde{e}'P\tilde{e} . \tag{4.52}$$

Again the estimate $\bar{\bar{\sigma}}^2$ is biased. To show this, we compute Ω_H with (4.29) and (4.35)

$$\Omega_H = (X(\tilde{\beta} - \hat{\beta}) + X\hat{\beta} - y)'P(X(\tilde{\beta} - \hat{\beta}) + X\hat{\beta} - y)$$
$$= (X\hat{\beta} - y)'P(X\hat{\beta} - y) + (\tilde{\beta} - \hat{\beta})'X'PX(\tilde{\beta} - \hat{\beta}) .$$

We substitute (4.29) in (4.44) and the result in the second term. We then obtain with (4.34)

$$\Omega_H = \Omega + R \quad \text{with}$$
$$R = (H\hat{\beta} - w)'(H(X'PX)^{-1}H')^{-1}(H\hat{\beta} - w) . \tag{4.53}$$

The matrix $(H(X'PX)^{-1}H')^{-1}$ is positive definite because of $\text{rank}H = r$. Thus, $R \geq 0$ so that by introducing the constraints $H\beta - w = 0$ the sum of squares Ω of the residuals in general increases. As was shown with (4.26) in connection with (4.13) and as will be shown with (4.145) and (4.197), the quadratic form R is used for the test of point null hypotheses. It is checked by this test, how much the constraint $H\beta - w = 0$, which is introduced by the point null hypothesis, causes the quadratic form R to increase.

The expected value $E(\Omega_H)$ is computed from (2.165) with (4.37) and with $E(H\hat{\beta} - w) = H\beta - w = 0$ because of (4.30) and $D(H\hat{\beta} - w) = \sigma^2 H(X'PX)^{-1}H'$ from (2.158) and (4.31) by

$$E(\Omega_H) = E(\Omega) + \sigma^2\text{tr}[(H(X'PX)^{-1}H')^{-1}H(X'PX)^{-1}H']$$
$$= \sigma^2(n - u + r) . \tag{4.54}$$

In contrary to (4.50) the unbiased estimate $\tilde{\sigma}^2$ of the variance factor σ^2 therefore follows with

$$\tilde{\sigma}^2 = \Omega_H/(n - u + r) \; . \tag{4.55}$$

Instead of estimating parameters in the linear model with constraints we may apply the model without constraints, if the constraints are introduced as observations with very small variances (KOCH 1999, p.176). The estimates in the linear model with constraints can be also derived as a limiting process of a sequence of Bayes estimates with an appropriate sequence of prior distributions (PILZ 1983, p.82). But the derivation by the method of least squares is simpler.

4.2.5 Robust Parameter Estimation

In the course of measurements some observations may be out of any reason grossly falsified. These observations are called *outliers*. Outliers may considerably influence the estimation of parameters. Outliers should therefore be eliminated from the observations. This is achieved by tests for outliers, if only a few outliers are present in observations which control each other, see for instance KOCH (1999, p.302).

If a larger number of outliers exists, the tests for outliers fail and one needs an estimation of the parameters which is insensitive to outliers. Such an estimation is called *robust*. The estimates of the parameters derived so far are not robust. They possess according to (4.17) or (4.28) the property that their weighted sum of squares of the residuals from (4.34) or (4.52) is minimal. The effect of an outlier is therefore not eliminated but distributed over the remaining observations.

Outliers change the probability distribution of the observations. Thus, the normal distribution should not be exclusively applied as done so far. HUBER (1964; 1981, p.71) proposed the density function $p(x|c)$ of the random variable X which is well suited for a robust estimation of the parameters

$$p(x|c) \propto e^{-x^2/2} \quad \text{for} \quad |x| \le c \tag{4.56}$$

$$p(x|c) \propto e^{-c|x|+c^2/2} \quad \text{for} \quad |x| > c \; . \tag{4.57}$$

The quantity c denotes a constant whose value depends on the portion of outliers in the data (HUBER 1981, p.87). In general $c = 1.5$ is chosen. The middle part (4.56) of the density function $p(x|c)$ is formed because of (2.166) by the standard normal distribution $N(0, 1)$, while the tails (4.57) of the density function $p(x|c)$ are represented by the Laplace distribution (2.191). At the tails of the distribution more probability mass is concentrated than at the tails of the normal distribution. Outliers are thus taken care of.

The robust estimation of the parameters is applied for the linear model (4.1) whose variance factor σ^2 is supposed to be known. Independent observations are assumed for simplicity, small modifications only are needed to work with dependent observations (YANG et al. 2002). Thus, we have because of (2.153) in (4.1)

$$\boldsymbol{P} = \operatorname{diag}(p_1, p_2, \ldots, p_n) \ . \tag{4.58}$$

If we define $\boldsymbol{X} = (\boldsymbol{x}_i')$ and $\boldsymbol{y} = (y_i)$, the observation equations are obtained from (4.1) and (4.3) as in the Example 2 to (4.15) by

$$\boldsymbol{x}_i'\boldsymbol{\beta} = E(y_i|\boldsymbol{\beta}) = y_i + e_i \quad \text{and} \quad V(y_i) = \sigma^2/p_i \ , \ i \in \{1, \ldots, n\} \ . \tag{4.59}$$

Instead of the error $e_i = \boldsymbol{x}_i'\boldsymbol{\beta} - y_i$ of the observation y_i the standardized error $\bar{e}_i$ is introduced

$$\bar{e}_i = \sqrt{p_i}(\boldsymbol{x}_i'\boldsymbol{\beta} - y_i)/\sigma \quad \text{with} \quad E(\bar{e}_i|\boldsymbol{\beta}) = 0 \ , \ V(\bar{e}_i|\boldsymbol{\beta}) = 1 \tag{4.60}$$

whose variance $V(\bar{e}_i|\boldsymbol{\beta})$ is equal to one because of (2.158). We therefore may assume that the density function $p(\bar{e}_i|\boldsymbol{\beta})$ is given by the density functions (4.56) and (4.57). By transforming $\bar{e}_i$ to y_i we obtain $p(y_i|\boldsymbol{\beta}) \propto p(\bar{e}_i|\boldsymbol{\beta})$. The observations $\boldsymbol{y}$ are independent by assumption. Thus, the likelihood function $p(\boldsymbol{y}|\boldsymbol{\beta})$ follows with (2.110) from

$$p(\boldsymbol{y}|\boldsymbol{\beta}) \propto \prod_{i=1}^{n} p(\bar{e}_i|\boldsymbol{\beta}) \ . \tag{4.61}$$

As in Chapter 4.2.1 the noninformative prior density function (2.216) is introduced as prior for the vector $\boldsymbol{\beta}$ of unknown parameters. It is determined by a constant. The posterior density function $p(\boldsymbol{\beta}|\boldsymbol{y})$ for $\boldsymbol{\beta}$ therefore follows with Bayes' theorem (2.122) immediately from the likelihood function (4.61) where $\boldsymbol{\beta}$ is variable and not fixed

$$p(\boldsymbol{\beta}|\boldsymbol{y}) \propto \prod_{i=1}^{n} p(\bar{e}_i|\boldsymbol{\beta}) \ . \tag{4.62}$$

The vector $\boldsymbol{\beta}$ of unknown parameters shall be estimated based on this posterior density function. The MAP estimate (3.30) will be applied. Depending on $\boldsymbol{\beta}$ not the maximum of $p(\boldsymbol{\beta}|\boldsymbol{y})$ but of $\ln p(\boldsymbol{\beta}|\boldsymbol{y})$ and the minimum of $-\ln p(\boldsymbol{\beta}|\boldsymbol{y})$, respectively, is sought as for (4.32)

$$-\ln p(\boldsymbol{\beta}|\boldsymbol{y}) \propto \sum_{i=1}^{n} -\ln p(\bar{e}_i|\boldsymbol{\beta}) \ . \tag{4.63}$$

If we introduce the score function $\rho(\bar{e}_i)$ with

$$\rho(\bar{e}_i) = -\ln p(\bar{e}_i|\boldsymbol{\beta}) \ , \tag{4.64}$$

we obtain from (4.56) and (4.57)

$$\rho(\bar{e}_i) \propto \bar{e}_i^2/2 \quad \text{for} \quad |\bar{e}_i| \leq c \tag{4.65}$$

$$\rho(\bar{e}_i) \propto c|\bar{e}_i| - c^2/2 \quad \text{for} \quad |\bar{e}_i| > c \ . \tag{4.66}$$

Thus, the MAP estimate is determined with (4.63) by

$$\sum_{i=1}^{n} \rho(\bar{e}_i) \to \min .$$ (4.67)

It leads, if the score function (4.65) only is applied, to the score function (4.28) of the method of least squares.

With the derivative $\psi(\bar{e}_i) = \partial\rho(\bar{e}_i)/\partial\bar{e}_i$ of the score function, with $\boldsymbol{\beta} = (\beta_l)$, $\boldsymbol{x}_i' = (x_{il})$ and with (4.60) we obtain

$$\frac{\partial}{\partial\beta_l}\rho(\bar{e}_i) = \psi(\bar{e}_i)\sqrt{p_i}x_{il}/\sigma \quad \text{for} \quad l \in \{1,\ldots,u\} .$$ (4.68)

The MAP estimate $\hat{\boldsymbol{\beta}}_M$ of $\boldsymbol{\beta}$, which is because of (4.62) identical with the maximum-likelihood estimate (3.33) and which for the robust estimation of parameters is therefore called *M-estimate*, follows with (4.67) by

$$\frac{1}{\sigma}\sum_{i=1}^{n}\sqrt{p_i}\psi(\hat{\bar{e}}_i)x_{il} = 0 \quad \text{for} \quad l \in \{1,\ldots,u\}$$ (4.69)

where $\hat{\bar{e}}_i$ denotes corresponding to (4.33) the standardized residual

$$\hat{\bar{e}}_i = \sqrt{p_i}(\boldsymbol{x}_i'\hat{\boldsymbol{\beta}}_M - y_i)/\sigma .$$ (4.70)

By introducing the weights

$$w_i = p_i\psi(\hat{\bar{e}}_i)/\hat{\bar{e}}_i$$ (4.71)

we find instead of (4.69)

$$\frac{1}{\sigma^2}\sum_{i=1}^{n}w_i(\boldsymbol{x}_i'\hat{\boldsymbol{\beta}}_M - y_i)x_{il} = 0$$

or with $\boldsymbol{W} = \text{diag}(w_1,\ldots,w_n)$

$$\boldsymbol{X}'\boldsymbol{W}\boldsymbol{X}\hat{\boldsymbol{\beta}}_M = \boldsymbol{X}'\boldsymbol{W}\boldsymbol{y} .$$ (4.72)

This system of equations has to be solved iteratively, since the weights w_i depend according to (4.71) on the standardized residuals $\hat{\bar{e}}_i$ which are determined because of (4.70) by the estimates $\hat{\boldsymbol{\beta}}_M$. The estimate (4.19) is chosen as first approximation $\hat{\boldsymbol{\beta}}_M^{(0)}$, thus

$$\hat{\boldsymbol{\beta}}_M^{(0)} = (\boldsymbol{X}'\boldsymbol{P}\boldsymbol{X})^{-1}\boldsymbol{X}'\boldsymbol{P}\boldsymbol{y} .$$ (4.73)

The $(m+1)$th iteration gives

$$\hat{\boldsymbol{\beta}}_M^{(m+1)} = (\boldsymbol{X}'\boldsymbol{W}^{(m)}\boldsymbol{X})^{-1}\boldsymbol{X}'\boldsymbol{W}^{(m)}\boldsymbol{y}$$ (4.74)

with

$$\boldsymbol{W}^{(m)} = \mathrm{diag}(w_1^{(m)}, \ldots, w_n^{(m)}) \ . \tag{4.75}$$

HUBER (1981, p.184) proves the convergence of the iterations.

To determine the weights $w_i^{(m)}$, the derivative $\psi(\hat{\bar{e}}_i)$ is computed with (4.65) and (4.66) by

$$\psi(\hat{\bar{e}}_i) \propto \hat{\bar{e}}_i \quad \text{for} \quad |\hat{\bar{e}}_i| \leq c \tag{4.76}$$

$$\psi(\hat{\bar{e}}_i) \propto c\hat{\bar{e}}_i/|\hat{\bar{e}}_i| \quad \text{for} \quad |\hat{\bar{e}}_i| > c \ . \tag{4.77}$$

The plausibility of the derivative of the only piecewise continuous differentiable function $|\hat{\bar{e}}_i|$ may be checked by $\partial\sqrt{|\hat{\bar{e}}_i|^2}/\partial\hat{\bar{e}}_i = \hat{\bar{e}}_i/|\hat{\bar{e}}_i|$. We obtain with the residual $\hat{e}_i = \boldsymbol{x}_i'\hat{\boldsymbol{\beta}}_M - y_i$ in (4.70) and with (4.71), since a constant need not be considered in the weights because of (4.72),

$$w_i^{(m)} = p_i \quad \text{for} \quad \sqrt{p_i}|\hat{e}_i^{(m)}| \leq c\sigma \tag{4.78}$$

$$w_i^{(m)} = c\sigma\sqrt{p_i}/|\hat{e}_i^{(m)}| \quad \text{for} \quad \sqrt{p_i}|\hat{e}_i^{(m)}| > c\sigma \ . \tag{4.79}$$

By iteratively reweighting the estimates $\hat{\boldsymbol{\beta}}_M^{(m+1)}$ according to (4.74) with the weights (4.78) and (4.79) we obtain a robust estimation, if the derivative $\psi(\hat{\bar{e}}_i)$, which is proportional to the so-called *influence function* (HAMPEL et al. 1986, p.40,101), is bounded. This is the case, since $\psi(\hat{\bar{e}}_i)$ for $|\hat{\bar{e}}_i| > c$ is constant.

Based on the posterior density function (4.62) for the unknown parameters from a robust estimation of parameters also confidence regions for the unknown parameters may be established and hypotheses may be tested, as explained in Chapters 3.3 and 3.4. Examples for determining confidence regions are given in Chapters 6.2.5 and 6.3.6. Detecting outliers by using prior information has been proposed by GUI et al. (2007).

Instead of combining the normal distribution (4.56) and the Laplace distribution (4.57) the Laplace distribution (2.191) only shall now be chosen for taking care of outliers. We obtain

$$p(x) \propto e^{-|x|} \tag{4.80}$$

and instead of (4.65) and (4.66)

$$\rho(\bar{e}_i) \propto |\bar{e}_i| \tag{4.81}$$

and from (4.67)

$$\sum_{i=1}^{n} |\bar{e}_i| \to \min \ . \tag{4.82}$$

Thus, the sum of the absolute values of the standardized errors or of the errors in case of observations with identical weights has to be minimized which leads to the L_1-*norm estimate*. The absolute value of the errors as loss function was already introduced in Chapter 3.2.2 and gave the median as estimate.

We obtain the derivative of the score function $\rho(\bar{e}_i)$ as in (4.77) with

$$\psi(\hat{\bar{e}}_i) \propto \hat{\bar{e}}_i/|\hat{\bar{e}}_i| \tag{4.83}$$

which is bounded so that the L_1-norm estimate is robust. We get instead of (4.78) and (4.79), since a constant may be omitted because of (4.72),

$$w_i^{(m)} = \sqrt{p_i}/|\hat{e}_i^{(m)}| \, . \tag{4.84}$$

By iteratively reweighting (4.74) with (4.84) the robust L_1-norm estimate is computed. However, one has to be aware that the L_1-norm estimate is not always unique and that nonunique solutions are not detected by the reweighting with (4.84), whereas the Simplex algorithm of linear programming is pointing them out, see for instance SPÄTH (1987, p.58).

Example: As in the example to (2.227) let an unknown quantity s be measured n times giving the observations y_i which shall be independent and have equal variances. The L_1-norm estimate from (4.69) leads with $p_i = 1, x_{il} = 1, \hat{\boldsymbol{\beta}}_M = \hat{s}_M$ and (4.83) to

$$\sum_{i=1}^{n} \frac{\hat{s}_M - y_i}{|\hat{s}_M - y_i|} = 0 \, .$$

The sum is equal to zero, if the difference $\hat{s}_M - y_i$ is as often positive as negative. This condition is fulfilled for the median $\hat{s}_{\mathrm{med}}$ of s from (3.25) and (3.26), thus $\hat{s}_{\mathrm{med}} = \hat{s}_M$.

The median is extremely robust because among 3 observations 1 gross error can be detected, among 5 observations 2 gross errors, among 7 observations 3 gross errors and so forth. In 4 observations 1 gross error may be identified, in 6 observations 2 gross errors and so on. The number of gross errors to be detected goes for n observations towards the maximum possible number of $n/2$, since with a larger number of outliers one cannot differentiate between the observations and the outliers. $\qquad\qquad\Delta$

The parameter estimation (4.74) does not react in a robust manner in case of outliers in *leverage points*. These are data which lie because of their geometry far away from the rest of the data so that they may have a considerable influence on the estimation of the parameters. Leverage points ask for a special approach to the robust estimation of parameters, see for instance ROUSSEEUW (1984), ROUSSEEUW and LEROY (1987), KOCH (1996),(1999, p.263), JUNHUAN (2005), XU (2005).

4.2.6 Informative Priors

In contrast to the preceding chapters where noninformative prior density functions for the vector $\boldsymbol{\beta}$ of unknown parameters were applied, it will now be assumed that prior information on $\boldsymbol{\beta}$ is available. Let the vector $E(\boldsymbol{\beta}) = \boldsymbol{\mu}$ of expected values for $\boldsymbol{\beta}$ and the positive definite covariance matrix $D(\boldsymbol{\beta}) = \sigma^2 \boldsymbol{\Sigma}$ be given and let the unknown parameters $\boldsymbol{\beta}$ be normally distributed so that with (2.196) the prior distribution follows

$$\boldsymbol{\beta} \sim N(\boldsymbol{\mu}, \sigma^2 \boldsymbol{\Sigma}) . \tag{4.85}$$

Let the observations $\boldsymbol{y}$ have the normal distribution (4.2) and let the variance factor σ^2 be given. The density function of the prior distribution (4.85) is then a conjugate prior, as the comparison with (2.225) and (2.226) shows, since the posterior density function for the unknown parameters $\boldsymbol{\beta}$ follows with (2.227) from the normal distribution

$$\boldsymbol{\beta}|\boldsymbol{y} \sim N\big(\boldsymbol{\mu}_0, \sigma^2 (\boldsymbol{X}'\boldsymbol{P}\boldsymbol{X} + \boldsymbol{\Sigma}^{-1})^{-1}\big) \tag{4.86}$$

with

$$\boldsymbol{\mu}_0 = (\boldsymbol{X}'\boldsymbol{P}\boldsymbol{X} + \boldsymbol{\Sigma}^{-1})^{-1}(\boldsymbol{X}'\boldsymbol{P}\boldsymbol{y} + \boldsymbol{\Sigma}^{-1}\boldsymbol{\mu}) .$$

The Bayes estimate $\hat{\boldsymbol{\beta}}_B$ of the unknown parameters $\boldsymbol{\beta}$ from (3.9) is obtained because of (2.196) by

$$\hat{\boldsymbol{\beta}}_B = (\boldsymbol{X}'\boldsymbol{P}\boldsymbol{X} + \boldsymbol{\Sigma}^{-1})^{-1}(\boldsymbol{X}'\boldsymbol{P}\boldsymbol{y} + \boldsymbol{\Sigma}^{-1}\boldsymbol{\mu}) \tag{4.87}$$

with the associated covariance matrix $D(\boldsymbol{\beta}|\boldsymbol{y})$ from (3.11)

$$D(\boldsymbol{\beta}|\boldsymbol{y}) = \sigma^2 (\boldsymbol{X}'\boldsymbol{P}\boldsymbol{X} + \boldsymbol{\Sigma}^{-1})^{-1} . \tag{4.88}$$

As can be seen with (2.195) the density function of the normal distribution (4.86) becomes maximal at $\boldsymbol{\mu}_0$. The MAP estimate $\hat{\boldsymbol{\beta}}_M$ of $\boldsymbol{\beta}$ according to (3.30) therefore leads to the estimate which is identical with (4.87)

$$\hat{\boldsymbol{\beta}}_M = (\boldsymbol{X}'\boldsymbol{P}\boldsymbol{X} + \boldsymbol{\Sigma}^{-1})^{-1}(\boldsymbol{X}'\boldsymbol{P}\boldsymbol{y} + \boldsymbol{\Sigma}^{-1}\boldsymbol{\mu}) . \tag{4.89}$$

The Bayes estimate $\hat{\boldsymbol{\beta}}_B$ from (4.87) and the MAP estimate $\hat{\boldsymbol{\beta}}_M$ from (4.89), respectively, may be also derived as estimate (4.29) of the method of least squares by introducing the prior information as an additional vector $\boldsymbol{\mu}$ of observations for $\boldsymbol{\beta}$ with the covariance matrix $\sigma^2 \boldsymbol{\Sigma}$. Thus, we set

$$\bar{\boldsymbol{X}} = \left| \begin{array}{c} \boldsymbol{X} \\ \boldsymbol{I} \end{array} \right|, \; \bar{\boldsymbol{y}} = \left| \begin{array}{c} \boldsymbol{y} \\ \boldsymbol{\mu} \end{array} \right|, \; \bar{\boldsymbol{P}} = \left| \begin{array}{cc} \boldsymbol{P} & \boldsymbol{0} \\ \boldsymbol{0} & \boldsymbol{\Sigma}^{-1} \end{array} \right|, \tag{4.90}$$

and apply (4.29) with $\bar{\boldsymbol{X}}, \bar{\boldsymbol{P}}$ and $\bar{\boldsymbol{y}}$ and obtain

$$\hat{\boldsymbol{\beta}} = (\boldsymbol{X}'\boldsymbol{P}\boldsymbol{X} + \boldsymbol{\Sigma}^{-1})^{-1}(\boldsymbol{X}'\boldsymbol{P}\boldsymbol{y} + \boldsymbol{\Sigma}^{-1}\boldsymbol{\mu}) . \tag{4.91}$$

If $\hat{\boldsymbol{\beta}}$ is an estimate of traditional statistics, we get $E(\hat{\boldsymbol{\beta}}) = \boldsymbol{\beta}$, since $E(\boldsymbol{\mu}) = \boldsymbol{\beta}$ has been supposed and the vector $\boldsymbol{y}$ of observations is assumed to be a random vector with $E(\boldsymbol{y}) = \boldsymbol{X\beta}$. An unbiased estimate in the sense of traditional statistics is therefore found.

By comparing the estimates (4.87), (4.89) or (4.91) under the assumption of prior information with the estimates (4.15), (4.19) or (4.29) without prior information, it becomes obvious that the estimate with prior information results as weighted mean (4.20) from the estimate without prior information and from the prior information. The weight matrices are obtained with $c = \sigma^2$ in (2.159) by $\boldsymbol{X'PX}$ from (4.16) and by $\boldsymbol{\Sigma}^{-1}$ from the prior information. Thus, we get

$$\hat{\boldsymbol{\beta}} = (\boldsymbol{X'PX} + \boldsymbol{\Sigma}^{-1})^{-1}(\boldsymbol{X'PX}(\boldsymbol{X'PX})^{-1}\boldsymbol{X'Py} + \boldsymbol{\Sigma}^{-1}\boldsymbol{\mu})\,.$$

Example 1: In Example 1 to (4.3) we will now assume that in addition to the observations for determining the unknown quantity s the prior information $E(s) = \mu$ and $V(s) = \sigma^2\sigma_s^2$ for s is available. With $\hat{\boldsymbol{\beta}}_B = \hat{\boldsymbol{\beta}}_M = \hat{\boldsymbol{\beta}} = \hat{s}, \boldsymbol{\mu} = \mu, \boldsymbol{\Sigma} = \sigma_s^2$ and $\boldsymbol{X'P} = |p_1,\ldots,p_n|$ we obtain from (4.87), (4.89) or (4.91)

$$\hat{s} = \frac{1}{\sum_{i=1}^n p_i + 1/\sigma_s^2}\Big(\sum_{i=1}^n p_i y_i + \mu/\sigma_s^2\Big)$$

with the variance $V(s|\boldsymbol{y})$ of s from (4.88)

$$V(s|\boldsymbol{y}) = \sigma^2\Big(\sum_{i=1}^n p_i + 1/\sigma_s^2\Big)^{-1}\,.$$

The variance of the weighted mean of the observations follows from (4.21) by $\sigma^2/\sum_{i=1}^n p_i$ and the variance of the prior information is $\sigma^2\sigma_s^2$. We obtain with $c = \sigma^2$ from (2.160) the weights $\sum_{i=1}^n p_i$ and $1/\sigma_s^2$. The estimate $\hat{s}$ therefore is the weighted mean of the weighted arithmetic mean of the observations and of the prior information. $\blacktriangle$

Applying the matrix identities (4.48) and (4.47) gives

$$(\boldsymbol{X'PX} + \boldsymbol{\Sigma}^{-1})^{-1}\boldsymbol{X'P} = \boldsymbol{\Sigma X'}(\boldsymbol{X\Sigma X'} + \boldsymbol{P}^{-1})^{-1} \tag{4.92}$$

$$(\boldsymbol{X'PX} + \boldsymbol{\Sigma}^{-1})^{-1} = \boldsymbol{\Sigma} - \boldsymbol{\Sigma X'}(\boldsymbol{X\Sigma X'} + \boldsymbol{P}^{-1})^{-1}\boldsymbol{X\Sigma} \tag{4.93}$$

by which the estimates $\hat{\boldsymbol{\beta}}_B, \hat{\boldsymbol{\beta}}_M$ and $\hat{\boldsymbol{\beta}}$, respectively, from (4.87), (4.89) and (4.91) are transformed to

$$\hat{\boldsymbol{\beta}}_B = \hat{\boldsymbol{\beta}}_M = \hat{\boldsymbol{\beta}} = \boldsymbol{\mu} + \boldsymbol{\Sigma X'}(\boldsymbol{X\Sigma X'} + \boldsymbol{P}^{-1})^{-1}(\boldsymbol{y} - \boldsymbol{X\mu})\,. \tag{4.94}$$

The covariance matrix $D(\boldsymbol{\beta}|\boldsymbol{y})$ from (4.88) follows with (4.93) by

$$D(\boldsymbol{\beta}|\boldsymbol{y}) = \sigma^2(\boldsymbol{\Sigma} - \boldsymbol{\Sigma X'}(\boldsymbol{X\Sigma X'} + \boldsymbol{P}^{-1})^{-1}\boldsymbol{X\Sigma})\,. \tag{4.95}$$

It can be recognized from these results which changes the prior information $\boldsymbol{\mu}$ undergoes in the estimate (4.94) and the prior information $\sigma^2\boldsymbol{\Sigma}$ in the covariance matrix (4.95) by adding the observation vector $\boldsymbol{y}$. If $n < u$, the dimensions of the matrices to be inverted in (4.94) and (4.95) are smaller than in (4.87) and (4.88) so that (4.94) and (4.95) have to be preferred to (4.87) and (4.88) for numerical computations.

Example 2: The $u \times 1$ vector $\boldsymbol{\beta}$ of unknown parameters shall be estimated by the $n_1 \times 1$ vector $\boldsymbol{y}_1$ of observations with the $n_1 \times n_1$ positive definite weight matrix $\boldsymbol{P}_1$ and by the independent $n_2 \times 1$ vector $\boldsymbol{y}_2$ of observations with the $n_2 \times n_2$ positive definite weight matrix $\boldsymbol{P}_2$. Let $\boldsymbol{X}_1$ and $\boldsymbol{X}_2$ denote the associated coefficient matrices of the linear model (4.1) or (4.3) which are determined by the linearization (4.8) if necessary. Let the coefficient matrix $\boldsymbol{X}_1$ have full column rank. Let the estimate $\hat{\boldsymbol{\beta}}_1$ of $\boldsymbol{\beta}$ from (4.15), (4.19) or (4.29) by the observation vector $\boldsymbol{y}_1$

$$\hat{\boldsymbol{\beta}}_1 = (\boldsymbol{X}_1'\boldsymbol{P}_1\boldsymbol{X}_1)^{-1}\boldsymbol{X}_1'\boldsymbol{P}_1\boldsymbol{y}_1$$

and the covariance matrix from (4.16)

$$D(\boldsymbol{\beta}_1|\boldsymbol{y}_1) = \sigma^2(\boldsymbol{X}_1'\boldsymbol{P}_1\boldsymbol{X}_1)^{-1}$$

serve as prior information to estimate $\boldsymbol{\beta}$ by $\boldsymbol{y}_2$. Thus, the prior distribution (4.85) follows with

$$\boldsymbol{\beta} \sim N(\hat{\boldsymbol{\beta}}_1, \sigma^2(\boldsymbol{X}_1'\boldsymbol{P}_1\boldsymbol{X}_1)^{-1}) \ .$$

The estimate $\hat{\boldsymbol{\beta}}$ of $\boldsymbol{\beta}$ is then obtained from (4.87), (4.89) or (4.91) by

$$\hat{\boldsymbol{\beta}} = (\boldsymbol{X}_1'\boldsymbol{P}_1\boldsymbol{X}_1 + \boldsymbol{X}_2'\boldsymbol{P}_2\boldsymbol{X}_2)^{-1}(\boldsymbol{X}_1'\boldsymbol{P}_1\boldsymbol{y}_1 + \boldsymbol{X}_2'\boldsymbol{P}_2\boldsymbol{y}_2)$$

with the covariance matrix from (4.88)

$$D(\boldsymbol{\beta}|\boldsymbol{y}_1, \boldsymbol{y}_2) = \sigma^2(\boldsymbol{X}_1'\boldsymbol{P}_1\boldsymbol{X}_1 + \boldsymbol{X}_2'\boldsymbol{P}_2\boldsymbol{X}_2)^{-1} \ .$$

This approach corresponds to the recursive application (2.134) of Bayes' theorem. An identical result is obtained, if the vectors $\boldsymbol{y}_1$ and $\boldsymbol{y}_2$ of observations are jointly analyzed to estimate $\boldsymbol{\beta}$, i.e. if (4.15), (4.19) or (4.29) are applied in the linear model

$$\begin{vmatrix} \boldsymbol{X}_1 \\ \boldsymbol{X}_2 \end{vmatrix} \boldsymbol{\beta} = E(\begin{vmatrix} \boldsymbol{y}_1 \\ \boldsymbol{y}_2 \end{vmatrix} |\boldsymbol{\beta}) \quad \text{with} \quad D(\begin{vmatrix} \boldsymbol{y}_1 \\ \boldsymbol{y}_2 \end{vmatrix} |\sigma^2) = \sigma^2 \begin{vmatrix} \boldsymbol{P}_1^{-1} & \boldsymbol{0} \\ \boldsymbol{0} & \boldsymbol{P}_2^{-1} \end{vmatrix} \ .$$

The alternative estimate (4.94) follows by

$$\hat{\boldsymbol{\beta}} = \hat{\boldsymbol{\beta}}_1 + (\boldsymbol{X}_1'\boldsymbol{P}_1\boldsymbol{X}_1)^{-1}\boldsymbol{X}_2'(\boldsymbol{X}_2(\boldsymbol{X}_1'\boldsymbol{P}_1\boldsymbol{X}_1)^{-1}\boldsymbol{X}_2' + \boldsymbol{P}_2^{-1})^{-1}$$
$$(\boldsymbol{y}_2 - \boldsymbol{X}_2\hat{\boldsymbol{\beta}}_1)$$

and the covariance matrix (4.95) by

$$D(\boldsymbol{\beta}|\boldsymbol{y}_1, \boldsymbol{y}_2) = \sigma^2\big((\boldsymbol{X}_1'\boldsymbol{P}_1\boldsymbol{X}_1)^{-1}$$
$$-(\boldsymbol{X}_1'\boldsymbol{P}_1\boldsymbol{X}_1)^{-1}\boldsymbol{X}_2'(\boldsymbol{X}_2(\boldsymbol{X}_1'\boldsymbol{P}_1\boldsymbol{X}_1)^{-1}\boldsymbol{X}_2'+\boldsymbol{P}_2^{-1})^{-1}\boldsymbol{X}_2(\boldsymbol{X}_1'\boldsymbol{P}_1\boldsymbol{X}_1)^{-1}\big) \ .$$

As was mentioned in connection with (4.94) and (4.95), the change of the prior information by adding observations is recognized from these results. If $\boldsymbol{\beta}$ and $\hat{\boldsymbol{\beta}}$ get the index m and the index 1 is replaced by $m-1$ and the index 2 by m, we obtain the formulas for a recursive estimate of $\boldsymbol{\beta}_m$ by $\hat{\boldsymbol{\beta}}_m$ from $\hat{\boldsymbol{\beta}}_{m-1}$. $\qquad\qquad\blacktriangle$

The $1-\alpha$ confidence hyperellipsoid for $\boldsymbol{\beta}$ is given corresponding to (4.22) with the posterior distribution (4.86) and with (4.91) by

$$(\boldsymbol{\beta} - \hat{\boldsymbol{\beta}})'(\boldsymbol{X}'\boldsymbol{P}\boldsymbol{X} + \boldsymbol{\Sigma}^{-1})(\boldsymbol{\beta} - \hat{\boldsymbol{\beta}})/\sigma^2 = \chi^2_{1-\alpha;u} \; . \tag{4.96}$$

In addition, hypotheses for $\boldsymbol{\beta}$ may be tested by means of the posterior distribution (4.86).

Example 3: The point null hypothesis (4.23)

$$H_0 : \boldsymbol{H}\boldsymbol{\beta} = \boldsymbol{w} \quad \text{versus} \quad H_1 : \boldsymbol{H}\boldsymbol{\beta} \neq \boldsymbol{w}$$

shall be tested by the confidence hyperellipsoid for $\boldsymbol{H}\boldsymbol{\beta}$ which follows corresponding to (4.25) by

$$(\boldsymbol{H}\boldsymbol{\beta} - \boldsymbol{H}\hat{\boldsymbol{\beta}})'(\boldsymbol{H}(\boldsymbol{X}'\boldsymbol{P}\boldsymbol{X} + \boldsymbol{\Sigma}^{-1})^{-1}\boldsymbol{H}')^{-1}(\boldsymbol{H}\boldsymbol{\beta} - \boldsymbol{H}\hat{\boldsymbol{\beta}})/\sigma^2 = \chi^2_{1-\alpha;r} \; . \tag{4.97}$$

By substituting $\boldsymbol{H} = |\boldsymbol{0}, \boldsymbol{I}, \boldsymbol{0}|$, for instance, the confidence hyperellipsoid for a subset of the unknown parameters in $\boldsymbol{\beta}$ is obtained.

The point null hypothesis is accepted as in (4.26), if

$$(\boldsymbol{H}\hat{\boldsymbol{\beta}} - \boldsymbol{w})'(\boldsymbol{H}(\boldsymbol{X}'\boldsymbol{P}\boldsymbol{X} + \boldsymbol{\Sigma}^{-1})^{-1}\boldsymbol{H}')^{-1}(\boldsymbol{H}\hat{\boldsymbol{\beta}} - \boldsymbol{w})/\sigma^2 < \chi^2_{1-\alpha;r} \; . \tag{4.98}$$

$$\blacktriangle$$

4.2.7 Kalman Filter

The unknown parameters of the linear models considered so far do not change with time. Parameters of dynamical systems which are functions of time shall now be introduced. The differential equations which govern, for instance, the orbits of artificial satellites or the currents of oceans establish such dynamical systems. The unknown parameters of the system, for instance, the position and the velocity of a satellite at a certain time, have to be estimated.

Let $\boldsymbol{\beta}_k$ be the $u \times 1$ random vector of unknown parameters at the time k. It is called *state vector*. Let it be linearly transformed by the $u \times u$ *transition matrix* $\boldsymbol{\phi}(k+1, k)$ into the unknown $u \times 1$ state vector $\boldsymbol{\beta}_{k+1}$ at the time $k+1$. The matrix $\boldsymbol{\phi}(k+1, k)$ is assumed to be known. A $u \times 1$ random vector $\boldsymbol{w}_k$ of disturbances which is independent of $\boldsymbol{\beta}_k$ is added. Let $E(\boldsymbol{w}_k) = \boldsymbol{0}$ and $D(\boldsymbol{w}_k) = \boldsymbol{\Omega}_k$ hold true and let the $u \times u$ positive definite covariance matrix $\boldsymbol{\Omega}_k$ be known. If N different moments of time are considered, the *linear dynamical system* is obtained

$$\boldsymbol{\beta}_{k+1} = \boldsymbol{\phi}(k+1, k)\boldsymbol{\beta}_k + \boldsymbol{w}_k$$

with

$$E(\boldsymbol{w}_k) = \boldsymbol{0} , \ D(\boldsymbol{w}_k) = \boldsymbol{\Omega}_k , \ k \in \{1,\ldots,N-1\} \tag{4.99}$$

after linearizing the generally nonlinear differential equations, see for instance JAZWINSKI (1970, p.273) and ARENT et al. (1992).

Observations are available which contain information on the unknown state vectors. Let the relations between the observations and the state vectors be given, if necessary after a linearization, by the linear model (4.1) which for the dynamical system reads

$$\boldsymbol{X}_k\boldsymbol{\beta}_k = E(\boldsymbol{y}_k|\boldsymbol{\beta}_k) \quad \text{with} \quad D(\boldsymbol{y}_k) = \sigma^2\boldsymbol{P}_k^{-1} . \tag{4.100}$$

Here, $\boldsymbol{X}_k$ denotes the $n \times u$ matrix of given coefficients with $\text{rank}\boldsymbol{X}_k = u$, $\boldsymbol{y}_k$ the $n \times 1$ random vector of observations and $\boldsymbol{P}_k$ the $n \times n$ positive definite weight matrix of the observations. The variance factor σ^2 is being assumed as given. The estimation of the parameters with unknown σ^2 can be found for instance in KOCH (1990, p.96) and WEST and HARRISON (1989, p.117).

Let the disturbances $\boldsymbol{w}_k$ and $\boldsymbol{w}_l$ and the observations $\boldsymbol{y}_k$ and $\boldsymbol{y}_l$ be independent for $k \neq l$. In addition, let $\boldsymbol{w}_k$ and $\boldsymbol{y}_k$ be independent and normally distributed, hence

$$\boldsymbol{w}_k \sim N(\boldsymbol{0},\boldsymbol{\Omega}_k) \tag{4.101}$$

and

$$\boldsymbol{y}_k|\boldsymbol{\beta}_k \sim N(\boldsymbol{X}_k\boldsymbol{\beta}_k,\sigma^2\boldsymbol{P}_k^{-1}) . \tag{4.102}$$

Based on the observations $\boldsymbol{y}_k$ at the time k the posterior density function of the unknown random vector $\boldsymbol{\beta}_k$ shall be determined. The prior information is given by the posterior distribution for $\boldsymbol{\beta}_{k-1}$ which is obtained by the observations $\boldsymbol{y}_{k-1}$ at the time $k-1$. The posterior distribution for $\boldsymbol{\beta}_{k-1}$ on the other hand uses the prior information which stems from $\boldsymbol{y}_{k-2}$ and so forth. Bayes' theorem is therefore applied recursively according to (2.134). This is possible, because the vectors of observations are independent by assumption.

Let the prior information for the state vector $\boldsymbol{\beta}_1$ at the first moment of time be introduced by the prior distribution

$$\boldsymbol{\beta}_1|\boldsymbol{y}_0 \sim N(\hat{\boldsymbol{\beta}}_{1,0},\sigma^2\boldsymbol{\Sigma}_{1,0}) \tag{4.103}$$

whose parameters, the vector $\hat{\boldsymbol{\beta}}_{1,0}$ and the covariance matrix $\sigma^2\boldsymbol{\Sigma}_{1,0}$, are given by the prior information. The first index in $\hat{\boldsymbol{\beta}}_{1,0}$ and $\boldsymbol{\Sigma}_{1,0}$ refers to the time for which $\boldsymbol{\beta}_1$ is defined and the second index to the time when the vector $\boldsymbol{y}_0$ of observations has been taken. No observations are available at the beginning so that the vector $\boldsymbol{y}_0$ is not used anymore in the sequel.

As a comparison with (2.225) and (2.226) shows, the density function of the prior distribution (4.103) in connection with the likelihood function

(4.102) for $k = 1$ is a conjugate prior. The posterior density function for $\boldsymbol{\beta}_1$ therefore results again from the normal distribution. By transforming $\boldsymbol{\beta}_1$ to $\boldsymbol{\beta}_2$ with (4.99) also $\boldsymbol{\beta}_2$ is normally distributed of the form (4.103) so that a prior distribution is obtained for $\boldsymbol{\beta}_2$. This will be shown immediately for the time k, since Bayes' theorem is recursively applied with (2.134). Let the prior distribution for $\boldsymbol{\beta}_k$ be given according to (2.134) as in (4.103) by

$$\boldsymbol{\beta}_k | \boldsymbol{y}_1, \ldots, \boldsymbol{y}_{k-1} \sim N(\hat{\boldsymbol{\beta}}_{k,k-1}, \sigma^2 \boldsymbol{\Sigma}_{k,k-1}) \ . \tag{4.104}$$

This conjugate prior leads according to (2.227) together with the likelihood function from (4.102) to the posterior distribution for $\boldsymbol{\beta}_k$

$$\boldsymbol{\beta}_k | \boldsymbol{y}_1, \ldots, \boldsymbol{y}_k \sim N\big(\boldsymbol{\mu}_0, \sigma^2 (\boldsymbol{X}_k' \boldsymbol{P}_k \boldsymbol{X}_k + \boldsymbol{\Sigma}_{k,k-1}^{-1})^{-1}\big) \tag{4.105}$$

with

$$\boldsymbol{\mu}_0 = (\boldsymbol{X}_k' \boldsymbol{P}_k \boldsymbol{X}_k + \boldsymbol{\Sigma}_{k,k-1}^{-1})^{-1} (\boldsymbol{X}_k' \boldsymbol{P}_k \boldsymbol{y}_k + \boldsymbol{\Sigma}_{k,k-1}^{-1} \hat{\boldsymbol{\beta}}_{k,k-1}) \ .$$

Equivalently to (4.87), (4.89) or (4.91), the estimate $\hat{\boldsymbol{\beta}}_{k,k}$ of $\boldsymbol{\beta}_k$ is then obtained by

$$\hat{\boldsymbol{\beta}}_{k,k} = (\boldsymbol{X}_k' \boldsymbol{P}_k \boldsymbol{X}_k + \boldsymbol{\Sigma}_{k,k-1}^{-1})^{-1} (\boldsymbol{X}_k' \boldsymbol{P}_k \boldsymbol{y}_k + \boldsymbol{\Sigma}_{k,k-1}^{-1} \hat{\boldsymbol{\beta}}_{k,k-1}) \tag{4.106}$$

and with (4.88) the covariance matrix $D(\boldsymbol{\beta}_k | \boldsymbol{y}_1, \ldots, \boldsymbol{y}_k) = \sigma^2 \boldsymbol{\Sigma}_{k,k}$ of $\boldsymbol{\beta}_k$ by

$$\boldsymbol{\Sigma}_{k,k} = (\boldsymbol{X}_k' \boldsymbol{P}_k \boldsymbol{X}_k + \boldsymbol{\Sigma}_{k,k-1}^{-1})^{-1} \ . \tag{4.107}$$

With the identities (4.92) and (4.93) the result (4.106) and with (4.93) the result (4.107) is transformed to

$$\hat{\boldsymbol{\beta}}_{k,k} = \hat{\boldsymbol{\beta}}_{k,k-1} + \boldsymbol{F}_k (\boldsymbol{y}_k - \boldsymbol{X}_k \hat{\boldsymbol{\beta}}_{k,k-1}) \tag{4.108}$$

with

$$\boldsymbol{F}_k = \boldsymbol{\Sigma}_{k,k-1} \boldsymbol{X}_k' (\boldsymbol{X}_k \boldsymbol{\Sigma}_{k,k-1} \boldsymbol{X}_k' + \boldsymbol{P}_k^{-1})^{-1} \tag{4.109}$$

and

$$\boldsymbol{\Sigma}_{k,k} = (\boldsymbol{I} - \boldsymbol{F}_k \boldsymbol{X}_k) \boldsymbol{\Sigma}_{k,k-1} \ . \tag{4.110}$$

Thus, the estimate $\hat{\boldsymbol{\beta}}_{k,k}$ and the covariance matrix $\sigma^2 \boldsymbol{\Sigma}_{k,k}$ is recursively computed from $\hat{\boldsymbol{\beta}}_{k,k-1}$ and $\sigma^2 \boldsymbol{\Sigma}_{k,k-1}$.

It remains to be shown, how the prior density distribution (4.104) for $\boldsymbol{\beta}_k$ results from the posterior distribution for $\boldsymbol{\beta}_{k-1}$. With (4.105) to (4.107) we find the posterior distribution

$$\boldsymbol{\beta}_{k-1} | \boldsymbol{y}_1, \ldots, \boldsymbol{y}_{k-1} \sim N(\hat{\boldsymbol{\beta}}_{k-1,k-1}, \sigma^2 \boldsymbol{\Sigma}_{k-1,k-1}) \ . \tag{4.111}$$

By the linear dynamical system (4.99) the state vector $\boldsymbol{\beta}_{k-1}$ is transformed to $\boldsymbol{\beta}_k$. The distribution for $\boldsymbol{\beta}_k$ is therefore obtained with (2.202), since $\boldsymbol{\beta}_{k-1}$ and $\boldsymbol{w}_{k-1}$ are independent by assumption,

$$\boldsymbol{\beta}_k|\boldsymbol{y}_1,\ldots,\boldsymbol{y}_{k-1} \sim N(\hat{\boldsymbol{\beta}}_{k,k-1},\sigma^2\boldsymbol{\Sigma}_{k,k-1}) \tag{4.112}$$

with

$$\hat{\boldsymbol{\beta}}_{k,k-1} = \boldsymbol{\phi}(k,k-1)\hat{\boldsymbol{\beta}}_{k-1,k-1} \tag{4.113}$$

and

$$\sigma^2\boldsymbol{\Sigma}_{k,k-1} = \sigma^2\boldsymbol{\phi}(k,k-1)\boldsymbol{\Sigma}_{k-1,k-1}\boldsymbol{\phi}'(k,k-1) + \boldsymbol{\Omega}_{k-1} \ . \tag{4.114}$$

The normal distribution (4.112) is identical with (4.104) so that the prior distribution for $\boldsymbol{\beta}_k$ has been found. It was therefore justified to start the derivation of the distributions immediately at the time k instead of at the first moment of time.

The recursive estimation (4.108) to (4.110) together with (4.113) and (4.114) establishes the *Kalman filter*, also called *Kalman-Bucy filter*. It computes new estimates of the state vector, as soon as new observations arrive. It may therefore be applied in real time.

If one assumes that the observations are not normally distributed but have the density functions (4.56) and (4.57), which take care of outliers in the observations, a robust Kalman filter is obtained (KOCH and YANG 1998B). For adaptive Kalman filters see for instance YANG and GAO (2006).

4.3 Linear Model with Unknown Variance Factor

Contrary to Chapter 4.2 the variance factor σ^2 of the linear model (4.1) is now assumed as unknown random variable. Noninformative priors are first introduced for the unknown parameters, then, informative priors in Chapter 4.3.2. Since the variance factor σ^2 is now an unknown parameter, it is replaced, as proposed for the conjugate normal-gamma distribution (2.234), by the unknown weight parameter τ from (2.219)

$$\tau = 1/\sigma^2 \ . \tag{4.115}$$

We then obtain as in (2.233) instead of (4.4) the likelihood function

$$p(\boldsymbol{y}|\boldsymbol{\beta},\tau) = (2\pi)^{-n/2}(\det \boldsymbol{P})^{1/2}\tau^{n/2} \exp\left[-\frac{\tau}{2}(\boldsymbol{y}-\boldsymbol{X}\boldsymbol{\beta})'\boldsymbol{P}(\boldsymbol{y}-\boldsymbol{X}\boldsymbol{\beta})\right] \ . \tag{4.116}$$

4.3.1 Noninformative Priors

The noninformative prior density function (2.216) which is determined by a constant is selected for the vector $\boldsymbol{\beta}$ of unknown parameters and the noninformative prior density function (2.220), which is proportional to $1/\tau$, for the weight parameter τ. Bayes' theorem (2.122) then gives with the likelihood function (4.116) the posterior density function $p(\boldsymbol{\beta}, \tau|\boldsymbol{y})$ with

$$p(\boldsymbol{\beta}, \tau|\boldsymbol{y}) \propto \tau^{n/2-1} \exp\left[-\frac{\tau}{2}(\boldsymbol{y} - \boldsymbol{X}\boldsymbol{\beta})'\boldsymbol{P}(\boldsymbol{y} - \boldsymbol{X}\boldsymbol{\beta})\right] \tag{4.117}$$

where only terms depending on $\boldsymbol{\beta}$ and τ need to be considered. The exponent in (4.117) may be transformed as in (2.237) and (4.12) by

$$\begin{aligned}
(\boldsymbol{y} - \boldsymbol{X}\boldsymbol{\beta})'\boldsymbol{P}(\boldsymbol{y} - \boldsymbol{X}\boldsymbol{\beta}) &= \boldsymbol{y}'\boldsymbol{P}\boldsymbol{y} - 2\boldsymbol{\beta}'\boldsymbol{X}'\boldsymbol{P}\boldsymbol{y} + \boldsymbol{\beta}'\boldsymbol{X}'\boldsymbol{P}\boldsymbol{X}\boldsymbol{\beta} \\
&= \boldsymbol{y}'\boldsymbol{P}\boldsymbol{y} - 2\boldsymbol{\mu}_0'\boldsymbol{X}'\boldsymbol{P}\boldsymbol{y} + \boldsymbol{\mu}_0'\boldsymbol{X}'\boldsymbol{P}\boldsymbol{X}\boldsymbol{\mu}_0 + (\boldsymbol{\beta} - \boldsymbol{\mu}_0)'\boldsymbol{X}'\boldsymbol{P}\boldsymbol{X}(\boldsymbol{\beta} - \boldsymbol{\mu}_0) \\
&= (\boldsymbol{y} - \boldsymbol{X}\boldsymbol{\mu}_0)'\boldsymbol{P}(\boldsymbol{y} - \boldsymbol{X}\boldsymbol{\mu}_0) + (\boldsymbol{\beta} - \boldsymbol{\mu}_0)'\boldsymbol{X}'\boldsymbol{P}\boldsymbol{X}(\boldsymbol{\beta} - \boldsymbol{\mu}_0)
\end{aligned} \tag{4.118}$$

with $\boldsymbol{\mu}_0$ from (4.13)

$$\boldsymbol{\mu}_0 = (\boldsymbol{X}'\boldsymbol{P}\boldsymbol{X})^{-1}\boldsymbol{X}'\boldsymbol{P}\boldsymbol{y} . \tag{4.119}$$

By substituting this result in (4.117) a comparison with (2.212) shows because of $n/2 - 1 = u/2 + (n - u)/2 - 1$ and because of (4.39) that the posterior density function (4.117) results from the normal-gamma distribution

$$\boldsymbol{\beta}, \tau|\boldsymbol{y} \sim NG(\boldsymbol{\mu}_0, (\boldsymbol{X}'\boldsymbol{P}\boldsymbol{X})^{-1}, (n - u)\hat{\sigma}^2/2, (n - u)/2) . \tag{4.120}$$

The posterior marginal distribution for the vector $\boldsymbol{\beta}$ of unknown parameters is then according to (2.213) determined by the multivariate t-distribution

$$\boldsymbol{\beta}|\boldsymbol{y} \sim t(\boldsymbol{\mu}_0, \hat{\sigma}^2(\boldsymbol{X}'\boldsymbol{P}\boldsymbol{X})^{-1}, n - u) \tag{4.121}$$

with the vector of expected values from (2.208)

$$E(\boldsymbol{\beta}|\boldsymbol{y}) = \boldsymbol{\mu}_0 . \tag{4.122}$$

The Bayes estimate $\hat{\boldsymbol{\beta}}_B$ of $\boldsymbol{\beta}$ therefore is obtained from (3.9) by

$$\hat{\boldsymbol{\beta}}_B = (\boldsymbol{X}'\boldsymbol{P}\boldsymbol{X})^{-1}\boldsymbol{X}'\boldsymbol{P}\boldsymbol{y} \tag{4.123}$$

and in agreement with this result the MAP estimate $\hat{\boldsymbol{\beta}}_M$ of $\boldsymbol{\beta}$ from (3.30), since the density function (2.204) of the multivariate t-distribution has a maximum at the point $\boldsymbol{\mu}_0$, thus

$$\hat{\boldsymbol{\beta}}_M = (\boldsymbol{X}'\boldsymbol{P}\boldsymbol{X})^{-1}\boldsymbol{X}'\boldsymbol{P}\boldsymbol{y} . \tag{4.124}$$

These two results are identical with the two estimates (4.15) and (4.19) and with the estimate (4.29) of the method of least squares. The estimates for the unknown parameters $\boldsymbol{\beta}$ are therefore identical independent of the variance factor σ^2 being known or unknown. Assumed are for both cases noninformative priors. With (4.165) a corresponding result is obtained for informative priors.

The covariance matrix $D(\boldsymbol{\beta}|\boldsymbol{y})$ if found from (4.121) with (2.208) and (3.11) by

$$D(\boldsymbol{\beta}|\boldsymbol{y}) = \frac{n-u}{n-u-2}\hat{\sigma}^2(\boldsymbol{X}'\boldsymbol{P}\boldsymbol{X})^{-1} \; . \tag{4.125}$$

Except for the factor $(n-u)/(n-u-2)$ which goes for large values of $n-u$ towards one, this is also a result of traditional statistics. It follows from (4.31) after substituting σ^2 by $\hat{\sigma}^2$ from (4.39).

The marginal distribution for the weight parameter τ obtained from the posterior distribution (4.120) is the gamma distribution because of (2.214)

$$\tau|\boldsymbol{y} \sim G((n-u)\hat{\sigma}^2/2, (n-u)/2) \; . \tag{4.126}$$

The inverted gamma distribution therefore results from (2.176) for the variance factor σ^2

$$\sigma^2|\boldsymbol{y} \sim IG((n-u)\hat{\sigma}^2/2, (n-u)/2) \tag{4.127}$$

with the expected value from (2.177)

$$E(\sigma^2|\boldsymbol{y}) = \frac{n-u}{n-u-2}\hat{\sigma}^2 \; . \tag{4.128}$$

We therefore get the Bayes estimate $\hat{\sigma}_B^2$ of σ^2 from (3.9) by

$$\hat{\sigma}_B^2 = \frac{n-u}{n-u-2}\hat{\sigma}^2 \; . \tag{4.129}$$

Except for the factor $(n-u)/(n-u-2)$ this is again a result of the traditional statistics, as shown with (4.39).

The variance $V(\sigma^2|\boldsymbol{y})$ of σ^2 follows with (2.177) and (3.11) by

$$V(\sigma^2|\boldsymbol{y}) = \frac{2(n-u)^2(\hat{\sigma}^2)^2}{(n-u-2)^2(n-u-4)} \; . \tag{4.130}$$

The corresponding result of traditional statistics is $V(\sigma^2) = 2(\hat{\sigma}^2)^2/(n-u)$, see for instance KOCH (1999, p.244).

The confidence region for the vector $\boldsymbol{\beta}$ of unknown parameters is obtained by the multivariate t-distribution (4.121). Its density function is according to (2.204) a monotonically decreasing function of the quadratic form $(\boldsymbol{\beta}-\boldsymbol{\mu}_0)'\boldsymbol{X}'\boldsymbol{P}\boldsymbol{X}(\boldsymbol{\beta}-\boldsymbol{\mu}_0)/\hat{\sigma}^2$. A hypersurface of equal density is therefore given with (4.29) and (4.119) by the relation

$$(\boldsymbol{\beta}-\hat{\boldsymbol{\beta}})'\boldsymbol{X}'\boldsymbol{P}\boldsymbol{X}(\boldsymbol{\beta}-\hat{\boldsymbol{\beta}})/\hat{\sigma}^2 = \text{const}$$

which has the shape of a hyperellipsoid. The quadratic form has because of (2.211) the F-distribution (2.182)

$$(\boldsymbol{\beta}-\hat{\boldsymbol{\beta}})'\boldsymbol{X}'\boldsymbol{P}\boldsymbol{X}(\boldsymbol{\beta}-\hat{\boldsymbol{\beta}})/(u\hat{\sigma}^2) \sim F(u, n-u) \; . \tag{4.131}$$

The $1 - \alpha$ confidence hyperellipsoid is therefore corresponding to (3.36) determined by

$$(\boldsymbol{\beta} - \hat{\boldsymbol{\beta}})' \boldsymbol{X}' \boldsymbol{P} \boldsymbol{X} (\boldsymbol{\beta} - \hat{\boldsymbol{\beta}})/(u\hat{\sigma}^2) = F_{1-\alpha;u,n-u} \tag{4.132}$$

where $F_{1-\alpha;u,n-u}$ denotes the upper α-percentage point of the F-distribution defined in (2.183). The confidence interval for the variance factor σ^2 follows from (4.127).

Example 1: We find for the unknown quantity s in the Example 1 to (4.3) with $\hat{s}$ from (4.20) according to (4.131) the distribution

$$(s - \hat{s})^2 \sum_{i=1}^{n} p_i/\hat{\sigma}^2 \sim F(1, n-1) \, . \tag{4.133}$$

With (4.21) we define $\hat{\sigma}_s^2$ by

$$\hat{\sigma}_s^2 = \hat{\sigma}^2 / \sum_{i=1}^{n} p_i \, . \tag{4.134}$$

It is the variance of the quantity s which is computed by the estimated variance factor $\hat{\sigma}^2$. With the definition (2.183) of the upper α-percentage point of the F-distribution in (4.132) we find

$$P((s - \hat{s})^2/\hat{\sigma}_s^2 < F_{1-\alpha;1,n-1}) = 1 - \alpha \, . \tag{4.135}$$

The quantity $t_{1-\alpha;n-1}$ of the t-distribution from (2.186) which is equivalent to the upper α-percentage point $F_{1-\alpha;1,n-1}$ of the F-distribution

$$t_{1-\alpha;n-1} = (F_{1-\alpha;1,n-1})^{1/2} \tag{4.136}$$

gives because of (2.187)

$$P(-t_{1-\alpha;n-1} < (s - \hat{s})/\hat{\sigma}_s < t_{1-\alpha;n-1}) = 1 - \alpha \, . \tag{4.137}$$

This finally leads to the $1 - \alpha$ confidence interval for s

$$P(\hat{s} - \hat{\sigma}_s t_{1-\alpha;n-1} < s < \hat{s} + \hat{\sigma}_s t_{1-\alpha;n-1}) = 1 - \alpha \, . \tag{4.138}$$

This confidence interval may also be derived from the posterior marginal distribution (4.121). With (4.20) and (4.134) the generalized t-distribution follows by

$$s|\boldsymbol{y} \sim t(\hat{s}, \hat{\sigma}_s^2, n-1) \tag{4.139}$$

and the t-distribution from the transformation (2.206) into the standard form (2.207)

$$(s - \hat{s})/\hat{\sigma}_s \sim t(n-1) \, . \tag{4.140}$$

Then, because of (2.187) the relation (4.137) is valid which leads to the confidence interval (4.138). $\triangle$

By the posterior distribution (4.121) for $\boldsymbol{\beta}$ and (4.127) for σ^2 also hypotheses for $\boldsymbol{\beta}$ and σ^2 may be tested as explained in Chapter 3.4.

Example 2: Let the point null hypothesis (4.23)

$$H_0 : \boldsymbol{H\beta} = \boldsymbol{w} \quad \text{versus} \quad H_1 : \boldsymbol{H\beta} \neq \boldsymbol{w} \tag{4.141}$$

be tested with (3.82) by the confidence hyperellipsoid for $\boldsymbol{H\beta}$. The distribution of the linear transformation $\boldsymbol{H\beta}$ follows from (2.210) with (4.121) by

$$\boldsymbol{H\beta}|\boldsymbol{y} \sim t(\boldsymbol{H\mu}_0, \hat{\sigma}^2 \boldsymbol{H}(\boldsymbol{X'PX})^{-1}\boldsymbol{H'}, n - u) \, . \tag{4.142}$$

Furthermore, we get with (2.211), (4.29) and (4.119)

$$(\boldsymbol{H\beta} - \boldsymbol{H\hat{\beta}})'(\boldsymbol{H}(\boldsymbol{X'PX})^{-1}\boldsymbol{H'})^{-1}(\boldsymbol{H\beta} - \boldsymbol{H\hat{\beta}})/(r\hat{\sigma}^2) \sim F(r, n-u) \, . \tag{4.143}$$

The $1 - \alpha$ confidence hyperellipsoid for $\boldsymbol{H\beta}$ is therefore obtained corresponding to (3.36) and (4.132) by

$$(\boldsymbol{H\beta} - \boldsymbol{H\hat{\beta}})'(\boldsymbol{H}(\boldsymbol{X'PX})^{-1}\boldsymbol{H'})^{-1}(\boldsymbol{H\beta} - \boldsymbol{H\hat{\beta}})/(r\hat{\sigma}^2) = F_{1-\alpha;r,n-u} \, . \tag{4.144}$$

It is identical with the confidence hyperellipsoid of traditional statistics, see for instance KOCH (1999, p.300). If we substitute $\boldsymbol{H} = |\boldsymbol{0}, \boldsymbol{I}, \boldsymbol{0}|$, for instance, the confidence hyperellipsoid for a subset of the unknown parameters in $\boldsymbol{\beta}$ follows.

Corresponding to (3.44) and (3.82) the point null hypothesis (4.141) is accepted, if

$$(\boldsymbol{H\hat{\beta}} - \boldsymbol{w})'(\boldsymbol{H}(\boldsymbol{X'PX})^{-1}\boldsymbol{H'})^{-1}(\boldsymbol{H\hat{\beta}} - \boldsymbol{w})/(r\hat{\sigma}^2) < F_{1-\alpha;r,n-u} \, . \tag{4.145}$$

This is the test procedure of traditional statistics for the hypothesis (4.141), see for instance KOCH (1999, p.280). $\triangle$

Example 3: The point null hypothesis for the unknown quantity s in Example 1 to (4.3)

$$H_0 : s = s_0 \quad \text{versus} \quad H_1 : s \neq s_0 \tag{4.146}$$

shall be tested. With $r = 1, \boldsymbol{w} = s_0, \boldsymbol{H} = 1$, $\hat{s}$ from (4.20) and $\hat{\sigma}_s^2$ from (4.134) the null hypothesis is accepted according to (4.145), if

$$(\hat{s} - s_0)^2/\hat{\sigma}_s^2 < F_{1-\alpha;1,n-1} \, . \tag{4.147}$$

$\triangle$

Example 4: To detect movements of buildings, the coordinates $\boldsymbol{\beta}_i$ of points connected with the buildings are determined by the measurements $\boldsymbol{y}_i$ with $i \in \{1, \ldots, p\}$ at p different times. The observation vectors $\boldsymbol{y}_i$ of the different times are independent and have the positive definite weight matrices $\boldsymbol{P}_i$. The associated matrices of coefficients are $\boldsymbol{X}_i$. If necessary, they are determined from (4.8) by a linearization. They are assumed to have full column rank. It is suspected that the positions of a subset of points whose coordinates are denoted by $\boldsymbol{\beta}_{fi}$ remain unchanged during the p times of the measurements, while the positions of the remaining points with the coordinates $\boldsymbol{\beta}_{ui}$ have changed. The coordinates $\boldsymbol{\beta}_{fi}$ shall appear first in the vector $\boldsymbol{\beta}_i$ of coordinates, thus $\boldsymbol{\beta}_i = |\boldsymbol{\beta}'_{fi}, \boldsymbol{\beta}'_{ui}|'$ with $i \in \{1, \dots, p\}$. The null hypothesis that the points with the coordinates $\boldsymbol{\beta}_{fi}$ did not move shall be tested against the alternative hypothesis that the points have moved.

The linear model (4.1) for estimating the unknown coordinates $\boldsymbol{\beta}_i$ is given by

$$
\left|\begin{array}{cccc}
\boldsymbol{X}_1 & \boldsymbol{0} & \dots & \boldsymbol{0} \\
\boldsymbol{0} & \boldsymbol{X}_2 & \dots & \boldsymbol{0} \\
\hdotsfor{4} \\
\boldsymbol{0} & \boldsymbol{0} & \dots & \boldsymbol{X}_p
\end{array}\right|
\left|\begin{array}{c}
\boldsymbol{\beta}_1 \\ \boldsymbol{\beta}_2 \\ \dots \\ \boldsymbol{\beta}_p
\end{array}\right|
= E\left(
\left|\begin{array}{c}
\boldsymbol{y}_1|\boldsymbol{\beta}_1 \\ \boldsymbol{y}_2|\boldsymbol{\beta}_2 \\ \dots \\ \boldsymbol{y}_p|\boldsymbol{\beta}_p
\end{array}\right|
\right)
$$

$$
\text{with}\quad
D\left(
\left|\begin{array}{c}
\boldsymbol{y}_1 \\ \boldsymbol{y}_2 \\ \dots \\ \boldsymbol{y}_p
\end{array}\right|
|\sigma^2\right) = \sigma^2
\left|\begin{array}{cccc}
\boldsymbol{P}_1^{-1} & \boldsymbol{0} & \dots & \boldsymbol{0} \\
\boldsymbol{0} & \boldsymbol{P}_2^{-1} & \dots & \boldsymbol{0} \\
\hdotsfor{4} \\
\boldsymbol{0} & \boldsymbol{0} & \dots & \boldsymbol{P}_p^{-1}
\end{array}\right| .
$$

According to (4.123) or (4.124) the estimate $\hat{\boldsymbol{\beta}}_i$ of $\boldsymbol{\beta}_i$ is obtained with

$$
\left|\begin{array}{cccc}
\boldsymbol{X}'_1 & \boldsymbol{0} & \dots & \boldsymbol{0} \\
\boldsymbol{0} & \boldsymbol{X}'_2 & \dots & \boldsymbol{0} \\
\hdotsfor{4} \\
\boldsymbol{0} & \boldsymbol{0} & \dots & \boldsymbol{X}'_p
\end{array}\right|
\left|\begin{array}{cccc}
\boldsymbol{P}_1 & \boldsymbol{0} & \dots & \boldsymbol{0} \\
\boldsymbol{0} & \boldsymbol{P}_2 & \dots & \boldsymbol{0} \\
\hdotsfor{4} \\
\boldsymbol{0} & \boldsymbol{0} & \dots & \boldsymbol{P}_p
\end{array}\right|
$$

$$
=
\left|\begin{array}{cccc}
\boldsymbol{X}'_1\boldsymbol{P}_1 & \boldsymbol{0} & \dots & \boldsymbol{0} \\
\boldsymbol{0} & \boldsymbol{X}'_2\boldsymbol{P}_2 & \dots & \boldsymbol{0} \\
\hdotsfor{4} \\
\boldsymbol{0} & \boldsymbol{0} & \dots & \boldsymbol{X}'_p\boldsymbol{P}_p
\end{array}\right|
$$

from the normal equations

$$
\left|\begin{array}{c}
\hat{\boldsymbol{\beta}}_1 \\ \hat{\boldsymbol{\beta}}_2 \\ \dots \\ \hat{\boldsymbol{\beta}}_p
\end{array}\right|
$$

$$
=
\left|\begin{array}{cccc}
(\boldsymbol{X}'_1\boldsymbol{P}_1\boldsymbol{X}_1)^{-1} & \boldsymbol{0} & \dots & \boldsymbol{0} \\
\boldsymbol{0} & (\boldsymbol{X}'_2\boldsymbol{P}_2\boldsymbol{X}_2)^{-1} & \dots & \boldsymbol{0} \\
\hdotsfor{4} \\
\boldsymbol{0} & \boldsymbol{0} & \dots & (\boldsymbol{X}'_p\boldsymbol{P}_p\boldsymbol{X}_p)^{-1}
\end{array}\right|
\left|\begin{array}{c}
\boldsymbol{X}'_1\boldsymbol{P}_1\boldsymbol{y}_1 \\ \boldsymbol{X}'_2\boldsymbol{P}_2\boldsymbol{y}_2 \\ \dots \\ \boldsymbol{X}'_p\boldsymbol{P}_p\boldsymbol{y}_p
\end{array}\right| ,
$$

thus

$$\hat{\boldsymbol{\beta}}_i = (\boldsymbol{X}'_i \boldsymbol{P}_i \boldsymbol{X}_i)^{-1} \boldsymbol{X}'_i \boldsymbol{P}_i \boldsymbol{y}_i \ .$$

The null hypothesis means that the coordinates $\boldsymbol{\beta}_{fi}$ between successive times of measurements remain unchanged. The point null hypothesis

$$H_0 : \begin{vmatrix} \boldsymbol{\beta}_{f1} - \boldsymbol{\beta}_{f2} \\ \boldsymbol{\beta}_{f2} - \boldsymbol{\beta}_{f3} \\ \cdots\cdots\cdots \\ \boldsymbol{\beta}_{f,p-1} - \boldsymbol{\beta}_{fp} \end{vmatrix} = \boldsymbol{0} \quad \text{versus} \quad H_1 : \begin{vmatrix} \boldsymbol{\beta}_{f1} - \boldsymbol{\beta}_{f2} \\ \boldsymbol{\beta}_{f2} - \boldsymbol{\beta}_{f3} \\ \cdots\cdots\cdots \\ \boldsymbol{\beta}_{f,p-1} - \boldsymbol{\beta}_{fp} \end{vmatrix} \neq \boldsymbol{0}$$

therefore shall be tested. We apply (4.145) so that first the matrix $\boldsymbol{H}$ needs to be specified. We obtain corresponding to the partitioning of $\boldsymbol{\beta}$ in

$$\boldsymbol{\beta} = |\boldsymbol{\beta}'_{f1}, \boldsymbol{\beta}'_{u1}, \boldsymbol{\beta}'_{f2}, \boldsymbol{\beta}'_{u2}, \ldots, \boldsymbol{\beta}'_{fp}, \boldsymbol{\beta}'_{up}|'$$

the matrix $\boldsymbol{H}$ with

$$\boldsymbol{H} = \begin{vmatrix} \boldsymbol{I} & \boldsymbol{0} & -\boldsymbol{I} & \boldsymbol{0} & \boldsymbol{0} & \boldsymbol{0} & \ldots & \boldsymbol{0} & \boldsymbol{0} & \boldsymbol{0} & \boldsymbol{0} \\ \boldsymbol{0} & \boldsymbol{0} & \boldsymbol{I} & \boldsymbol{0} & -\boldsymbol{I} & \boldsymbol{0} & \ldots & \boldsymbol{0} & \boldsymbol{0} & \boldsymbol{0} & \boldsymbol{0} \\ \cdots & \cdots & \cdots & \cdots & \cdots & \cdots & \cdots & \cdots & \cdots & \cdots & \cdots \\ \boldsymbol{0} & \boldsymbol{0} & \boldsymbol{0} & \boldsymbol{0} & \boldsymbol{0} & \boldsymbol{0} & \ldots & \boldsymbol{I} & \boldsymbol{0} & -\boldsymbol{I} & \boldsymbol{0} \end{vmatrix} ,$$

by which we get

$$\boldsymbol{H}\boldsymbol{\beta} = \begin{vmatrix} \boldsymbol{\beta}_{f1} - \boldsymbol{\beta}_{f2} \\ \boldsymbol{\beta}_{f2} - \boldsymbol{\beta}_{f3} \\ \cdots\cdots\cdots \\ \boldsymbol{\beta}_{f,p-1} - \boldsymbol{\beta}_{fp} \end{vmatrix} .$$

If the inverse of the matrix of normal equations given above is split according to the partitioning $\boldsymbol{\beta}_i = |\boldsymbol{\beta}'_{fi}, \boldsymbol{\beta}'_{ui}|'$ into the part $(\boldsymbol{X}'_i \boldsymbol{P}_i \boldsymbol{X}_i)^{-1}_f$, which belongs to $\boldsymbol{\beta}_{fi}$, and into the part belonging to $\boldsymbol{\beta}_{ui}$, which is of no further interest and which is denoted in the following equation by ——, we obtain for the product $\boldsymbol{H}(\boldsymbol{X}'\boldsymbol{P}\boldsymbol{X})^{-1}$ in (4.145)

$$\boldsymbol{H}(\boldsymbol{X}'\boldsymbol{P}\boldsymbol{X})^{-1}$$

$$= \begin{vmatrix} (\boldsymbol{X}'_1 \boldsymbol{P}_1 \boldsymbol{X}_1)^{-1}_f & \text{——} & -(\boldsymbol{X}'_2 \boldsymbol{P}_2 \boldsymbol{X}_2)^{-1}_f & \text{——} & \boldsymbol{0} & \boldsymbol{0} & \ldots \\ \boldsymbol{0} & \boldsymbol{0} & (\boldsymbol{X}'_2 \boldsymbol{P}_2 \boldsymbol{X}_2)^{-1}_f & \text{——} & -(\boldsymbol{X}'_3 \boldsymbol{P}_3 \boldsymbol{X}_3)^{-1}_f & \text{——} & \ldots \\ \cdots & \cdots & \cdots & \cdots & \cdots & \cdots & \cdots \end{vmatrix}$$

and for the product $(\boldsymbol{H}(\boldsymbol{X}'\boldsymbol{P}\boldsymbol{X})^{-1}\boldsymbol{H}')^{-1}$

$$(\boldsymbol{H}(\boldsymbol{X}'\boldsymbol{P}\boldsymbol{X})^{-1}\boldsymbol{H}')^{-1} = \left| \begin{array}{cccc} (\boldsymbol{X}_1'\boldsymbol{P}_1\boldsymbol{X}_1)_f^{-1} + (\boldsymbol{X}_2'\boldsymbol{P}_2\boldsymbol{X}_2)_f^{-1} & & & \\ -(\boldsymbol{X}_2'\boldsymbol{P}_2\boldsymbol{X}_2)_f^{-1} & & & \\ \hdotsfor{4} \\ -(\boldsymbol{X}_2'\boldsymbol{P}_2\boldsymbol{X}_2)_f^{-1} & \mathbf{0} & \mathbf{0} & \cdots \\ (\boldsymbol{X}_2'\boldsymbol{P}_2\boldsymbol{X}_2)_f^{-1} + (\boldsymbol{X}_3'\boldsymbol{P}_3\boldsymbol{X}_3)_f^{-1} & -(\boldsymbol{X}_3'\boldsymbol{P}_3\boldsymbol{X}_3)_f^{-1} & \mathbf{0} & \cdots \\ \hdotsfor{4} \end{array} \right|^{-1} \cdot$$

This matrix contains for larger dimensions many zeros which should be considered when inverting it, see for instance GEORGE and LIU (1981). Furthermore, the estimate $\hat{\sigma}^2$ of the variance factor σ^2 needs to be computed in (4.145). With (4.34) and (4.39) we obtain

$$\hat{\sigma}^2 = \frac{1}{n-u} \sum_{i=1}^{p} (\boldsymbol{X}_i\hat{\boldsymbol{\beta}}_i - \boldsymbol{y}_i)' \boldsymbol{P}_i (\boldsymbol{X}_i\hat{\boldsymbol{\beta}}_i - \boldsymbol{y}_i)$$

where n denotes the number of all observations and u the number of all unknown coordinates. Finally, r in (4.145) denotes the number of hypotheses, that is the number of all coordinate differences in H_0. Thus, all quantities are given for the test of the point null hypothesis according to (4.145). Δ

4.3.2 Informative Priors

To estimate the parameters of the linear model (4.1), prior information shall now be available as in Chapter 4.2.6 for the vector $\boldsymbol{\beta}$ of unknown parameters. It is given by the vector $E(\boldsymbol{\beta}) = \boldsymbol{\mu}$ of expected values and by the covariance matrix $D(\boldsymbol{\beta})$ of $\boldsymbol{\beta}$. The variance factor σ^2 is also unknown. Prior information is therefore assumed by the expected value $E(\sigma^2) = \sigma_p^2$ and by the variance $V(\sigma^2) = V_{\sigma^2}$ of σ^2. Thus, the covariance matrix $D(\boldsymbol{\beta})$ of $\boldsymbol{\beta}$ known by prior information is introduced by $D(\boldsymbol{\beta}) = \sigma_p^2\boldsymbol{\Sigma}$. A conjugate prior shall be applied. The normal-gamma distribution (2.234) is therefore chosen for the parameter vector $\boldsymbol{\beta}$ and the weight parameter τ with $\tau = 1/\sigma^2$ from (4.115)

$$\boldsymbol{\beta}, \tau \sim NG(\boldsymbol{\mu}, \boldsymbol{V}, b, p) \ . \tag{4.148}$$

Its parameters are determined by the prior-information, as will be shown in the following.

The vector $\boldsymbol{\beta}$ has with (2.213) because of (4.148) as prior marginal distribution the multivariate t-distribution

$$\boldsymbol{\beta} \sim t(\boldsymbol{\mu}, b\boldsymbol{V}/p, 2p) \tag{4.149}$$

with the vector of expected values from (2.208)

$$E(\boldsymbol{\beta}) = \boldsymbol{\mu} \tag{4.150}$$

in agreement with the prior information. The covariance matrix $D(\boldsymbol{\beta})$ of $\boldsymbol{\beta}$ is obtained with (2.208) and with the prior information by

$$D(\boldsymbol{\beta}) = \frac{b}{p-1}\boldsymbol{V} = \sigma_p^2\boldsymbol{\Sigma} \ . \tag{4.151}$$

The prior marginal distribution for τ follows with (2.214) by $\tau \sim G(b,p)$. The variance factor $\sigma^2 = 1/\tau$ therefore possesses with (2.176) the inverted gamma distribution

$$\sigma^2 \sim IG(b,p) \tag{4.152}$$

with the expected value and the variance from (2.177) and from the prior information

$$E(\sigma^2) = \frac{b}{p-1} = \sigma_p^2 \tag{4.153}$$

and

$$V(\sigma^2) = b^2/[(p-1)^2(p-2)] = V_{\sigma^2} \ . \tag{4.154}$$

We obtain with (4.151) and (4.153)

$$\boldsymbol{V} = \boldsymbol{\Sigma} \tag{4.155}$$

and from (4.154)

$$p = (\sigma_p^2)^2/V_{\sigma^2} + 2 \tag{4.156}$$

and finally from (4.153) and (4.156)

$$b = [(\sigma_p^2)^2/V_{\sigma^2} + 1]\sigma_p^2 \ . \tag{4.157}$$

Thus, the parameters of the distribution (4.148) are determined by the prior information.

By comparing the likelihood function (4.116) and the prior distribution (4.148) with (2.233) and (2.234) it becomes obvious from (2.235) that the posterior distribution for $\boldsymbol{\beta}$ and τ is the normal-gamma distribution

$$\boldsymbol{\beta}, \tau|\boldsymbol{y} \sim NG(\boldsymbol{\mu}_0, \boldsymbol{V}_0, b_0, p_0) \tag{4.158}$$

with the parameters from (2.236), (4.150) and (4.155) to (4.157)

$$\begin{aligned}
\boldsymbol{\mu}_0 &= (\boldsymbol{X}'\boldsymbol{P}\boldsymbol{X} + \boldsymbol{\Sigma}^{-1})^{-1}(\boldsymbol{X}'\boldsymbol{P}\boldsymbol{y} + \boldsymbol{\Sigma}^{-1}\boldsymbol{\mu}) \\
\boldsymbol{V}_0 &= (\boldsymbol{X}'\boldsymbol{P}\boldsymbol{X} + \boldsymbol{\Sigma}^{-1})^{-1} \\
b_0 &= \{2[(\sigma_p^2)^2/V_{\sigma^2} + 1]\sigma_p^2 \\
&\quad + (\boldsymbol{\mu} - \boldsymbol{\mu}_0)'\boldsymbol{\Sigma}^{-1}(\boldsymbol{\mu} - \boldsymbol{\mu}_0) + (\boldsymbol{y} - \boldsymbol{X}\boldsymbol{\mu}_0)'\boldsymbol{P}(\boldsymbol{y} - \boldsymbol{X}\boldsymbol{\mu}_0)\}/2 \\
p_0 &= (n + 2(\sigma_p^2)^2/V_{\sigma^2} + 4)/2 \ .
\end{aligned} \tag{4.159}$$

The posterior marginal distribution for $\boldsymbol{\beta}$ is with (2.213) and (4.158) the multivariate t-distribution

$$\boldsymbol{\beta}|\boldsymbol{y} \sim t(\boldsymbol{\mu}_0, b_0 \boldsymbol{V}_0/p_0, 2p_0) \tag{4.160}$$

with the vector of expected values for $\boldsymbol{\beta}$ from (2.208)

$$E(\boldsymbol{\beta}|\boldsymbol{y}) = \boldsymbol{\mu}_0 \ . \tag{4.161}$$

The Bayes estimate $\hat{\boldsymbol{\beta}}_B$ of $\boldsymbol{\beta}$ therefore follows from (3.9) and (4.159) by

$$\hat{\boldsymbol{\beta}}_B = (\boldsymbol{X}'\boldsymbol{P}\boldsymbol{X} + \boldsymbol{\Sigma}^{-1})^{-1}(\boldsymbol{X}'\boldsymbol{P}\boldsymbol{y} + \boldsymbol{\Sigma}^{-1}\boldsymbol{\mu}) \tag{4.162}$$

and in agreement with this result the MAP estimate $\hat{\boldsymbol{\beta}}_M$ of $\boldsymbol{\beta}$ from (3.30) corresponding to (4.124) by

$$\hat{\boldsymbol{\beta}}_M = (\boldsymbol{X}'\boldsymbol{P}\boldsymbol{X} + \boldsymbol{\Sigma}^{-1})^{-1}(\boldsymbol{X}'\boldsymbol{P}\boldsymbol{y} + \boldsymbol{\Sigma}^{-1}\boldsymbol{\mu}) \ . \tag{4.163}$$

The Bayes estimate $\hat{\boldsymbol{\beta}}_B$ and the MAP estimate $\hat{\boldsymbol{\beta}}_M$, respectively, may be also derived as estimate (4.29) of the method of least squares. If we define as in (4.90)

$$\bar{\boldsymbol{X}} = \left|\begin{array}{c} \boldsymbol{X} \\ \boldsymbol{I} \end{array}\right|, \ \bar{\boldsymbol{y}} = \left|\begin{array}{c} \boldsymbol{y} \\ \boldsymbol{\mu} \end{array}\right|, \ \bar{\boldsymbol{P}} = \left|\begin{array}{cc} \boldsymbol{P} & \boldsymbol{0} \\ \boldsymbol{0} & \boldsymbol{\Sigma}^{-1} \end{array}\right| \tag{4.164}$$

and compute the estimate $\hat{\boldsymbol{\beta}}$ of $\boldsymbol{\beta}$ by substituting $\bar{\boldsymbol{X}}, \bar{\boldsymbol{P}}$ and $\bar{\boldsymbol{y}}$ in (4.29), we obtain in agreement with (4.162) and (4.163)

$$\hat{\boldsymbol{\beta}} = (\boldsymbol{X}'\boldsymbol{P}\boldsymbol{X} + \boldsymbol{\Sigma}^{-1})^{-1}(\boldsymbol{X}'\boldsymbol{P}\boldsymbol{y} + \boldsymbol{\Sigma}^{-1}\boldsymbol{\mu}) \ . \tag{4.165}$$

The estimate $\hat{\boldsymbol{\beta}}_B, \hat{\boldsymbol{\beta}}_M$ or $\hat{\boldsymbol{\beta}}$ agrees with the estimate from (4.87), (4.89) or (4.91). Independent of the variance factor σ^2 being known or unknown the estimates of the unknown parameters $\boldsymbol{\beta}$ are identical. Informative priors were assumed. This result was already mentioned in connection with the corresponding result (4.124) for noninformative priors.

The marginal distribution of the posterior distribution (4.158) for the weight parameter τ is according to (2.214) the gamma distribution. The variance factor $\sigma^2 = 1/\tau$ has because of (2.176) the inverted gamma distribution

$$\sigma^2|\boldsymbol{y} \sim IG(b_0, p_0) \tag{4.166}$$

with the expected value from (2.177)

$$E(\sigma^2|\boldsymbol{y}) = b_0/(p_0 - 1) \ . \tag{4.167}$$

Hence, the Bayes estimate $\hat{\sigma}_B^2$ of σ^2 is obtained from (3.9) with (4.159) and (4.165) by

$$\hat{\sigma}_B^2 = (n + 2(\sigma_p^2)^2/V_{\sigma^2} + 2)^{-1}\big\{2[(\sigma_p^2)^2/V_{\sigma^2} + 1]\sigma_p^2$$
$$+ (\boldsymbol{\mu} - \hat{\boldsymbol{\beta}})'\boldsymbol{\Sigma}^{-1}(\boldsymbol{\mu} - \hat{\boldsymbol{\beta}}) + (\boldsymbol{y} - \boldsymbol{X}\hat{\boldsymbol{\beta}})'\boldsymbol{P}(\boldsymbol{y} - \boldsymbol{X}\hat{\boldsymbol{\beta}})\big\} \quad (4.168)$$

and the variance $V(\sigma^2|\boldsymbol{y})$ of σ^2 from (2.177) by

$$V(\sigma^2|\boldsymbol{y}) = \frac{2(\hat{\sigma}_B^2)^2}{n + 2(\sigma_p^2)^2/V_{\sigma^2}} \; . \tag{4.169}$$

The covariance matrix $D(\boldsymbol{\beta}|\boldsymbol{y})$ of $\boldsymbol{\beta}$ follows with $\hat{\sigma}_B^2$ from (4.168) and with (2.208), (4.159) and (4.160) by

$$D(\boldsymbol{\beta}|\boldsymbol{y}) = \hat{\sigma}_B^2(\boldsymbol{X}'\boldsymbol{P}\boldsymbol{X} + \boldsymbol{\Sigma}^{-1})^{-1} \; . \tag{4.170}$$

Example 1: We now assume for the Example 1 to (4.3) that in addition to the measurements $\boldsymbol{y} = |y_1, y_2, \ldots, y_n|'$ with the variances $V(y_i) = \sigma^2/p_i$ for determining the quantity s prior information for s is available by the expected value $E(s) = \mu$ and the variance $V(s) = \sigma_p^2\sigma_s^2$. Furthermore, prior information for the variance factor σ^2 is given by the expected value $E(\sigma^2) = \sigma_p^2$ and the variance $V(\sigma^2) = V_{\sigma^2}$. With $\hat{\boldsymbol{\beta}}_B = \hat{\boldsymbol{\beta}}_M = \hat{\boldsymbol{\beta}} = \hat{s}, \boldsymbol{\mu} = \mu, \boldsymbol{\Sigma} = \sigma_s^2, \boldsymbol{P} = \text{diag}(p_1, \ldots, p_n)$ and $\boldsymbol{X}'\boldsymbol{P} = |p_1, \ldots, p_n|$ we then obtain from (4.162), (4.163) or (4.165) or from (4.87), (4.89) or (4.91) the estimate $\hat{s}$ with the variance $V(s|\boldsymbol{y})$ of the Example 1 to (4.91).

The Bayes estimate $\hat{\sigma}_B^2$ of the variance factor σ^2 follows from (4.168) by

$$\hat{\sigma}_B^2 = (n + 2(\sigma_p^2)^2/V_{\sigma^2} + 2)^{-1}\big\{2[(\sigma_p^2)^2/V_{\sigma^2} + 1]\sigma_p^2$$
$$+ (\hat{s} - \mu)^2/\sigma_s^2 + \sum_{i=1}^{n} p_i(\hat{s} - y_i)^2\big\} \quad (4.171)$$

with the variance $V(\sigma^2|\boldsymbol{y})$ from (4.169). Δ

The confidence region for the vector $\boldsymbol{\beta}$ of unknown parameters is obtained by the multivariate t-distribution (4.160). Its density function is because of (2.204) a monotonically decreasing function of the quadratic form $p_0(\boldsymbol{\beta} - \boldsymbol{\mu}_0)'\boldsymbol{V}_0^{-1}(\boldsymbol{\beta} - \boldsymbol{\mu}_0)/b_0$. A hyperellipsoid of equal density is therefore found by

$$p_0(\boldsymbol{\beta} - \boldsymbol{\mu}_0)'\boldsymbol{V}_0^{-1}(\boldsymbol{\beta} - \boldsymbol{\mu}_0)/b_0 = \text{const} \; .$$

The quadratic form has with (2.211) the F-distribution (2.182)

$$p_0(\boldsymbol{\beta} - \boldsymbol{\mu}_0)'\boldsymbol{V}_0^{-1}(\boldsymbol{\beta} - \boldsymbol{\mu}_0)/(ub_0) \sim F(u, 2p_0) \; .$$

Thus, the $1 - \alpha$ confidence hyperellipsoid is determined with (4.159), (4.165) and (4.168) corresponding to (3.36) by

$$(n + 2(\sigma_p^2)^2/V_{\sigma^2} + 4)(\boldsymbol{\beta} - \hat{\boldsymbol{\beta}})'(\boldsymbol{X}'\boldsymbol{P}\boldsymbol{X} + \boldsymbol{\Sigma}^{-1})(\boldsymbol{\beta} - \hat{\boldsymbol{\beta}})/$$
$$[u(n + 2(\sigma_p^2)^2/V_{\sigma^2} + 2)\hat{\sigma}_B^2] = F_{1-\alpha;u,n+2(\sigma_p^2)^2/V_{\sigma^2}+4} \quad (4.172)$$

where $F_{1-\alpha;u,n+2(\sigma_p^2)^2/V_{\sigma^2}+4}$ denotes the upper α-percentage point defined by (2.183) of the F-distribution with u and $n+2(\sigma_p^2)^2/V_{\sigma^2}+4$ as parameters. The confidence interval for the variance factor σ^2 is obtained from (4.166).

Based on the posterior distributions (4.160) for $\boldsymbol{\beta}$ and (4.166) for σ^2 hypotheses for $\boldsymbol{\beta}$ and σ^2 may be tested, as described in Chapter 3.4.

Example 2: Let the point null hypothesis (4.23)

$$H_0 : \boldsymbol{H\beta} = \boldsymbol{w} \quad \text{versus} \quad H_1 : \boldsymbol{H\beta} \neq \boldsymbol{w} \tag{4.173}$$

be tested with (3.82) by means of the confidence hyperellipsoid for $\boldsymbol{H\beta}$. We obtain with (4.160) the distribution of the linear transformation $\boldsymbol{H\beta}$ from (2.210) by

$$\boldsymbol{H\beta}|\boldsymbol{y} \sim t(\boldsymbol{H\mu}_0, b_0\boldsymbol{HV}_0\boldsymbol{H}'/p_0, 2p_0) \ . \tag{4.174}$$

Hence, we find from (2.211)

$$p_0(\boldsymbol{H\beta} - \boldsymbol{H\mu}_0)'(\boldsymbol{HV}_0\boldsymbol{H}')^{-1}(\boldsymbol{H\beta} - \boldsymbol{H\mu}_0)/(rb_0) \sim F(r, 2p_0) \ .$$

This result leads to the confidence hyperellipsoid for $\boldsymbol{H\beta}$ as in (4.144). Corresponding to (4.145) and (4.172) the point null hypothesis (4.173) is accepted, if

$$(n+2(\sigma_p^2)^2/V_{\sigma^2}+4)(\boldsymbol{H\hat{\beta}}-\boldsymbol{w})'(\boldsymbol{H}(\boldsymbol{X}'\boldsymbol{PX}+\boldsymbol{\Sigma}^{-1})^{-1}\boldsymbol{H}')^{-1}(\boldsymbol{H\hat{\beta}}-\boldsymbol{w})/$$
$$[r(n+2(\sigma_p^2)^2/V_{\sigma^2}+2)\hat{\sigma}_B^2] < F_{1-\alpha;r,n+2(\sigma_p^2)^2/V_{\sigma^2}+4} \ . \tag{4.175}$$
$$\Delta$$

4.4 Linear Model not of Full Rank

So far, it has been assumed that the coefficient matrix $\boldsymbol{X}$ in the linear model (4.1) has full column rank. This supposition is, for instance, not fulfilled for the analysis of variance and covariance, see for instance KOCH (1999, p.200 and 207). A coefficient matrix not of full rank results also from observations which determine the coordinates of points of a so-called free network, since the observations do not contain information on the position and the orientation of the network, see for instance KOCH (1999, p.187). The linear model *not of full rank*

$$\boldsymbol{X\beta} = E(\boldsymbol{y}|\boldsymbol{\beta}) \quad \text{with} \quad \text{rank}\boldsymbol{X} = q < u \quad \text{and} \quad D(\boldsymbol{y}|\sigma^2) = \sigma^2\boldsymbol{P}^{-1} \tag{4.176}$$

therefore needs to be considered, too, where $\boldsymbol{X}, \boldsymbol{\beta}, \boldsymbol{y}$ and $\boldsymbol{P}$ are defined as in (4.1). By allowing a rank deficiency for $\boldsymbol{X}$ the likelihood functions (4.4) and (4.116) do not change. When estimating the vector $\boldsymbol{\beta}$ of unknown parameters, however, we have to be aware that because of rank$\boldsymbol{X} = q$ also rank$(\boldsymbol{X}'\boldsymbol{PX}) = q$ is valid. The matrix $\boldsymbol{X}'\boldsymbol{PX}$ of normal equations is therefore singular. Its inverse does not exist, a symmetric reflexive generalized inverse $(\boldsymbol{X}'\boldsymbol{P}\boldsymbol{X})_{rs}^{-}$ of the matrix $\boldsymbol{X}'\boldsymbol{P}\boldsymbol{X}$ is therefore introduced. It is chosen to be symmetric, since $\boldsymbol{X}'\boldsymbol{P}\boldsymbol{X}$ is symmetric, and it is chosen to be reflexive because of

$$\mathrm{rank}(\boldsymbol{X}'\boldsymbol{P}\boldsymbol{X}) = \mathrm{rank}(\boldsymbol{X}'\boldsymbol{P}\boldsymbol{X})_{rs}^{-} = q \;. \tag{4.177}$$

In addition we have

$$\boldsymbol{X}'\boldsymbol{P}\boldsymbol{X}(\boldsymbol{X}'\boldsymbol{P}\boldsymbol{X})_{rs}^{-}\boldsymbol{X}' = \boldsymbol{X}' \tag{4.178}$$

and

$$\left|\begin{array}{cc} \boldsymbol{X}'\boldsymbol{P}\boldsymbol{X} & \boldsymbol{B}' \\ \boldsymbol{B} & \boldsymbol{0} \end{array}\right|^{-1} = \left|\begin{array}{cc} (\boldsymbol{X}'\boldsymbol{P}\boldsymbol{X})_{rs}^{-} & \cdots \\ \cdots & \boldsymbol{0} \end{array}\right|$$

where a $(u-q)\times u$ matrix $\boldsymbol{B}$ is selected such that the matrix on the left-hand side becomes regular so that it can be inverted to compute $(\boldsymbol{X}'\boldsymbol{P}\boldsymbol{X})_{rs}^{-}$, see for instance KOCH (1999, p.51,52,60).

When estimating the unknown parameters $\boldsymbol{\beta}$ and the variance factor σ^2 one has to distinguish between the case of noninformative and informative prior information.

4.4.1   Noninformative Priors

It will be first assumed that the variance factor σ^2 is known. If we set instead of (4.13)

$$\boldsymbol{\mu}_0 = (\boldsymbol{X}'\boldsymbol{P}\boldsymbol{X})_{rs}^{-}\boldsymbol{X}'\boldsymbol{P}\boldsymbol{y} \;, \tag{4.179}$$

the transformation (4.12) is valid because of (4.178). The posterior density function for the vector $\boldsymbol{\beta}$ of unknown parameters is therefore obtained with (4.11) from the normal distribution by

$$p(\boldsymbol{\beta}|\boldsymbol{y}) \propto e^{-\frac{1}{2\sigma^2}(\boldsymbol{\beta}-\boldsymbol{\mu}_0)'\boldsymbol{X}'\boldsymbol{P}\boldsymbol{X}(\boldsymbol{\beta}-\boldsymbol{\mu}_0)} \;. \tag{4.180}$$

Because of $\det(\boldsymbol{X}'\boldsymbol{P}\boldsymbol{X})_{rs}^{-} = 0$ from (4.177) this density function cannot be normalized which follows from (2.195). If we restrict ourselves, however, to q components in the vector $\boldsymbol{\beta}$ and if we choose the corresponding elements in $(\boldsymbol{X}'\boldsymbol{P}\boldsymbol{X})_{rs}^{-}$, the resulting matrix is regular because of (4.177) and the density function of the normal distribution can be normalized. With $\boldsymbol{\beta} = (\beta_i), \boldsymbol{\mu}_0 = (\mu_i)$, $\boldsymbol{\Sigma} = (\sigma_{ij}) = (\boldsymbol{X}'\boldsymbol{P}\boldsymbol{X})_{rs}^{-}$ and $k - j + 1 = q$ we define

$$\boldsymbol{\beta}_{j\cdot\cdot k} = (\beta_l), \boldsymbol{\mu}_{0,j\cdot\cdot k} = (\mu_l) \quad\text{and}\quad \boldsymbol{\Sigma}_{j\cdot\cdot k} = (\sigma_{lm})$$
$$\text{with}\quad l, m \in \{j, j+1, \ldots, k\} \tag{4.181}$$

and obtain because of

$$\mathrm{rank}\boldsymbol{\Sigma}_{j\cdot\cdot k} = q \tag{4.182}$$

from (4.180) the normal distribution

$$\boldsymbol{\beta}_{j\cdot\cdot k}|\boldsymbol{y} \sim N(\boldsymbol{\mu}_{0,j\cdot\cdot k}, \sigma^2\boldsymbol{\Sigma}_{j\cdot\cdot k}) \tag{4.183}$$

which can be normalized.

The q components of the vector $\boldsymbol{\beta}$ may be arbitrarily chosen. Thus, the Bayes estimate $\hat{\boldsymbol{\beta}}_B$, the MAP estimate $\hat{\boldsymbol{\beta}}_M$ and also the estimate $\hat{\boldsymbol{\beta}}$ of the method of least squares of the unknown parameters $\boldsymbol{\beta}$ follow instead of (4.15), (4.19) and (4.29) by

$$\hat{\boldsymbol{\beta}} = \hat{\boldsymbol{\beta}}_B = \hat{\boldsymbol{\beta}}_M = (\boldsymbol{X}'\boldsymbol{P}\boldsymbol{X})^-_{rs}\boldsymbol{X}'\boldsymbol{P}\boldsymbol{y} \tag{4.184}$$

and the covariance matrix instead of (4.16) by

$$D(\boldsymbol{\beta}|\boldsymbol{y}) = \sigma^2(\boldsymbol{X}'\boldsymbol{P}\boldsymbol{X})^-_{rs} . \tag{4.185}$$

In traditional statistics we find for the model (4.176) not of full rank, see for instance KOCH (1999, p.183),

$$E(\Omega) = \sigma^2(n - q) \tag{4.186}$$

instead of (4.37). The unbiased estimate $\hat{\sigma}^2$ of the variance factor σ^2 therefore follows with

$$\hat{\sigma}^2 = \frac{1}{n - q}\Omega \tag{4.187}$$

instead of (4.39). Correspondingly, we obtain for the linear model not of full rank with constraints instead of (4.55)

$$\tilde{\sigma}^2 = \Omega_H/(n - q + r) . \tag{4.188}$$

When testing the point null hypothesis (4.23) one has to be aware that

$$\text{rank}\boldsymbol{H}(\boldsymbol{X}'\boldsymbol{P}\boldsymbol{X})^-_{rs}\boldsymbol{X}'\boldsymbol{P}\boldsymbol{X} = r \tag{4.189}$$

must be fulfilled. The matrix $\boldsymbol{H}(\boldsymbol{X}'\boldsymbol{P}\boldsymbol{X})^-_{rs}\boldsymbol{H}'$ is then regular (KOCH 1999, p.196). Hence, because of (4.26) the point null hypothesis is accepted, if

$$(\boldsymbol{H}\hat{\boldsymbol{\beta}} - \boldsymbol{w})'(\boldsymbol{H}(\boldsymbol{X}'\boldsymbol{P}\boldsymbol{X})^-_{rs}\boldsymbol{H}')^{-1}(\boldsymbol{H}\hat{\boldsymbol{\beta}} - \boldsymbol{w})/\sigma^2 < \chi^2_{1-\alpha;r} . \tag{4.190}$$

If the variance factor σ^2 is unknown, the transformation (4.118) is also valid with $\boldsymbol{\mu}_0$ from (4.179) because of (4.178). The normal-gamma distribution therefore follows with (4.181) and (4.187) instead of (4.120) by

$$\boldsymbol{\beta}_{j\cdot\cdot k}, \tau|\boldsymbol{y} \sim NG(\boldsymbol{\mu}_{0,j\cdot\cdot k}, \boldsymbol{\Sigma}_{j\cdot\cdot k}, (n - q)\hat{\sigma}^2/2, (n - q)/2) . \tag{4.191}$$

The marginal distribution for the vector $\boldsymbol{\beta}$ of unknown parameters is then obtained corresponding to (4.121) by

$$\boldsymbol{\beta}_{j\cdot\cdot k}|\boldsymbol{y} \sim t(\boldsymbol{\mu}_{0,j\cdot\cdot k}, \hat{\sigma}^2\boldsymbol{\Sigma}_{j\cdot\cdot k}, n - q) . \tag{4.192}$$

The q components of the vector $\boldsymbol{\beta}$ may be arbitrarily chosen. The estimates therefore follow instead of (4.123), (4.124) and (4.29) by

$$\hat{\boldsymbol{\beta}} = \hat{\boldsymbol{\beta}}_B = \hat{\boldsymbol{\beta}}_M = (\boldsymbol{X}'\boldsymbol{P}\boldsymbol{X})_{rs}^{-}\boldsymbol{X}'\boldsymbol{P}\boldsymbol{y} \tag{4.193}$$

in agreement with (4.184). The covariance matrix $D(\boldsymbol{\beta}|\boldsymbol{y})$ is obtained from (4.125) with (4.187) by

$$D(\boldsymbol{\beta}|\boldsymbol{y}) = \frac{n-q}{n-q-2}\hat{\sigma}^2(\boldsymbol{X}'\boldsymbol{P}\boldsymbol{X})_{rs}^{-} . \tag{4.194}$$

Finally, we find instead of (4.129)

$$\hat{\sigma}_B^2 = \frac{n-q}{n-q-2}\hat{\sigma}^2 \tag{4.195}$$

and instead of (4.130)

$$V(\sigma^2|\boldsymbol{y}) = \frac{2(n-q)^2(\hat{\sigma}^2)^2}{(n-q-2)^2(n-q-4)} . \tag{4.196}$$

With observing (4.189) the point null hypothesis (4.141) has to be accepted corresponding to (4.145), if

$$(\boldsymbol{H}\hat{\boldsymbol{\beta}} - \boldsymbol{w})'(\boldsymbol{H}(\boldsymbol{X}'\boldsymbol{P}\boldsymbol{X})_{rs}^{-}\boldsymbol{H}')^{-1}(\boldsymbol{H}\hat{\boldsymbol{\beta}} - \boldsymbol{w})/(r\hat{\sigma}^2) < F_{1-\alpha;r,n-q} . \tag{4.197}$$

4.4.2   Informative Priors

Prior information for the vector $\boldsymbol{\beta}$ of unknown parameters is introduced by the vector $E(\boldsymbol{\beta}) = \boldsymbol{\mu}$ of expected values of $\boldsymbol{\beta}$ and in case of a known variance factor σ^2 by the covariance matrix $D(\boldsymbol{\beta}) = \sigma^2\boldsymbol{\Sigma}$ of $\boldsymbol{\beta}$, as was described in Chapter 4.2.6, and in case of an unknown σ^2 by $D(\boldsymbol{\beta}) = \sigma_p^2\boldsymbol{\Sigma}$, as was explained in Chapter 4.3.2. If the matrix $\boldsymbol{\Sigma}$ is positive definite which was assumed so far, then

$$\mathrm{rank}(\boldsymbol{X}'\boldsymbol{P}\boldsymbol{X} + \boldsymbol{\Sigma}^{-1}) = u \tag{4.198}$$

holds true, since $\boldsymbol{X}'\boldsymbol{P}\boldsymbol{X}$ is positive semidefinite. The matrix $\boldsymbol{X}'\boldsymbol{P}\boldsymbol{X} + \boldsymbol{\Sigma}^{-1}$ in (4.86) and in (4.159) is then regular so that under the assumption of the positive definite matrix $\boldsymbol{\Sigma}$ all results of Chapter 4.2.6 and 4.3.2 of the model with full rank are valid in case of a known or unknown variance factor σ^2.

We will now consider a singular matrix $\boldsymbol{\Sigma}$. It appears, if the prior information for the vector $\boldsymbol{\beta}$ of unknown parameters is introduced by already available observations $\boldsymbol{y}_p$ with the associated weight matrix $\boldsymbol{P}_p$ and the coefficient matrix $\boldsymbol{X}_p$ of the linear model. The covariance matrix $D(\boldsymbol{\beta}|\boldsymbol{y}_p)$ of $\boldsymbol{\beta}$ then follows with the matrix $\boldsymbol{X}_p'\boldsymbol{P}_p\boldsymbol{X}_p$ of normal equations in the model of full rank from (4.16) by

$$D(\boldsymbol{\beta}|\boldsymbol{y}_p) = \sigma^2(\boldsymbol{X}_p'\boldsymbol{P}_p\boldsymbol{X}_p)^{-1} .$$

The prior information for the vector $\boldsymbol{\beta}$, which is introduced by the covariance matrix $\sigma^2\boldsymbol{\Sigma}$ or $\sigma_p^2\boldsymbol{\Sigma}$, is therefore expressed by the inverse $(\boldsymbol{X}_p'\boldsymbol{P}_p\boldsymbol{X}_p)^{-1}$ of the matrix of normal equations, thus

$$\boldsymbol{\Sigma} = (\boldsymbol{X}_p'\boldsymbol{P}_p\boldsymbol{X}_p)^{-1} \quad \text{or} \quad \boldsymbol{\Sigma}^{-1} = \boldsymbol{X}_p'\boldsymbol{P}_p\boldsymbol{X}_p \ . \tag{4.199}$$

By defining the prior information in this manner the matrix $\boldsymbol{\Sigma}^{-1}$ results, which in the model (4.176) not of full rank becomes singular because of

$$\text{rank}(\boldsymbol{X}_p'\boldsymbol{P}_p\boldsymbol{X}_p) = q \tag{4.200}$$

from (4.177). To derive the estimates for this model from the estimates of the model of full rank, $\boldsymbol{\Sigma}^{-1}$ is replaced by $\boldsymbol{X}_p'\boldsymbol{P}_p\boldsymbol{X}_p$. Furthermore, it holds

$$\text{rank}(\boldsymbol{X}'\boldsymbol{P}\boldsymbol{X} + \boldsymbol{X}_p'\boldsymbol{P}_p\boldsymbol{X}_p) = q \ . \tag{4.201}$$

Let the variance factor σ^2 first be known. We set corresponding to (4.86)

$$\boldsymbol{\mu}_0 = (\boldsymbol{X}'\boldsymbol{P}\boldsymbol{X} + \boldsymbol{X}_p'\boldsymbol{P}_p\boldsymbol{X}_p)_{rs}^-(\boldsymbol{X}'\boldsymbol{P}\boldsymbol{y} + \boldsymbol{X}_p'\boldsymbol{P}_p\boldsymbol{X}_p\boldsymbol{\mu}) \ , \tag{4.202}$$

where $(\boldsymbol{X}'\boldsymbol{P}\boldsymbol{X} + \boldsymbol{X}_p'\boldsymbol{P}_p\boldsymbol{X}_p)_{rs}^-$ again means a symmetric reflexive generalized inverse of $\boldsymbol{X}'\boldsymbol{P}\boldsymbol{X} + \boldsymbol{X}_p'\boldsymbol{P}_p\boldsymbol{X}_p$. It fulfills after substituting $\boldsymbol{X}'$ by $|\boldsymbol{X}', \boldsymbol{X}_p'|$ and $\boldsymbol{P}$ by $\text{diag}(\boldsymbol{P}, \boldsymbol{P}_p)$ in (4.178)

$$(\boldsymbol{X}'\boldsymbol{P}\boldsymbol{X} + \boldsymbol{X}_p'\boldsymbol{P}_p\boldsymbol{X}_p)(\boldsymbol{X}'\boldsymbol{P}\boldsymbol{X} + \boldsymbol{X}_p'\boldsymbol{P}_p\boldsymbol{X}_p)_{rs}^-|\boldsymbol{X}', \boldsymbol{X}_p'| = |\boldsymbol{X}', \boldsymbol{X}_p'| \ .$$

With this result we may transform like in (2.228). The posterior density function for the unknown parameters $\boldsymbol{\beta}$ therefore follows as the density function of the normal distribution by

$$p(\boldsymbol{\beta}|\boldsymbol{y}) \propto \exp\left\{-\frac{1}{2\sigma^2}[(\boldsymbol{\beta}-\boldsymbol{\mu}_0)'(\boldsymbol{X}'\boldsymbol{P}\boldsymbol{X} + \boldsymbol{X}_p'\boldsymbol{P}_p\boldsymbol{X}_p)_{rs}^-(\boldsymbol{\beta}-\boldsymbol{\mu}_0)]\right\} \ . \tag{4.203}$$

This density function cannot be normalized like the density function (4.180). If we restrict ourselves, however, to q components of the vector $\boldsymbol{\beta}$, we obtain like in (4.181) to (4.183) the normal distribution which can be normalized. The Bayes estimate $\hat{\boldsymbol{\beta}}_B$, the MAP estimate $\hat{\boldsymbol{\beta}}_M$ and the estimate $\hat{\boldsymbol{\beta}}$ of the method of least squares are obtained instead of (4.87), (4.89) and (4.91) by

$$\hat{\boldsymbol{\beta}} = \hat{\boldsymbol{\beta}}_B = \hat{\boldsymbol{\beta}}_M = (\boldsymbol{X}'\boldsymbol{P}\boldsymbol{X} + \boldsymbol{X}_p'\boldsymbol{P}_p\boldsymbol{X}_p)_{rs}^-(\boldsymbol{X}'\boldsymbol{P}\boldsymbol{y} + \boldsymbol{X}_p'\boldsymbol{P}_p\boldsymbol{X}_p\boldsymbol{\mu}) \ . \tag{4.204}$$

The covariance matrix $D(\boldsymbol{\beta}|\boldsymbol{y})$ of $\boldsymbol{\beta}$ follows instead of (4.88) with

$$D(\boldsymbol{\beta}|\boldsymbol{y}) = \sigma^2(\boldsymbol{X}'\boldsymbol{P}\boldsymbol{X} + \boldsymbol{X}_p'\boldsymbol{P}_p\boldsymbol{X}_p)_{rs}^- \ . \tag{4.205}$$

Provided that (4.189) holds, the point null hypothesis (4.23) is accepted because of (4.98), if

$$(\boldsymbol{H}\hat{\boldsymbol{\beta}}-\boldsymbol{w})'(\boldsymbol{H}(\boldsymbol{X}'\boldsymbol{P}\boldsymbol{X} + \boldsymbol{X}_p'\boldsymbol{P}_p\boldsymbol{X}_p)_{rs}^-\boldsymbol{H}')^{-1}(\boldsymbol{H}\hat{\boldsymbol{\beta}}-\boldsymbol{w})/\sigma^2 < \chi^2_{1-\alpha,r} \ . \tag{4.206}$$

If the variance factor σ^2 is unknown, the prior information $E(\sigma^2) = \sigma_p^2$ and $V(\sigma^2) = V_{\sigma^2}$ for σ^2 is introduced, as explained in Chapter 4.3.2 . The prior information for the vector $\boldsymbol{\beta}$ of unknown parameters which is obtained by the covariance matrix $D(\boldsymbol{\beta}) = \sigma_p^2 \boldsymbol{\Sigma}$ of $\boldsymbol{\beta}$ is then given according to (4.199) by the matrix

$$\boldsymbol{\Sigma}^{-1} = \boldsymbol{X}_p' \boldsymbol{P}_p \boldsymbol{X}_p \; . \tag{4.207}$$

Thus, $\boldsymbol{\mu}_0$ and $\boldsymbol{V}_0$ in (4.159) follow for the model not of full rank with

$$\boldsymbol{\mu}_0 = (\boldsymbol{X}'\boldsymbol{P}\boldsymbol{X} + \boldsymbol{X}_p'\boldsymbol{P}_p\boldsymbol{X}_p)_{rs}^-(\boldsymbol{X}'\boldsymbol{P}\boldsymbol{y} + \boldsymbol{X}_p'\boldsymbol{P}_p\boldsymbol{X}_p\boldsymbol{\mu})$$
$$\boldsymbol{V}_0 = (\boldsymbol{X}'\boldsymbol{P}\boldsymbol{X} + \boldsymbol{X}_p'\boldsymbol{P}_p\boldsymbol{X}_p)_{rs}^- \; . \tag{4.208}$$

If again we restrict ourselves to q components in the vector $\boldsymbol{\beta}$, we obtain instead of (4.158) and (4.160) the normal-gamma distribution and the multivariate t-distribution which both can be normalized. The Bayes estimate $\hat{\boldsymbol{\beta}}_B$, the MAP estimate $\hat{\boldsymbol{\beta}}_M$ and the estimate $\hat{\boldsymbol{\beta}}$ of the method of least squares of $\boldsymbol{\beta}$ are therefore obtained instead of (4.162), (4.163) and (4.165) by

$$\hat{\boldsymbol{\beta}} = \hat{\boldsymbol{\beta}}_B = \hat{\boldsymbol{\beta}}_M = (\boldsymbol{X}'\boldsymbol{P}\boldsymbol{X} + \boldsymbol{X}_p'\boldsymbol{P}_p\boldsymbol{X}_p)_{rs}^-(\boldsymbol{X}'\boldsymbol{P}\boldsymbol{y} + \boldsymbol{X}_p'\boldsymbol{P}_p\boldsymbol{X}_p\boldsymbol{\mu}) \; . \tag{4.209}$$

This result agrees with (4.204).

The Bayes estimate $\hat{\sigma}_B^2$ of σ^2 follows instead of (4.168) with

$$\hat{\sigma}_B^2 = (n + 2(\sigma_p^2)^2/V_{\sigma^2} + 2)^{-1}\big\{2[(\sigma_p^2)^2/V_{\sigma^2} + 1]\sigma_p^2$$
$$+ (\boldsymbol{\mu} - \hat{\boldsymbol{\beta}})'\boldsymbol{X}_p'\boldsymbol{P}_p\boldsymbol{X}_p(\boldsymbol{\mu} - \hat{\boldsymbol{\beta}}) + (\boldsymbol{y} - \boldsymbol{X}\hat{\boldsymbol{\beta}})'\boldsymbol{P}(\boldsymbol{y} - \boldsymbol{X}\hat{\boldsymbol{\beta}})\big\} \tag{4.210}$$

and the covariance matrix $D(\boldsymbol{\beta}|\boldsymbol{y})$ of $\boldsymbol{\beta}$ instead of (4.170) with

$$D(\boldsymbol{\beta}|\boldsymbol{y}) = \hat{\sigma}_B^2(\boldsymbol{X}'\boldsymbol{P}\boldsymbol{X} + \boldsymbol{X}_p'\boldsymbol{P}_p\boldsymbol{X}_p)_{rs}^- \; . \tag{4.211}$$

Finally, under the assumption of (4.189) the point null hypothesis (4.173) is accepted because of (4.175), if

$$(n + 2(\sigma_p^2)^2/V_{\sigma^2} + 4)(\boldsymbol{H}\hat{\boldsymbol{\beta}} - \boldsymbol{w})'(\boldsymbol{H}(\boldsymbol{X}'\boldsymbol{P}\boldsymbol{X} + \boldsymbol{X}_p'\boldsymbol{P}_p\boldsymbol{X}_p)_{rs}^-\boldsymbol{H}')^{-1}$$
$$(\boldsymbol{H}\hat{\boldsymbol{\beta}} - \boldsymbol{w})/[r(n + 2(\sigma_p^2)^2/V_{\sigma^2} + 2)\hat{\sigma}_B^2] < F_{1-\alpha;r,n+2(\sigma_p^2)^2/V_{\sigma^2}+4} \; . \tag{4.212}$$

Example: Observations $\boldsymbol{y}$ are given which have to be analyzed in the model (4.176) not of full rank. Let the prior information on the vector $\boldsymbol{\beta}$ of unknown parameters be introduced by the observations $\boldsymbol{y}_p$ with the weight matrix $\boldsymbol{P}_p$. They have been taken at an earlier time and are analyzed with the associated coefficient matrix $\boldsymbol{X}_p$ also in the model (4.176). The vector $\boldsymbol{\mu}$ of expected values from the prior information for the vector $\boldsymbol{\beta}$ of unknown parameters in (4.204) or (4.209) therefore follows with (4.184) or (4.193) by

$$\boldsymbol{\mu} = (\boldsymbol{X}_p'\boldsymbol{P}_p\boldsymbol{X}_p)_{rs}^-\boldsymbol{X}_p'\boldsymbol{P}_p\boldsymbol{y}_p \; . \tag{4.213}$$

The prior information for $\boldsymbol{\beta}$ which is introduced by the covariance matrix $D(\boldsymbol{\beta})$ shall be determined with (4.199). By substituting (4.213) in (4.204) or (4.209) we then obtain because of (4.178)

$$\hat{\boldsymbol{\beta}} = \hat{\boldsymbol{\beta}}_B = \hat{\boldsymbol{\beta}}_M = (\boldsymbol{X}'\boldsymbol{P}\boldsymbol{X} + \boldsymbol{X}'_p\boldsymbol{P}_p\boldsymbol{X}_p)^-_{rs}(\boldsymbol{X}'\boldsymbol{P}\boldsymbol{y} + \boldsymbol{X}'_p\boldsymbol{P}_p\boldsymbol{y}_p) \ . \quad (4.214)$$

This estimate with prior information for $\boldsymbol{\beta}$, which is obtained by the vector $\boldsymbol{y}_p$ of observations, is identical with the estimate obtained from (4.184) or (4.193), if the observations $\boldsymbol{y}$ and $\boldsymbol{y}_p$ are assumed as being independent and if they are jointly analyzed. $\quad\Delta$

5 Special Models and Applications

In the following two special linear models are presented, the model of prediction and filtering and the model with unknown variance and covariance components. The model with variance components is augmented by a variance component for prior information so that the regularization parameter of a type Tykhonov regularization can be estimated by the ratio of two variance components. The reconstruction and the smoothing of digital three-dimensional images and pattern recognition in two-dimensional images are covered. Finally, Bayesian networks are presented for making decisions in systems with uncertainties.

5.1 Prediction and Filtering

As was explained in Chapter 2.2, measurements are the results of random experiments, the results therefore vary. The disturbances should be removed from the measurements. This is also true, if measurements are taken as functions of time or position. Measurements depending on time belong to the stochastic processes and depending on position to the random fields, see for instance KOCH and SCHMIDT (1994) and MEIER and KELLER (1990). One does not only want to remove the disturbances of the measurements which is called *filtering*, but also to forecast the measurements at times or at positions when or where no measurements were taken. This task is called *prediction*. WIENER (1949) solved the problem of prediction and filtering by minimizing the expected value of the quadratic error of the prediction and the filtering, see for instance KOCH and SCHMIDT (1994, p.233). The prediction and filtering was not only applied for the interpolation of measurements but by HEITZ (1968) and MORITZ (1969) also for estimating unknown parameters. It was generalized by KRARUP (1969) and MORITZ (1973, 1980) to the so-called *collocation* for which it is assumed that the measurements are represented by *signals* as linear functions of unknown parameters and by a *trend* which allows to assume the expected value zero for the signals.

In traditional statistics the prediction and filtering is developed from the mixed model which contains the unknown parameters as constants and as random variables. The unknown parameters of the trend represent the fixed quantities of the prediction and filtering, while the unknown parameters of the signals are the random parameters, see for instance KOCH (1999, p.221). The joint appearance of fixed parameters and of random parameters in the model of prediction and filtering of the traditional statistics is not easily interpreted. Difficulties also arise when to decide which parameters are constant and which parameters are random. This becomes apparent for applications in physical

geodesy where frequently the coordinates of points at the surface of the earth
are defined as fixed parameters and the parameters of the earth's gravity field
as random parameters.

The model of prediction and filtering is much clearer interpreted by the
linear model of Bayesian statistics which defines the unknown parameters as
random variables, as was explained in Chapter 2.2.8. One does not have to
distinguish between fixed parameters and random parameters. One differen-
tiates between the unknown parameters by their prior information only. This
means for the model of prediction and filtering that no prior information is
assumed for the parameters of the trend, while the parameters of the signals
are introduced with the prior information that the expected values are equal
to zero and that the variances and covariances are given. As will be shown
in the following chapter, the estimates of the parameters are then obtained
which are identical with the estimates of the model of prediction and filtering
of the traditional statistics, see for instance KOCH (1990, p.117).

If prior information is assumed only for the unknown parameters of the
signals, an improper prior density function is obtained for the unknown parameters. It is simpler to introduce prior information for the parameters of
the trend as well as for the parameters of the signals. This leads to a special
linear model which is also well suited for practical applications. It readily
gives, as shown in Chapter 4, the estimates, the confidence regions and the
hypothesis tests not only for the unknown parameters of the trend but also
for the unknown parameters of the signal. Afterwards, the prior information
may be restricted (KOCH 1994). By this manner prediction and filtering is
covered in the following.

The method of prediction in geostatistics is called kriging and Bayesian
kriging if Bayesian statistics is applied, see for instance CRESSIE (1991,
p.170), MENZ and PILZ (1994), PILZ and WEBER (1998). The latter corresponds to the procedure derived here.

5.1.1   Model of Prediction and Filtering as Special Linear Model

The linear model (4.1) is introduced in the form

$$\bar{\boldsymbol{X}}\bar{\boldsymbol{\beta}} = E(\boldsymbol{y}|\bar{\boldsymbol{\beta}}) \quad \text{with} \quad D(\boldsymbol{y}|\sigma^2) = \sigma^2 \bar{\boldsymbol{P}}^{-1} \tag{5.1}$$

or in the alternative formulation (4.3)

$$\bar{\boldsymbol{X}}\bar{\boldsymbol{\beta}} = \boldsymbol{y} + \boldsymbol{e} \quad \text{with} \quad E(\boldsymbol{e}|\bar{\boldsymbol{\beta}}) = \boldsymbol{0}$$

$$\text{and} \quad D(\boldsymbol{e}|\bar{\boldsymbol{\beta}}, \sigma^2) = D(\boldsymbol{y}|\sigma^2) = \sigma^2 \bar{\boldsymbol{P}}^{-1} \; . \tag{5.2}$$

If we substitute

$$\bar{\boldsymbol{X}} = |\boldsymbol{X}, \boldsymbol{Z}| \, , \; \bar{\boldsymbol{\beta}} = \left| \begin{array}{c} \beta \\ \gamma \end{array} \right| , \; \bar{\boldsymbol{P}}^{-1} = \boldsymbol{\Sigma}_{ee} \, , \tag{5.3}$$

the *model of prediction and filtering,* also called *mixed model,* is obtained

$$\boldsymbol{X\beta} + \boldsymbol{Z\gamma} = E(\boldsymbol{y}|\boldsymbol{\beta},\boldsymbol{\gamma}) \quad \text{with} \quad D(\boldsymbol{y}|\sigma^2) = \sigma^2\boldsymbol{\Sigma}_{ee} \tag{5.4}$$

or in the alternative formulation

$$\boldsymbol{X\beta} + \boldsymbol{Z\gamma} = \boldsymbol{y} + \boldsymbol{e} \quad \text{with} \quad E(\boldsymbol{e}|\boldsymbol{\beta},\boldsymbol{\gamma}) = \boldsymbol{0}$$
$$\text{and} \quad D(\boldsymbol{e}|\boldsymbol{\beta},\boldsymbol{\gamma},\sigma^2) = D(\boldsymbol{y}|\sigma^2) = \sigma^2\boldsymbol{\Sigma}_{ee} \, . \tag{5.5}$$

Here, $\boldsymbol{X}$ means an $n \times u$ and $\boldsymbol{Z}$ an $n \times r$ matrix of known coefficients, $\boldsymbol{\beta}$ the $u \times 1$ vector of unknown parameters of the trend, $\boldsymbol{\gamma}$ the $r \times 1$ vector of unknown signals, $\boldsymbol{y}$ the $n \times 1$ vector of observations, $\boldsymbol{e}$ the $n \times 1$ vector of their errors, σ^2 the unknown variance factor and $\boldsymbol{\Sigma}_{ee}$ the known positive definite $n \times n$ matrix, which multiplied by σ^2 gives the covariance matrix of the observations and the errors, respectively. Thus, the observations $\boldsymbol{y}$ and their errors $\boldsymbol{e}$ are represented with (5.5) by the linear function $\boldsymbol{Z\gamma}$ of the unknown signals $\boldsymbol{\gamma}$ and by the linear function $\boldsymbol{X\beta}$ of the unknown parameters $\boldsymbol{\beta}$ of the trend.

The model of prediction and filtering is therefore defined as a special linear model. As prior information for the unknown parameters $\bar{\boldsymbol{\beta}}$ and the unknown variance factor σ^2 in (5.1) the expected values of $\bar{\boldsymbol{\beta}}$ and σ^2 and the variances and covariances of $\bar{\boldsymbol{\beta}}$ and σ^2 are introduced as in Chapter 4.3.2

$$E(\bar{\boldsymbol{\beta}}) = E\left(\begin{vmatrix} \boldsymbol{\beta} \\ \boldsymbol{\gamma} \end{vmatrix}\right) = \bar{\boldsymbol{\mu}} = \begin{vmatrix} \boldsymbol{\mu} \\ \boldsymbol{0} \end{vmatrix}, \ D(\bar{\boldsymbol{\beta}}) = D\left(\begin{vmatrix} \boldsymbol{\beta} \\ \boldsymbol{\gamma} \end{vmatrix}\right) = \sigma_p^2\bar{\boldsymbol{\Sigma}}$$

$$= \sigma_p^2 \begin{vmatrix} \boldsymbol{\Sigma}_\beta & \boldsymbol{0} \\ \boldsymbol{0} & \boldsymbol{\Sigma}_{\gamma\gamma} \end{vmatrix}, \ E(\sigma^2) = \sigma_p^2 \, , \ V(\sigma^2) = V_{\sigma^2} \, . \tag{5.6}$$

As was already mentioned in Chapter 5.1, it is feasible because of the trend, to use the prior information $E(\boldsymbol{\gamma}) = \boldsymbol{0}$ for the unknown signals $\boldsymbol{\gamma}$. If one substitutes the matrices and vectors (5.3) and the prior information (5.6) in the distribution (4.158) so that it becomes valid for the model (5.1), the posterior distribution for the unknown parameters $\boldsymbol{\beta}$ and $\boldsymbol{\gamma}$ and for the unknown weight parameter τ with $\tau = 1/\sigma^2$ follows as normal-gamma distribution by

$$\begin{vmatrix} \boldsymbol{\beta} \\ \boldsymbol{\gamma} \end{vmatrix}, \ \tau|\boldsymbol{y} \sim NG(\begin{vmatrix} \boldsymbol{\beta}_0 \\ \boldsymbol{\gamma}_0 \end{vmatrix}, \boldsymbol{V}_0, b_0, p_0) \, . \tag{5.7}$$

With

$$\bar{\boldsymbol{X}}'\bar{\boldsymbol{P}}\bar{\boldsymbol{X}} + \bar{\boldsymbol{\Sigma}}^{-1} = \begin{vmatrix} \boldsymbol{X}' \\ \boldsymbol{Z}' \end{vmatrix} \boldsymbol{\Sigma}_{ee}^{-1}|\boldsymbol{X},\boldsymbol{Z}| + \begin{vmatrix} \boldsymbol{\Sigma}_\beta^{-1} & \boldsymbol{0} \\ \boldsymbol{0} & \boldsymbol{\Sigma}_{\gamma\gamma}^{-1} \end{vmatrix}$$

the parameters in (5.7) are obtained from (4.159) by

$$\begin{vmatrix} \boldsymbol{\beta}_0 \\ \boldsymbol{\gamma}_0 \end{vmatrix} = \boldsymbol{V}_0 \begin{vmatrix} \boldsymbol{X}'\boldsymbol{\Sigma}_{ee}^{-1}\boldsymbol{y} + \boldsymbol{\Sigma}_{\beta}^{-1}\boldsymbol{\mu} \\ \boldsymbol{Z}'\boldsymbol{\Sigma}_{ee}^{-1}\boldsymbol{y} \end{vmatrix} \tag{5.8}$$

$$\boldsymbol{V}_0 = \begin{vmatrix} \boldsymbol{X}'\boldsymbol{\Sigma}_{ee}^{-1}\boldsymbol{X} + \boldsymbol{\Sigma}_{\beta}^{-1} & \boldsymbol{X}'\boldsymbol{\Sigma}_{ee}^{-1}\boldsymbol{Z} \\ \boldsymbol{Z}'\boldsymbol{\Sigma}_{ee}^{-1}\boldsymbol{X} & \boldsymbol{Z}'\boldsymbol{\Sigma}_{ee}^{-1}\boldsymbol{Z} + \boldsymbol{\Sigma}_{\gamma\gamma}^{-1} \end{vmatrix}^{-1} \tag{5.9}$$

$$\begin{aligned} b_0 &= \big\{2\big[(\sigma_p^2)^2/V_{\sigma^2} + 1\big]\sigma_p^2 \\ &\quad + (\boldsymbol{\mu} - \boldsymbol{\beta}_0)'\boldsymbol{\Sigma}_{\beta}^{-1}(\boldsymbol{\mu} - \boldsymbol{\beta}_0) + \boldsymbol{\gamma}_0'\boldsymbol{\Sigma}_{\gamma\gamma}^{-1}\boldsymbol{\gamma}_0 \\ &\quad + (\boldsymbol{y} - \boldsymbol{X}\boldsymbol{\beta}_0 - \boldsymbol{Z}\boldsymbol{\gamma}_0)'\boldsymbol{\Sigma}_{ee}^{-1}(\boldsymbol{y} - \boldsymbol{X}\boldsymbol{\beta}_0 - \boldsymbol{Z}\boldsymbol{\gamma}_0)\big\}/2 \tag{5.10} \\ p_0 &= \big(n + 2(\sigma_p^2)^2/V_{\sigma^2} + 4\big)/2 \,. \tag{5.11} \end{aligned}$$

The posterior marginal distribution for $\boldsymbol{\beta}$ and $\boldsymbol{\gamma}$ follows with (5.7) from (4.160) as the multivariate t-distribution

$$\begin{vmatrix} \boldsymbol{\beta} \\ \boldsymbol{\gamma} \end{vmatrix}\Big|\boldsymbol{y} \sim t\big(\begin{vmatrix} \boldsymbol{\beta}_0 \\ \boldsymbol{\gamma}_0 \end{vmatrix}, \; b_0\boldsymbol{V}_0/p_0, 2p_0\big) \,. \tag{5.12}$$

The Bayes estimates $\hat{\boldsymbol{\beta}}_B$ and $\hat{\boldsymbol{\gamma}}_B$, the MAP estimates $\hat{\boldsymbol{\beta}}_M$ and $\hat{\boldsymbol{\gamma}}_M$ and the estimates $\hat{\boldsymbol{\beta}}$ and $\hat{\boldsymbol{\gamma}}$ of the method of least squares of the unknown parameters $\boldsymbol{\beta}$ and $\boldsymbol{\gamma}$ are therefore obtained with (4.162), (4.163) and (4.165) from (5.8) by

$$\hat{\boldsymbol{\beta}} = \hat{\boldsymbol{\beta}}_B = \hat{\boldsymbol{\beta}}_M = \boldsymbol{\beta}_0 \quad \text{and} \quad \hat{\boldsymbol{\gamma}} = \hat{\boldsymbol{\gamma}}_B = \hat{\boldsymbol{\gamma}}_M = \boldsymbol{\gamma}_0 \,. \tag{5.13}$$

After multiplying both sides of (5.8) from the left by $\boldsymbol{V}_0^{-1}$ in (5.9) the second equation leads to

$$\hat{\boldsymbol{\gamma}} = (\boldsymbol{Z}'\boldsymbol{\Sigma}_{ee}^{-1}\boldsymbol{Z} + \boldsymbol{\Sigma}_{\gamma\gamma}^{-1})^{-1}\boldsymbol{Z}'\boldsymbol{\Sigma}_{ee}^{-1}(\boldsymbol{y} - \boldsymbol{X}\hat{\boldsymbol{\beta}}) \tag{5.14}$$

and the identity (4.48) to

$$\hat{\boldsymbol{\gamma}} = \boldsymbol{\Sigma}_{\gamma\gamma}\boldsymbol{Z}'(\boldsymbol{Z}\boldsymbol{\Sigma}_{\gamma\gamma}\boldsymbol{Z}' + \boldsymbol{\Sigma}_{ee})^{-1}(\boldsymbol{y} - \boldsymbol{X}\hat{\boldsymbol{\beta}}) \,. \tag{5.15}$$

If (5.14) is substituted in the first equation following from (5.8), we obtain

$$\begin{aligned} \big\{\boldsymbol{X}'[\boldsymbol{\Sigma}_{ee}^{-1} &- \boldsymbol{\Sigma}_{ee}^{-1}\boldsymbol{Z}(\boldsymbol{Z}'\boldsymbol{\Sigma}_{ee}^{-1}\boldsymbol{Z} + \boldsymbol{\Sigma}_{\gamma\gamma}^{-1})^{-1}\boldsymbol{Z}'\boldsymbol{\Sigma}_{ee}^{-1}]\boldsymbol{X} + \boldsymbol{\Sigma}_{\beta}^{-1}\big\}\hat{\boldsymbol{\beta}} \\ &= \boldsymbol{X}'[\boldsymbol{\Sigma}_{ee}^{-1} - \boldsymbol{\Sigma}_{ee}^{-1}\boldsymbol{Z}(\boldsymbol{Z}'\boldsymbol{\Sigma}_{ee}^{-1}\boldsymbol{Z} + \boldsymbol{\Sigma}_{\gamma\gamma}^{-1})^{-1}\boldsymbol{Z}'\boldsymbol{\Sigma}_{ee}^{-1}]\boldsymbol{y} + \boldsymbol{\Sigma}_{\beta}^{-1}\boldsymbol{\mu} \,. \tag{5.16} \end{aligned}$$

The identity (4.47) is applied to the matrices in brackets leading to

$$\begin{aligned} \hat{\boldsymbol{\beta}} = {}&(\boldsymbol{X}'(\boldsymbol{Z}\boldsymbol{\Sigma}_{\gamma\gamma}\boldsymbol{Z}' + \boldsymbol{\Sigma}_{ee})^{-1}\boldsymbol{X} + \boldsymbol{\Sigma}_{\beta}^{-1})^{-1} \\ &(\boldsymbol{X}'(\boldsymbol{Z}\boldsymbol{\Sigma}_{\gamma\gamma}\boldsymbol{Z}' + \boldsymbol{\Sigma}_{ee})^{-1}\boldsymbol{y} + \boldsymbol{\Sigma}_{\beta}^{-1}\boldsymbol{\mu}) \,. \tag{5.17} \end{aligned}$$

The posterior marginal distributions valid for $\boldsymbol{\beta}$ and for $\boldsymbol{\gamma}$ are obtained with (2.209) from (5.12) by computing the inverse on the right-hand side of (5.9) with (4.46) and for $\boldsymbol{\beta}$ in addition by the identity (4.47) so that the two multivariate t-distributions for $\boldsymbol{\beta}$ and $\boldsymbol{\gamma}$ are obtained by

$$\boldsymbol{\beta}|\boldsymbol{y} \sim t(\hat{\boldsymbol{\beta}}, b_0(\boldsymbol{X}'\boldsymbol{\Sigma}_{ee}^{-1}\boldsymbol{X} + \boldsymbol{\Sigma}_{\beta}^{-1} - \boldsymbol{X}'\boldsymbol{\Sigma}_{ee}^{-1}\boldsymbol{Z}$$
$$(\boldsymbol{Z}'\boldsymbol{\Sigma}_{ee}^{-1}\boldsymbol{Z} + \boldsymbol{\Sigma}_{\gamma\gamma}^{-1})^{-1}\boldsymbol{Z}'\boldsymbol{\Sigma}_{ee}^{-1}\boldsymbol{X})^{-1}/p_0, 2p_0) \qquad (5.18)$$

$$\boldsymbol{\gamma}|\boldsymbol{y} \sim t(\hat{\boldsymbol{\gamma}}, b_0(\boldsymbol{Z}'\boldsymbol{\Sigma}_{ee}^{-1}\boldsymbol{Z} + \boldsymbol{\Sigma}_{\gamma\gamma}^{-1} - \boldsymbol{Z}'\boldsymbol{\Sigma}_{ee}^{-1}\boldsymbol{X}$$
$$(\boldsymbol{X}'\boldsymbol{\Sigma}_{ee}^{-1}\boldsymbol{X} + \boldsymbol{\Sigma}_{\beta}^{-1})^{-1}\boldsymbol{X}'\boldsymbol{\Sigma}_{ee}^{-1}\boldsymbol{Z})^{-1}/p_0, 2p_0) \ . \qquad (5.19)$$

The two multivariate t-distributions are simplified by the indentity (4.47) and follow with

$$\boldsymbol{\beta}|\boldsymbol{y} \sim t(\hat{\boldsymbol{\beta}}, b_0(\boldsymbol{X}'(\boldsymbol{Z}\boldsymbol{\Sigma}_{\gamma\gamma}\boldsymbol{Z}' + \boldsymbol{\Sigma}_{ee})^{-1}\boldsymbol{X} + \boldsymbol{\Sigma}_{\beta}^{-1})^{-1}/p_0, 2p_0) \qquad (5.20)$$

$$\boldsymbol{\gamma}|\boldsymbol{y} \sim t(\hat{\boldsymbol{\gamma}}, b_0(\boldsymbol{Z}'(\boldsymbol{X}\boldsymbol{\Sigma}_{\beta}\boldsymbol{X}' + \boldsymbol{\Sigma}_{ee})^{-1}\boldsymbol{Z} + \boldsymbol{\Sigma}_{\gamma\gamma}^{-1})^{-1}/p_0, 2p_0) \ . \qquad (5.21)$$

With these results confidence regions for $\boldsymbol{\beta}$ and $\boldsymbol{\gamma}$ may be computed or hypotheses for $\boldsymbol{\beta}$ and $\boldsymbol{\gamma}$ may be tested. The covariance matrices for $\boldsymbol{\beta}$ and $\boldsymbol{\gamma}$ are obtained with (2.208) and (3.11) by

$$D(\boldsymbol{\beta}|\boldsymbol{y}) = \frac{b_0}{p_0 - 1}(\boldsymbol{X}'(\boldsymbol{Z}\boldsymbol{\Sigma}_{\gamma\gamma}\boldsymbol{Z}' + \boldsymbol{\Sigma}_{ee})^{-1}\boldsymbol{X} + \boldsymbol{\Sigma}_{\beta}^{-1})^{-1} \qquad (5.22)$$

$$D(\boldsymbol{\gamma}|\boldsymbol{y}) = \frac{b_0}{p_0 - 1}(\boldsymbol{Z}'(\boldsymbol{X}\boldsymbol{\Sigma}_{\beta}\boldsymbol{X}' + \boldsymbol{\Sigma}_{ee})^{-1}\boldsymbol{Z} + \boldsymbol{\Sigma}_{\gamma\gamma}^{-1})^{-1} \qquad (5.23)$$

where b_0 and p_0 are determined by (5.10) and (5.11). An alternative representation of the posterior marginal distribution (5.21) for $\boldsymbol{\gamma}$, which is needed in the following chapter, is obtained by applying again the identity (4.47)

$$\boldsymbol{\gamma}|\boldsymbol{y} \sim t(\hat{\boldsymbol{\gamma}}, b_0(\boldsymbol{\Sigma}_{\gamma\gamma} - \boldsymbol{\Sigma}_{\gamma\gamma}\boldsymbol{Z}'(\boldsymbol{Z}\boldsymbol{\Sigma}_{\gamma\gamma}\boldsymbol{Z}' + \boldsymbol{\Sigma}_{ee} + \boldsymbol{X}\boldsymbol{\Sigma}_{\beta}\boldsymbol{X}')^{-1}\boldsymbol{Z}\boldsymbol{\Sigma}_{\gamma\gamma})/$$
$$p_0, 2p_0). \qquad (5.24)$$

The $n \times 1$ vector $\boldsymbol{y}_f$ of filtered observations is found from (5.5) by

$$\boldsymbol{y}_f = \boldsymbol{y} + \boldsymbol{e} = \boldsymbol{X}\boldsymbol{\beta} + \boldsymbol{Z}\boldsymbol{\gamma} \qquad (5.25)$$

with its distribution as linear function $\boldsymbol{X}\boldsymbol{\beta} + \boldsymbol{Z}\boldsymbol{\gamma}$ of $\boldsymbol{\beta}$ and $\boldsymbol{\gamma}$ from (2.210), (5.12) and (5.13)

$$|\boldsymbol{X}, \boldsymbol{Z}| \begin{vmatrix} \boldsymbol{\beta} \\ \boldsymbol{\gamma} \end{vmatrix} |\boldsymbol{y} \sim t(|\boldsymbol{X}, \boldsymbol{Z}| \begin{vmatrix} \hat{\boldsymbol{\beta}} \\ \hat{\boldsymbol{\gamma}} \end{vmatrix}, b_0|\boldsymbol{X}, \boldsymbol{Z}|\boldsymbol{V}_0 \begin{vmatrix} \boldsymbol{X}' \\ \boldsymbol{Z}' \end{vmatrix}/p_0, 2p_0) \ . \qquad (5.26)$$

The estimate $\hat{\boldsymbol{y}}_f$ of the filtered observations $\boldsymbol{y}_f$ is therefore obtained by

$$\hat{\boldsymbol{y}}_f = \boldsymbol{X}\hat{\boldsymbol{\beta}} + \boldsymbol{Z}\hat{\boldsymbol{\gamma}} \ . \qquad (5.27)$$

Correspondingly, the $q \times 1$ vector $\boldsymbol{y}_p$ of predicted observations follows with

$$\boldsymbol{y}_p = \boldsymbol{X}^*\boldsymbol{\beta} + \boldsymbol{Z}^*\boldsymbol{\gamma} \tag{5.28}$$

where $\boldsymbol{X}^*$ and $\boldsymbol{Z}^*$ denote the $q \times u$ and $q \times r$ matrices of coefficients which express the trend and the signal at the positions of the predicted observations. The distribution of $\boldsymbol{y}_p$ is found from (5.26) by substituting $\boldsymbol{X} = \boldsymbol{X}^*$ and $\boldsymbol{Z} = \boldsymbol{Z}^*$ and the estimate $\hat{\boldsymbol{y}}_p$ of $\boldsymbol{y}_p$ by

$$\hat{\boldsymbol{y}}_p = \boldsymbol{X}^*\hat{\boldsymbol{\beta}} + \boldsymbol{Z}^*\hat{\boldsymbol{\gamma}} \; . \tag{5.29}$$

As was already mentioned in Chapter 5.1, the case is also dealt with that prior information is only introduced for the unknown parameters $\boldsymbol{\gamma}$ of the signals. Thus, we set in (5.6)

$$E(\boldsymbol{\beta}) = \boldsymbol{\mu} = \boldsymbol{0} \quad \text{and} \quad \boldsymbol{\Sigma}_\beta^{-1} = \boldsymbol{0} \tag{5.30}$$

which means the expectations zero with very large variances. The joint posterior distribution for $\boldsymbol{\beta}, \boldsymbol{\gamma}$ and τ is then obtained from (5.7) by substituting (5.30) in (5.8) to (5.10) and the posterior marginal distribution for $\boldsymbol{\beta}$ and $\boldsymbol{\gamma}$ by substituting in (5.12). The estimates $\hat{\boldsymbol{\beta}}$ and $\hat{\boldsymbol{\gamma}}$ of $\boldsymbol{\beta}$ and $\boldsymbol{\gamma}$ then follow from (5.15) and (5.17) by

$$\hat{\boldsymbol{\beta}} = (\boldsymbol{X}'(\boldsymbol{Z}\boldsymbol{\Sigma}_{\gamma\gamma}\boldsymbol{Z}' + \boldsymbol{\Sigma}_{ee})^{-1}\boldsymbol{X})^{-1}\boldsymbol{X}'(\boldsymbol{Z}\boldsymbol{\Sigma}_{\gamma\gamma}\boldsymbol{Z}' + \boldsymbol{\Sigma}_{ee})^{-1}\boldsymbol{y} \tag{5.31}$$

$$\hat{\boldsymbol{\gamma}} = \boldsymbol{\Sigma}_{\gamma\gamma}\boldsymbol{Z}'(\boldsymbol{Z}\boldsymbol{\Sigma}_{\gamma\gamma}\boldsymbol{Z}' + \boldsymbol{\Sigma}_{ee})^{-1}(\boldsymbol{y} - \boldsymbol{X}\hat{\boldsymbol{\beta}}) \; . \tag{5.32}$$

These are the estimates of the unknown parameters $\boldsymbol{\beta}$ and $\boldsymbol{\gamma}$ of the model of prediction and filtering of traditional statistics, see for instance KOCH (1999, p.223).

We find the posterior marginal distribution for $\boldsymbol{\beta}$ by substituting (5.30) in (5.20). However, (5.21) is not valid as posterior marginal distribution for $\boldsymbol{\gamma}$, since $\boldsymbol{\Sigma}_\beta$ does not exist because of (5.30). The distribution follows from (5.19) after substituting (5.30). This marginal distribution shall be presented in a form which is needed in the following chapter. The matrix identity (4.47) gives

$$\boldsymbol{A} = (\boldsymbol{Z}'\boldsymbol{\Sigma}_{ee}^{-1}\boldsymbol{Z} + \boldsymbol{\Sigma}_{\gamma\gamma}^{-1})^{-1}$$
$$= \boldsymbol{\Sigma}_{\gamma\gamma} - \boldsymbol{\Sigma}_{\gamma\gamma}\boldsymbol{Z}'(\boldsymbol{Z}\boldsymbol{\Sigma}_{\gamma\gamma}\boldsymbol{Z}' + \boldsymbol{\Sigma}_{ee})^{-1}\boldsymbol{Z}\boldsymbol{\Sigma}_{\gamma\gamma} \tag{5.33}$$

and furthermore we obtain with the same identity instead of (5.19)

$$\boldsymbol{\gamma}|\boldsymbol{y} \sim t(\hat{\boldsymbol{\gamma}}, b_0(\boldsymbol{A} + \boldsymbol{A}\boldsymbol{Z}'\boldsymbol{\Sigma}_{ee}^{-1}\boldsymbol{X}(\boldsymbol{X}'\boldsymbol{\Sigma}_{ee}^{-1}\boldsymbol{X}$$
$$- \boldsymbol{X}'\boldsymbol{\Sigma}_{ee}^{-1}\boldsymbol{Z}\boldsymbol{A}\boldsymbol{Z}'\boldsymbol{\Sigma}_{ee}^{-1}\boldsymbol{X})^{-1}\boldsymbol{X}'\boldsymbol{\Sigma}_{ee}^{-1}\boldsymbol{Z}\boldsymbol{A})/p_0, 2p_0) \; . \tag{5.34}$$

The Bayes estimate of the variance factor σ^2 in the model (5.4) or (5.5) of prediction and filtering is obtained with (3.9) and (4.167) by

$$\hat{\sigma}_B^2 = b_0/(p_0 - 1) \tag{5.35}$$

where b_0 and p_0 result from (5.10) and (5.11). The variance $V(\sigma^2|\boldsymbol{y})$ of σ^2 follows from (4.169) by

$$V(\sigma^2|\boldsymbol{y}) = \frac{2(\hat{\sigma}_B^2)^2}{n + 2(\sigma_p^2)^2/V_{\sigma^2}} \; . \tag{5.36}$$

5.1.2 Special Model of Prediction and Filtering

It is often assumed for the model (5.4) or (5.5) of prediction and filtering that the products of matrices

$$\boldsymbol{\Sigma}_{ss} = \boldsymbol{Z}\boldsymbol{\Sigma}_{\gamma\gamma}\boldsymbol{Z}' \quad \text{and} \quad \boldsymbol{\Sigma}_{\gamma y} = \boldsymbol{\Sigma}_{\gamma\gamma}\boldsymbol{Z}' \quad \text{with} \quad \boldsymbol{\Sigma}_{y\gamma} = \boldsymbol{\Sigma}'_{\gamma y} \tag{5.37}$$

are given. This can be explained by interpreting because of (2.158) the matrix $\sigma^2\boldsymbol{\Sigma}_{ss}$ as covariance matrix of the signal $\boldsymbol{Z}\boldsymbol{\gamma}$ and the matrix $\sigma^2\boldsymbol{\Sigma}_{\gamma y}$ as covariance matrix of $\boldsymbol{\gamma}$ and $\boldsymbol{y}$, since we obtain with (2.164) and (5.6)

$$C(\boldsymbol{\gamma},|\boldsymbol{X},\boldsymbol{Z}| \begin{vmatrix} \boldsymbol{\beta} \\ \boldsymbol{\gamma} \end{vmatrix}) = |\boldsymbol{0},\boldsymbol{\Sigma}_{\gamma\gamma}| \begin{vmatrix} \boldsymbol{X}' \\ \boldsymbol{Z}' \end{vmatrix} = \boldsymbol{\Sigma}_{\gamma\gamma}\boldsymbol{Z}' \; .$$

The sum $\sigma^2(\boldsymbol{\Sigma}_{ss}+\boldsymbol{\Sigma}_{ee})$ may, furthermore, be interpreted in traditional statistics as the covariance matrix of the observations $\boldsymbol{y}$, because $\boldsymbol{\beta}$ contains fixed parameters.

By substituting (5.37) in (5.20) and in (5.24) the posterior marginal distributions for $\boldsymbol{\beta}$ and $\boldsymbol{\gamma}$ follow with

$$\boldsymbol{\beta}|\boldsymbol{y} \sim t(\hat{\boldsymbol{\beta}}, b_0(\boldsymbol{X}'(\boldsymbol{\Sigma}_{ss} + \boldsymbol{\Sigma}_{ee})^{-1}\boldsymbol{X} + \boldsymbol{\Sigma}_\beta^{-1})^{-1}/p_0, 2p_0) \tag{5.38}$$

$$\boldsymbol{\gamma}|\boldsymbol{y} \sim t(\hat{\boldsymbol{\gamma}}, b_0(\boldsymbol{\Sigma}_{\gamma\gamma} - \boldsymbol{\Sigma}_{\gamma y}(\boldsymbol{\Sigma}_{ss} + \boldsymbol{\Sigma}_{ee} + \boldsymbol{X}\boldsymbol{\Sigma}_\beta\boldsymbol{X}')^{-1}\boldsymbol{\Sigma}_{y\gamma})/p_0, 2p_0) \tag{5.39}$$

and the estimates $\hat{\boldsymbol{\beta}}$ and $\hat{\boldsymbol{\gamma}}$ from (5.15) and (5.17) with

$$\hat{\boldsymbol{\beta}} = (\boldsymbol{X}'(\boldsymbol{\Sigma}_{ss} + \boldsymbol{\Sigma}_{ee})^{-1}\boldsymbol{X} + \boldsymbol{\Sigma}_\beta^{-1})^{-1}(\boldsymbol{X}'(\boldsymbol{\Sigma}_{ss} + \boldsymbol{\Sigma}_{ee})^{-1}\boldsymbol{y} + \boldsymbol{\Sigma}_\beta^{-1}\boldsymbol{\mu}) \tag{5.40}$$

$$\hat{\boldsymbol{\gamma}} = \boldsymbol{\Sigma}_{\gamma y}(\boldsymbol{\Sigma}_{ss} + \boldsymbol{\Sigma}_{ee})^{-1}(\boldsymbol{y} - \boldsymbol{X}\hat{\boldsymbol{\beta}}) \; . \tag{5.41}$$

If the signals are directly measured by the observations $\boldsymbol{y}$, we set

$$\boldsymbol{s} = \boldsymbol{Z}\boldsymbol{\gamma} \tag{5.42}$$

where $\boldsymbol{s}$ denotes the $n \times 1$ vector of signals. By substituting in (5.5) the special model of prediction and filtering is obtained

$$\boldsymbol{X}\boldsymbol{\beta} + \boldsymbol{s} = \boldsymbol{y} + \boldsymbol{e} \; . \tag{5.43}$$

The posterior distribution for $\boldsymbol{s}$ follows as a linear function of $\boldsymbol{Z}\boldsymbol{\gamma}$ with (2.210) and (5.37) from (5.39)

$$\boldsymbol{s}|\boldsymbol{y} \sim t(\hat{\boldsymbol{s}}, b_0(\boldsymbol{\Sigma}_{ss} - \boldsymbol{\Sigma}_{ss}(\boldsymbol{\Sigma}_{ss} + \boldsymbol{\Sigma}_{ee} + \boldsymbol{X}\boldsymbol{\Sigma}_\beta\boldsymbol{X}')^{-1}\boldsymbol{\Sigma}_{ss})/p_0, 2p_0) \tag{5.44}$$

where the estimate $\hat{s} = Z\hat{\gamma}$ of s is obtained from (5.41)

$$\hat{s} = \Sigma_{ss}(\Sigma_{ss} + \Sigma_{ee})^{-1}(y - X\hat{\beta}) , \tag{5.45}$$

the parameter b_0 from (5.10) with (5.13), (5.15) and (5.37)

$$\begin{aligned} b_0 = {} & \{2[(\sigma_p^2)^2/V_{\sigma^2} + 1]\sigma_p^2 + (\mu - \hat{\beta})'\Sigma_\beta^{-1}(\mu - \hat{\beta}) \\ & + (y - X\hat{\beta})'(\Sigma_{ss} + \Sigma_{ee})^{-1}\Sigma_{ss}(\Sigma_{ss} + \Sigma_{ee})^{-1}(y - X\hat{\beta}) \\ & + (y - X\hat{\beta} - \hat{s})'\Sigma_{ee}^{-1}(y - X\hat{\beta} - \hat{s})\}/2 \end{aligned} \tag{5.46}$$

and the parameter p_0 from (5.11).

The vector y_f of filtered observations is found by substituting (5.42) in (5.25), its distribution follows from (5.26) and its estimate from (5.27). The predicted observations y_p result from (5.28)

$$y_p = X^*\beta + s^* \quad \text{with} \quad s^* = Z^*\gamma . \tag{5.47}$$

In addition to Σ_{ss} and $\Sigma_{\gamma y}$ in (5.37) the following matrices are often given

$$\Sigma_{s^*s^*} = Z^*\Sigma_{\gamma\gamma}Z^{*\prime} \quad \text{and} \quad \Sigma_{s^*s} = Z^*\Sigma_{\gamma\gamma}Z'$$
$$\text{with} \quad \Sigma_{ss^*} = \Sigma'_{s^*s} \tag{5.48}$$

where $\sigma^2\Sigma_{s^*s^*}$ because of (2.158) is interpreted as covariance matrix of the predicted signal and $\sigma^2\Sigma_{s^*s}$ because of (2.164) as covariance matrix of the predicted signal and the original signal. The posterior distribution for s^* follows with (2.210), (5.47) and (5.48) from (5.39)

$$s^*|y \sim t(\hat{s^*}, b_0(\Sigma_{s^*s^*} - \Sigma_{s^*s}(\Sigma_{ss} + \Sigma_{ee} + X\Sigma_\beta X')^{-1}\Sigma_{ss^*})/p_0, 2p_0) \tag{5.49}$$

with the estimate $\hat{s^*}$ of s^* from (5.41)

$$\hat{s^*} = \Sigma_{s^*s}(\Sigma_{ss} + \Sigma_{ee})^{-1}(y - X\hat{\beta}) . \tag{5.50}$$

The estimate $\hat{y}_p$ of the predicted signal is obtained from (5.29) with (5.40), (5.47) and (5.50) by

$$\hat{y}_p = X^*\hat{\beta} + \hat{s^*} . \tag{5.51}$$

If the prior information is restricted such that with (5.30) prior information is only introduced for the parameters γ, the posterior marginal distribution for β is found from (5.38) and the posterior marginal distribution for γ from (5.33) and (5.34) with (5.37) by

$$\begin{aligned} \gamma|y \sim t\big(\hat{\gamma}, b_0\big[& \Sigma_{\gamma\gamma} - \Sigma_{\gamma y}(\Sigma_{ss} + \Sigma_{ee})^{-1}\Sigma_{y\gamma} \\ & + (\Sigma_{\gamma y} - \Sigma_{\gamma y}(\Sigma_{ss} + \Sigma_{ee})^{-1}\Sigma_{ss})\Sigma_{ee}^{-1}X(X'\Sigma_{ee}^{-1}X \\ & - X'\Sigma_{ee}^{-1}(\Sigma_{ss} - \Sigma_{ss}(\Sigma_{ss} + \Sigma_{ee})^{-1}\Sigma_{ss})\Sigma_{ee}^{-1}X)^{-1} \\ & X'\Sigma_{ee}^{-1}(\Sigma_{y\gamma} - \Sigma_{ss}(\Sigma_{ss} + \Sigma_{ee})^{-1}\Sigma_{y\gamma})\big]/p_0, 2p_0\big) . \end{aligned} \tag{5.52}$$

The distributions for the signal s and the predicted signal s^* follow with this distribution as (5.44) and (5.49) from (5.39).

An application of the special model of prediction and filtering for the deformation analysis is given, for instance, by KOCH and PAPO (2003).

Example: Let temperatures T_i be measured at certain times t_i with $i \in \{1, \ldots, n\}$. They are collected in the vector $\boldsymbol{y} = (T_i)$ of observations. Let the measurements be independent and have identical but unknown variances σ^2. The covariance matrix $D(\boldsymbol{y}|\sigma^2)$ of the observations $\boldsymbol{y}$ in model (5.4) and (5.5) is therefore determined by

$$D(\boldsymbol{y}|\sigma^2) = \sigma^2 \boldsymbol{\Sigma}_{ee} \quad \text{and} \quad \boldsymbol{\Sigma}_{ee} = \boldsymbol{I} \ . \tag{5.53}$$

Let the measurements $\boldsymbol{y}$ be represented according to (5.43) by the vector $\boldsymbol{s}$ of the signals, which are the deviations of the temperatures from the trend, and by the trend $\boldsymbol{X\beta}$, which results from a polynomial expansion. The matrix $\boldsymbol{X}$ of coefficients is therefore built up depending on the times t_i by a polynomial of degree $u - 1$ such that u unknown parameters appear in $\boldsymbol{\beta}$. One gets $\boldsymbol{X} = |\boldsymbol{x}_1, \ldots, \boldsymbol{x}_n|'$ where $\boldsymbol{x}_i'$ denotes the ith row of $\boldsymbol{X}$ with

$$\boldsymbol{x}_i' = |1, t_i, t_i^2, \ldots, t_i^{u-1}| \ , \ i \in \{1, \ldots, n\} \ . \tag{5.54}$$

The temperatures T_j^* at given times t_j^* have to be predicted by $\boldsymbol{y}_p = (T_j^*)$ and $j \in \{1, \ldots, q\}$ according to (5.47). In addition, the $1 - \alpha$ confidence interval for the component s_j^* of the predicted signal $\boldsymbol{s}^*$ with $\boldsymbol{s}^* = (s_j^*)$ and $j \in \{1, \ldots, q\}$ has to be determined.

Let the elements σ_{ik} of the matrix $\boldsymbol{\Sigma}_{ss}$ in (5.37) with $\boldsymbol{\Sigma}_{ss} = (\sigma_{ik})$ and $i, k \in \{1, \ldots, n\}$, which define with $\sigma^2 \boldsymbol{\Sigma}_{ss}$ the covariance matrix of the signals, be given by the covariance function $\sigma(t_i - t_k)$, for instance by

$$\sigma_{ik} = \sigma(t_i - t_k) = ab^{-(t_i - t_k)^2} \tag{5.55}$$

where t_i and t_k denote the times of the measurements and a and b constants. The elements $\sigma_{j^*l^*}$ of the matrix $\boldsymbol{\Sigma}_{s^*s^*}$ with $\boldsymbol{\Sigma}_{s^*s^*} = (\sigma_{j^*l^*})$ and $j^*, l^* \in \{1, \ldots, q\}$ and the elements σ_{j^*i} of the matrix $\boldsymbol{\Sigma}_{s^*s}$ in (5.48) with $\boldsymbol{\Sigma}_{s^*s} = (\sigma_{j^*i})$ shall be determined by the same covariance function, thus

$$\sigma_{j^*l^*} = \sigma(t_j^* - t_l^*) = ab^{-(t_j^* - t_l^*)^2} \tag{5.56}$$

and

$$\sigma_{j^*i} = \sigma(t_j^* - t_i) = ab^{-(t_j^* - t_i)^2} \ . \tag{5.57}$$

The times t_j^* and t_l^* indicate the times of the signals to be predicted and $\sigma^2 \boldsymbol{\Sigma}_{s^*s^*}$ the covariance matrix of the predicted signals and $\sigma^2 \boldsymbol{\Sigma}_{s^*s}$ the covariance matrix of the predicted and the original signals.

To predict the temperatures T_j^* at the given times t_j^* with $j \in \{1, \ldots, q\}$, the estimate $\hat{\boldsymbol{y}}_p$ of $\boldsymbol{y}_p = (T_j^*)$ has to be computed with (5.51) from

$$\hat{\boldsymbol{y}}_p = \boldsymbol{X}^* \hat{\boldsymbol{\beta}} + \hat{\boldsymbol{s}}^* \ .$$

The coefficient matrix $\boldsymbol{X}^*$ follows corresponding to (5.54) with $\boldsymbol{X}^* = |\boldsymbol{x}_1^*, \ldots, \boldsymbol{x}_q^*|'$ and

$$\boldsymbol{x}_j^{*\prime} = |1, t_j^*, t_j^{*2}, \ldots, t_j^{*(u-1)}| \ , \quad j \in \{1, \ldots, q\} \ . \tag{5.58}$$

The estimates $\hat{\boldsymbol{\beta}}$ of the trend parameters are obtained from (5.40) with (5.53) and with $\boldsymbol{\mu}$ and $\boldsymbol{\Sigma}_\beta$ which according to (5.6) come from prior information on $\boldsymbol{\beta}$. The estimates $\hat{\boldsymbol{s}}^*$ of the predicted signals are computed from (5.50) and the estimates of the predicted observations are obtained with $\hat{\boldsymbol{y}}_p = (\hat{T}_j^*)$.

To find the confidence interval for the component s_j^* of the predicted signals $\boldsymbol{s}^*$ with $\boldsymbol{s}^* = (s_j^*)$, the posterior marginal distribution for s_j^* is formed with (2.209) from (5.49). With $\hat{\boldsymbol{s}}^* = (\hat{s}_j^*)$ the generalized t-distribution

$$s_j^*|\boldsymbol{y} \sim t(\hat{s}_j^*, 1/f, 2p_0) \tag{5.59}$$

is obtained with

$$1/f = b_0(\boldsymbol{\Sigma}_{s^*s^*} - \boldsymbol{\Sigma}_{s^*s}(\boldsymbol{\Sigma}_{ss} + \boldsymbol{I} + \boldsymbol{X}\boldsymbol{\Sigma}_\beta\boldsymbol{X}')^{-1}\boldsymbol{\Sigma}_{ss^*})_{jj}/p_0 \ , \tag{5.60}$$

with b_0 from (5.46) and p_0 from (5.11). By the transformation (2.206) of the generalized t-distribution to the standard form the t-distributed variable follows with (2.207) by

$$(s_j^* - \hat{s}_j^*)\sqrt{f} \sim t(2p_0) \ . \tag{5.61}$$

We therefore obtain with (2.187) the relation

$$P(-t_{1-\alpha;2p_0} < (s_j^* - \hat{s}_j^*)\sqrt{f} < t_{1-\alpha;2p_0}) = 1 - \alpha \tag{5.62}$$

where $t_{1-\alpha;2p_0}$ because of (2.186) denotes the quantity of the t-distribution which is equivalent to the upper α-percentage point $F_{1-\alpha;1,2p_0}$ of the F-distribution. Thus, the $1 - \alpha$ confidence interval for s_j^* follows from (5.62) with

$$P(\hat{s}_j^* - t_{1-\alpha;2p_0}/\sqrt{f} < s_j^* < \hat{s}_j^* + t_{1-\alpha;2p_0}/\sqrt{f}) = 1 - \alpha \ . \tag{5.63}$$

The Bayes estimate $\hat{\sigma}_B^2$ of the unknown variance factor σ^2 in (5.53) and its variance $V(\sigma|\boldsymbol{y})$ are obtained from (5.35) and (5.36) with b_0 from (5.46) and p_0 from (5.11). Δ

5.2 Variance and Covariance Components

If the variance factor σ^2 in the linear model (4.1) is unknown, it is estimated by (4.39), (4.129) and in case of prior information by (4.168). Thus, the covariance matrix of the observations $\boldsymbol{y}$ in (4.1) may be expressed by means of the estimated variance factor σ^2. Often data have to be analyzed which contain different types of observations, for instance, angle and distance measurements or terrestrial and satellite observations. The variance factors belonging to these subsets of observations are called variance components. In general, they are unknown and have to be estimated. If factors for the covariances of different subsets of observations have to be introduced, they are called covariance components which also need to be estimated.

5.2.1 Model and Likelihood Function

Let the $n \times 1$ random vector $\boldsymbol{e}$ of errors of the observations in the linear model (4.3) be represented by

$$
\boldsymbol{e} = |\boldsymbol{U}_1, \ldots, \boldsymbol{U}_l| \begin{vmatrix} \boldsymbol{\gamma}_1 \\ \ldots \\ \boldsymbol{\gamma}_l \end{vmatrix} \tag{5.64}
$$

where the $n \times r_i$ matrices $\boldsymbol{U}_i$ contain constants and $\boldsymbol{\gamma}_i$ denote unknown and unobservable $r_i \times 1$ random vectors with $E(\boldsymbol{\gamma}_i) = \boldsymbol{0}$ and $C(\boldsymbol{\gamma}_i, \boldsymbol{\gamma}_j) = \sigma_{ij} \boldsymbol{R}_{ij}$ with $\sigma_{ij} = \sigma_{ji}$ and $\boldsymbol{R}_{ji} = \boldsymbol{R}'_{ij}$ for $i, j \in \{1, \ldots, l\}$. The covariance matrix $D(\boldsymbol{e}|\boldsymbol{\beta}, \boldsymbol{\sigma})$ of the vector $\boldsymbol{e}$ then follows with (2.158) by

$$
\begin{aligned}
D(\boldsymbol{e}|\boldsymbol{\beta}, \boldsymbol{\sigma}) = {} & \sigma_1^2 \boldsymbol{U}_1 \boldsymbol{R}_{11} \boldsymbol{U}'_1 + \sigma_{12}(\boldsymbol{U}_1 \boldsymbol{R}_{12} \boldsymbol{U}'_2 + \boldsymbol{U}_2 \boldsymbol{R}_{21} \boldsymbol{U}'_1) \\
& + \sigma_{13}(\boldsymbol{U}_1 \boldsymbol{R}_{13} \boldsymbol{U}'_3 + \boldsymbol{U}_3 \boldsymbol{R}_{31} \boldsymbol{U}'_1) + \ldots + \sigma_l^2 \boldsymbol{U}_l \boldsymbol{R}_{ll} \boldsymbol{U}'_l .
\end{aligned} \tag{5.65}
$$

The $k \times 1$ vector $\boldsymbol{\sigma}$ contains the components

$$
\boldsymbol{\sigma} = |\sigma_1^2, \sigma_{12}, \sigma_{13}, \ldots, \sigma_{1l}, \sigma_2^2, \ldots, \sigma_l^2|' \tag{5.66}
$$

with $k = l(l+1)/2$. Furthermore, we set

$$
\begin{aligned}
& \boldsymbol{V}_1 = \boldsymbol{U}_1 \boldsymbol{R}_{11} \boldsymbol{U}'_1 \, , \quad \boldsymbol{V}_2 = \boldsymbol{U}_1 \boldsymbol{R}_{12} \boldsymbol{U}'_2 + \boldsymbol{U}_2 \boldsymbol{R}_{21} \boldsymbol{U}'_1, \ldots, \\
& \boldsymbol{V}_k = \boldsymbol{U}_l \boldsymbol{R}_{ll} \boldsymbol{U}'_l
\end{aligned} \tag{5.67}
$$

and obtain the covariance matrix $D(\boldsymbol{y}|\boldsymbol{\sigma}) = \boldsymbol{\Sigma}$ of the observations $\boldsymbol{y}$ with $D(\boldsymbol{y}|\boldsymbol{\sigma}) = D(\boldsymbol{e}|\boldsymbol{\beta}, \boldsymbol{\sigma})$ because of (2.158) and (4.3) by

$$
D(\boldsymbol{y}|\boldsymbol{\sigma}) = \boldsymbol{\Sigma} = \sigma_1^2 \boldsymbol{V}_1 + \sigma_{12} \boldsymbol{V}_2 + \ldots + \sigma_l^2 \boldsymbol{V}_k . \tag{5.68}
$$

We will represent the covariance matrix of the observations $\boldsymbol{y}$ in the linear model (4.1) by this manner.

Hence, let $\boldsymbol{X}$ be an $n \times u$ matrix of given coefficients with full column rank$\boldsymbol{X} = u$, $\boldsymbol{\beta}$ a $u \times 1$ random vector of unknown parameters, $\boldsymbol{y}$ an $n \times 1$ random vector of observations, whose $n \times n$ covariance matrix $D(\boldsymbol{y}|\boldsymbol{\sigma}) = \boldsymbol{\Sigma}$ is positive definite, and $\boldsymbol{e}$ the $n \times 1$ vector of errors of $\boldsymbol{y}$. We then call

$$\boldsymbol{X}\boldsymbol{\beta} = E(\boldsymbol{y}|\boldsymbol{\beta}) = \boldsymbol{y} + \boldsymbol{e}$$
$$\text{with} \quad D(\boldsymbol{y}|\boldsymbol{\sigma}) = \boldsymbol{\Sigma} = \sigma_1^2 \boldsymbol{V}_1 + \sigma_{12} \boldsymbol{V}_2 + \ldots + \sigma_l^2 \boldsymbol{V}_k \quad (5.69)$$

the linear model with unknown *variance components* σ_i^2 and unknown *covariance components* σ_{ij} for $i \in \{1, \ldots, l\}, i < j \le l$ and $l \le k \le l(l+1)/2$. The variance and covariance components are unknown random variables and according to (5.66) components of the $k \times 1$ random vector $\boldsymbol{\sigma}$. The $n \times n$ matrices $\boldsymbol{V}_m$ with $m \in \{1, \ldots, k\}$ are known and symmetric.

Example: To determine the coordinates of points, let distances be measured collected in $\boldsymbol{y}_1$ and angles be measured collected in $\boldsymbol{y}_2$. Let the measurements $\boldsymbol{y}_1$ and $\boldsymbol{y}_2$ be independent and let the weight matrix of $\boldsymbol{y}_1$ be $\boldsymbol{P}_1$ and the one of $\boldsymbol{y}_2$ be $\boldsymbol{P}_2$ from (2.159). Let the constant in (2.159) be defined for the distance measurements $\boldsymbol{y}_1$ by the unknown variance component σ_1^2 and for the angle measurements $\boldsymbol{y}_2$ by the unknown variance component σ_2^2. The covariance matrix of the observations $\boldsymbol{y}_1$ and $\boldsymbol{y}_2$ is then obtained from (5.69) by

$$D(\left|\begin{array}{c} \boldsymbol{y}_1 \\ \boldsymbol{y}_2 \end{array}\right| |\sigma_1^2, \sigma_2^2) = \sigma_1^2 \left|\begin{array}{cc} \boldsymbol{P}_1^{-1} & \boldsymbol{0} \\ \boldsymbol{0} & \boldsymbol{0} \end{array}\right| + \sigma_2^2 \left|\begin{array}{cc} \boldsymbol{0} & \boldsymbol{0} \\ \boldsymbol{0} & \boldsymbol{P}_2^{-1} \end{array}\right| .$$

If the measurements $\boldsymbol{y}_1$ and $\boldsymbol{y}_2$ are dependent and if the covariance matrix $C(\boldsymbol{y}_1, \boldsymbol{y}_2) = \sigma_{12} \boldsymbol{R}_{12}$ of $\boldsymbol{y}_1$ and $\boldsymbol{y}_2$ defined by (2.161) is known except for the factor σ_{12} which is introduced as unknown covariance component σ_{12}, the covariance matrix of the observations $\boldsymbol{y}_1$ and $\boldsymbol{y}_2$ is obtained with (5.69) and $\boldsymbol{R}_{21} = \boldsymbol{R}'_{12}$ by

$$D(\left|\begin{array}{c} \boldsymbol{y}_1 \\ \boldsymbol{y}_2 \end{array}\right| |\sigma_1^2, \sigma_{12}, \sigma_2^2) = \sigma_1^2 \left|\begin{array}{cc} \boldsymbol{P}_1^{-1} & \boldsymbol{0} \\ \boldsymbol{0} & \boldsymbol{0} \end{array}\right| + \sigma_{12} \left|\begin{array}{cc} \boldsymbol{0} & \boldsymbol{R}_{12} \\ \boldsymbol{R}_{21} & \boldsymbol{0} \end{array}\right|$$
$$+ \sigma_2^2 \left|\begin{array}{cc} \boldsymbol{0} & \boldsymbol{0} \\ \boldsymbol{0} & \boldsymbol{P}_2^{-1} \end{array}\right| .$$

$$\Delta$$

Out of reasons explained in Chapter 2.4.1 the observations $\boldsymbol{y}$ are assumed to be normally distributed. We therefore obtain with (2.195), (2.196) and (5.69)

$$\boldsymbol{y}|\boldsymbol{\beta}, \boldsymbol{\sigma} \sim N(\boldsymbol{X}\boldsymbol{\beta}, \boldsymbol{\Sigma}) \tag{5.70}$$

and

$$p(\boldsymbol{y}|\boldsymbol{\beta}, \boldsymbol{\sigma}) = \frac{1}{(2\pi)^{n/2}(\det \boldsymbol{\Sigma})^{1/2}} e^{-\frac{1}{2}(\boldsymbol{y} - \boldsymbol{X}\boldsymbol{\beta})'\boldsymbol{\Sigma}^{-1}(\boldsymbol{y} - \boldsymbol{X}\boldsymbol{\beta})} . \tag{5.71}$$

Only the vector $\boldsymbol{\sigma}$ of unknown variance and covariance components is of interest. We therefore introduce for the vector $\boldsymbol{\beta}$ of unknown parameters the noninformative prior density function (2.216), which is determined by a constant, and eliminate the vector $\boldsymbol{\beta}$ from the density function (5.71) by an integration. Thus, the marginal density function for $\boldsymbol{\sigma}$ is computed with (2.89). We transform with

$$\hat{\boldsymbol{\beta}} = (\boldsymbol{X}'\boldsymbol{\Sigma}^{-1}\boldsymbol{X})^{-1}\boldsymbol{X}'\boldsymbol{\Sigma}^{-1}\boldsymbol{y} \tag{5.72}$$

the exponent in (5.71) corresponding to (4.118) and obtain

$$(\boldsymbol{y} - \boldsymbol{X}\boldsymbol{\beta})'\boldsymbol{\Sigma}^{-1}(\boldsymbol{y} - \boldsymbol{X}\boldsymbol{\beta}) = (\boldsymbol{y} - \boldsymbol{X}\hat{\boldsymbol{\beta}})'\boldsymbol{\Sigma}^{-1}(\boldsymbol{y} - \boldsymbol{X}\hat{\boldsymbol{\beta}})$$
$$+ (\boldsymbol{\beta} - \hat{\boldsymbol{\beta}})'\boldsymbol{X}'\boldsymbol{\Sigma}^{-1}\boldsymbol{X}(\boldsymbol{\beta} - \hat{\boldsymbol{\beta}}) \, . \tag{5.73}$$

Substituting this result in (5.71) and integrating over $\boldsymbol{\beta}$ gives

$$\int_{-\infty}^{\infty} \cdots \int_{-\infty}^{\infty} \exp\left[-\frac{1}{2}(\boldsymbol{\beta} - \hat{\boldsymbol{\beta}})'\boldsymbol{X}'\boldsymbol{\Sigma}^{-1}\boldsymbol{X}(\boldsymbol{\beta} - \hat{\boldsymbol{\beta}})\right] d\boldsymbol{\beta}$$
$$= (2\pi)^{u/2}(\det \boldsymbol{X}'\boldsymbol{\Sigma}^{-1}\boldsymbol{X})^{-1/2} \, , \tag{5.74}$$

since the normal distribution fulfills the second condition in (2.74), as already mentioned in connection with (2.195). If we set in addition

$$(\boldsymbol{y} - \boldsymbol{X}\hat{\boldsymbol{\beta}})'\boldsymbol{\Sigma}^{-1}(\boldsymbol{y} - \boldsymbol{X}\hat{\boldsymbol{\beta}}) = \boldsymbol{y}'\boldsymbol{W}\boldsymbol{y} \tag{5.75}$$

with

$$\boldsymbol{W} = \boldsymbol{\Sigma}^{-1} - \boldsymbol{\Sigma}^{-1}\boldsymbol{X}(\boldsymbol{X}'\boldsymbol{\Sigma}^{-1}\boldsymbol{X})^{-1}\boldsymbol{X}'\boldsymbol{\Sigma}^{-1} \, , \tag{5.76}$$

the likelihood function follows from (5.71) with

$$p(\boldsymbol{y}|\boldsymbol{\sigma}) \propto (\det \boldsymbol{\Sigma} \det \boldsymbol{X}'\boldsymbol{\Sigma}^{-1}\boldsymbol{X})^{-1/2} \exp(-\boldsymbol{y}'\boldsymbol{W}\boldsymbol{y}/2) \, . \tag{5.77}$$

It is only dependent on the vector $\boldsymbol{\sigma}$ of unknown variance and covariance components. Constants have not been considered.

The likelihood function (5.77) is also obtained, if by a linear transformation of the observation vector $\boldsymbol{y}$ the density function (5.71) is decomposed into the factors L_1 and L_2 where L_1 is identical with (5.77) (Koch 1987) and therefore only dependent on $\boldsymbol{\sigma}$. L_2 is a function of $\boldsymbol{\beta}$ and $\boldsymbol{\sigma}$ and leads with the maximum likelihood estimate (3.33) to the conventional estimate (5.72) of $\boldsymbol{\beta}$, if $\boldsymbol{\sigma}$ is assumed as known. Using (5.77) the maximum likelihood method gives the following estimate of $\boldsymbol{\sigma}$ (Koch 1986) which is identical with the MINQUE estimate and with the locally best invariant quadratic unbiased estimate (Koch 1999, p.229). Instead of $\boldsymbol{\sigma}$ the vector $\bar{\boldsymbol{\sigma}}$ is iteratively estimated by $\hat{\bar{\boldsymbol{\sigma}}}$ with

$$\hat{\bar{\boldsymbol{\sigma}}} = \bar{\boldsymbol{S}}^{-1}\bar{\boldsymbol{q}} \tag{5.78}$$

and

$$\bar{\boldsymbol{\sigma}} = |\ldots,\bar{\sigma}_i^2,\ldots,\bar{\sigma}_{ij},\ldots|' \; , \; \boldsymbol{\sigma} = |\ldots,\alpha_i^2\bar{\sigma}_i^2,\ldots,\alpha_{ij}\bar{\sigma}_{ij},\ldots|'$$

$$\text{for} \quad i \in \{1,\ldots,l\}, \; i < j \leq l$$

$$\bar{\boldsymbol{S}} = (\text{tr}\bar{\boldsymbol{W}}\bar{\boldsymbol{V}}_m\bar{\boldsymbol{W}}\bar{\boldsymbol{V}}_n) \;\; \text{for} \;\; m,n \in \{1,\ldots,k\}\,, \; l \leq k \leq l(l+1)/2$$

$$\bar{\boldsymbol{q}} = (\boldsymbol{y}'\bar{\boldsymbol{W}}\bar{\boldsymbol{V}}_m\bar{\boldsymbol{W}}\boldsymbol{y}) = (\hat{\boldsymbol{e}}'\boldsymbol{\Sigma}_0^{-1}\bar{\boldsymbol{V}}_m\boldsymbol{\Sigma}_0^{-1}\hat{\boldsymbol{e}})$$

$$\boldsymbol{V}_m = \bar{\boldsymbol{V}}_m/\alpha_i^2 \;\; \text{or} \;\; \boldsymbol{V}_m = \bar{\boldsymbol{V}}_m/\alpha_{ij}$$

$$\bar{\boldsymbol{W}} = \boldsymbol{\Sigma}_0^{-1} - \boldsymbol{\Sigma}_0^{-1}\boldsymbol{X}(\boldsymbol{X}'\boldsymbol{\Sigma}_0^{-1}\boldsymbol{X})^{-1}\boldsymbol{X}'\boldsymbol{\Sigma}_0^{-1}$$

$$\boldsymbol{\Sigma}_0 = \sum_{m=1}^{k} \bar{\boldsymbol{V}}_m, \; \bar{\boldsymbol{W}} = \bar{\boldsymbol{W}}\boldsymbol{\Sigma}_0\bar{\boldsymbol{W}}, \; \hat{\boldsymbol{e}} = \boldsymbol{X}\hat{\boldsymbol{\beta}} - \boldsymbol{y} \,. \tag{5.79}$$

Thus, the variance and covariance components σ_i^2 and σ_{ij} are divided for the estimation by their approximate values α_i^2 and α_{ij} to obtain the components $\bar{\sigma}_i^2$ and $\bar{\sigma}_{ij}$. They can be assumed of having values close to one. The approximate values are absorbed in the matrices $\boldsymbol{V}_m$ which then become $\bar{\boldsymbol{V}}_m$, i.e. we obtain in (5.69) for instance $\sigma_1^2\boldsymbol{V}_1 = (\sigma_1^2/\alpha_1^2)\alpha_1^2\boldsymbol{V}_1 = \bar{\sigma}_1^2\bar{\boldsymbol{V}}_1$ with $\sigma_1^2/\alpha_1^2 = \bar{\sigma}_1^2$ and $\alpha_1^2\boldsymbol{V}_1 = \bar{\boldsymbol{V}}_1$.

The estimates $\hat{\bar{\sigma}}_i^2$ and $\hat{\bar{\sigma}}_{ij}$ from (5.78) therefore depend on the approximate values α_i^2 and α_{ij}. Thus, they are iteratively computed by introducing $\alpha_i^2\hat{\bar{\sigma}}_i^2$ and $\alpha_{ij}\hat{\bar{\sigma}}_{ij}$ as new approximate values, until at the point of convergence the estimates reproduce themselves, that is until

$$\hat{\bar{\boldsymbol{\sigma}}} = |1,1,\ldots,1|' \tag{5.80}$$

is obtained. If this result is substituted in (5.78), we obtain with (5.79)

$$\sum_{j=1}^{k} \text{tr}(\bar{\boldsymbol{W}}\bar{\boldsymbol{V}}_i\bar{\boldsymbol{W}}\bar{\boldsymbol{V}}_j) = \text{tr}(\bar{\boldsymbol{W}}\bar{\boldsymbol{V}}_i) \tag{5.81}$$

and therefore the estimate

$$\hat{\bar{\boldsymbol{\sigma}}} = \bar{\boldsymbol{H}}\bar{\boldsymbol{q}} \tag{5.82}$$

with

$$\bar{\boldsymbol{H}} = \big(\text{diag}[1/\text{tr}(\bar{\boldsymbol{W}}\bar{\boldsymbol{V}}_i)]\big)\,, \; i \in \{1,\ldots,l\} \tag{5.83}$$

(FÖRSTNER 1979; KOCH 1999, p.234). Iteratively applied it gives results which agree with (5.78).

If the likelihood function (5.77) is used together with a noninformative prior, to derive the posterior density function for $\boldsymbol{\sigma}$ with Bayes' theorem (2.122), the estimate based on this posterior distribution will only slightly differ from (5.78). This will be shown with (5.118) for the variance components. In addition to the estimation of $\boldsymbol{\sigma}$, confidence regions for $\boldsymbol{\sigma}$ may be established and hypotheses for $\boldsymbol{\sigma}$ may be tested by means of the posterior distribution.

5.2.2 Noninformative Priors

Noninformative priors shall be introduced for the vector $\boldsymbol{\sigma}$ of unknown variance and covariance components. By the formula of JEFFREYS for deriving noninformative priors, which was mentioned in connection with (2.221), one obtains (KOCH 1987; KOCH 1990, p.126)

$$p(\boldsymbol{\sigma}) \propto (\det \boldsymbol{S})^{1/2} \tag{5.84}$$

with

$$\boldsymbol{S} = (\mathrm{tr}\boldsymbol{W}\boldsymbol{V}_i\boldsymbol{W}\boldsymbol{V}_j) \quad \text{for} \quad i,j \in \{1,\ldots,k\} \ . \tag{5.85}$$

This prior density function leads together with the likelihood function (5.77) because of Bayes' theorem (2.122) to the posterior density function $p(\boldsymbol{\sigma}|\boldsymbol{y})$ for $\boldsymbol{\sigma}$, where for simplifying the notation the condition C referring to the background information has been omitted, as was already mentioned at the beginning of Chapter 4.2.1,

$$p(\boldsymbol{\sigma}|\boldsymbol{y}) \propto (\det \boldsymbol{S})^{1/2}(\det \boldsymbol{\Sigma} \det \boldsymbol{X}'\boldsymbol{\Sigma}^{-1}\boldsymbol{X})^{-1/2} \exp(-\boldsymbol{y}'\boldsymbol{W}\boldsymbol{y}/2) \ . \tag{5.86}$$

This density function is because of $\boldsymbol{S}$ from (5.85), $\boldsymbol{\Sigma}$ from (5.69) and $\boldsymbol{W}$ from (5.76) a function of the variance and covariance components σ_i^2 and σ_{ij}.

Integrals over the posterior density function (5.86) for computing the Bayes estimate of $\boldsymbol{\sigma}$, for establishing confidence regions for $\boldsymbol{\sigma}$ or for testing hypotheses for $\boldsymbol{\sigma}$ could not be solved analytically. The numerical methods treated in Chapter 6 therefore have to be applied.

5.2.3 Informative Priors

It will be now assumed that prior information is available for the unknown variance and covariance components σ_i^2 and σ_{ij}. As described in Chapter 4.3.2 for the variance factor σ^2 of the linear model, the prior information will be given by the expected values and the variances

$$E(\sigma_i^2) = \mu_{\sigma_i} \ \text{and} \ \ V(\sigma_i^2) = V_{\sigma_i} \ \ \text{for} \ \ i \in \{1,\ldots,l\}$$
$$E(\sigma_{ij}) = \mu_{\sigma_{ij}} \ \text{and} \ \ V(\sigma_{ij}) = V_{\sigma_{ij}} \ \ \text{for} \ \ i < j \le l \ , \ l \le k \le l(l+1)/2 \ . \tag{5.87}$$

Variance components σ_i^2 must take on like the variance factor σ^2 only positive values. In the linear model σ^2 is replaced by the weight factor τ with $\tau = 1/\sigma^2$ from (4.115), and the gamma distribution is assumed for τ, if as conjugate prior the normal-gamma distribution (4.148) is applied. The inverted gamma distribution (2.176) then follows for σ^2. The inverted gamma distribution is therefore also chosen as prior distribution for the variance component σ_i^2, see for instance KOCH (1990, p.132). If the l variance components

σ_i^2 are assumed to be independent, the prior density function $p(\sigma_i^2)_l$ for the l components follows from (2.110) and (2.176) without the constants by

$$p(\sigma_i^2)_l \propto \prod_{i=1}^{l} \left(\frac{1}{\sigma_i^2}\right)^{p_i+1} e^{-b_i/\sigma_i^2} \ . \tag{5.88}$$

The product has to be taken over the l variance components σ_i^2, and b_i and p_i are the parameters of the distributions of the individual variance components. They are obtained because of (2.177) from the prior information (5.87) by

$$p_i = \mu_{\sigma_i}^2/V_{\sigma_i} + 2 \ , \ \ b_i = (p_i - 1)\mu_{\sigma_i} \ . \tag{5.89}$$

The truncated normal distribution mentioned in Chapter 2.6.2 may also be introduced as prior distribution for the variance components σ_i^2 instead of the inverted gamma distribution (KOCH 1990, p.133).

The covariance components σ_{ij} take on positive as well as negative values. The density function (2.166) of the normal distribution is therefore chosen as prior density function. If the $k - l$ covariance components σ_{ij} are assumed as independent, the prior density function $p(\sigma_{ij})_{kl}$ for the $k - l$ components is obtained with (2.110) and (5.87) by

$$p(\sigma_{ij})_{kl} \propto \prod_{1}^{k-l} \exp\left[-\frac{1}{2V_{\sigma_{ij}}}(\sigma_{ij} - \mu_{\sigma_{ij}})^2 \right] \ . \tag{5.90}$$

The product has to be taken over the $k - l$ covariance components σ_{ij}.

The posterior density function $p(\boldsymbol{\sigma}|\boldsymbol{y})$ for the vector $\boldsymbol{\sigma}$ of unknown variance and covariance components follows with Bayes' theorem (2.122) and with (5.77), (5.88) and (5.90) by

$$p(\boldsymbol{\sigma}|\boldsymbol{y}) \propto \prod_{i=1}^{l} \left(\frac{1}{\sigma_i^2}\right)^{p_i+1} e^{-b_i/\sigma_i^2} \prod_{1}^{k-l} \exp\left[-\frac{1}{2V_{\sigma_{ij}}}(\sigma_{ij} - \mu_{\sigma_{ij}})^2 \right]$$

$$(\det \boldsymbol{\Sigma} \det \boldsymbol{X}'\boldsymbol{\Sigma}^{-1}\boldsymbol{X})^{-1/2} \exp(-\boldsymbol{y}'\boldsymbol{W}\boldsymbol{y}/2) \ . \tag{5.91}$$

Integrals over this density function for computing the Bayes estimate of $\boldsymbol{\sigma}$, for establishing confidence regions for $\boldsymbol{\sigma}$ or for testing hypotheses for $\boldsymbol{\sigma}$ could not be found analytically. Thus, the numerical methods of Chapter 6 have to be used.

5.2.4   Variance Components

The linear model (5.69) with the unknown variance and covariance components simplifies, if the variance components σ_i^2 only with $i \in \{1, \ldots, k\}$ are considered. We get

$$\boldsymbol{X}\boldsymbol{\beta} = E(\boldsymbol{y}|\boldsymbol{\beta}) = \boldsymbol{y} + \boldsymbol{e} \tag{5.92}$$

with

$$X = \begin{vmatrix} X_1 \\ X_2 \\ \dots \\ X_k \end{vmatrix}, \; y = \begin{vmatrix} y_1 \\ y_2 \\ \dots \\ y_k \end{vmatrix}, \; e = \begin{vmatrix} e_1 \\ e_2 \\ \dots \\ e_k \end{vmatrix},$$

$$D(y|\sigma) = \Sigma = \sigma_1^2 V_1 + \sigma_2^2 V_2 + \dots + \sigma_k^2 V_k \ ,$$

$$\sigma = |\sigma_1^2, \sigma_2^2, \dots, \sigma_k^2|' \ ,$$

$$V_i = \begin{vmatrix} 0 & \dots & 0 & \dots & 0 \\ \hdotsfor{5} \\ 0 & \dots & P_i^{-1} & \dots & 0 \\ \hdotsfor{5} \\ 0 & \dots & 0 & \dots & 0 \end{vmatrix} . \tag{5.93}$$

Thus, the unknown variance components σ_i^2 are related to the independent vectors y_i of observations whose weight matrices P_i from (2.159) are known.

As mentioned for (5.78) the variance components are iteratively estimated. In the following we assume that we do not iterate to compute $\bar{\sigma}$ by (5.78) but the vector σ of variance components itself, which leads to a simpler notation. By substituting $\alpha_i^2 = \sigma_i^2$ in (5.79) we obtain

$$V_i = \bar{V}_i/\sigma_i^2 \quad \text{and by setting} \quad P_i^{-1} = \bar{P}_i^{-1}/\sigma_i^2 \tag{5.94}$$

we find

$$\Sigma_0 = \begin{vmatrix} \bar{P}_1^{-1} & 0 & \dots & 0 \\ 0 & \bar{P}_2^{-1} & \dots & 0 \\ \hdotsfor{4} \\ 0 & 0 & \dots & \bar{P}_k^{-1} \end{vmatrix}, \quad \Sigma_0^{-1} = \begin{vmatrix} \bar{P}_1 & 0 & \dots & 0 \\ 0 & \bar{P}_2 & \dots & 0 \\ \hdotsfor{4} \\ 0 & 0 & \dots & \bar{P}_k \end{vmatrix} \tag{5.95}$$

as well as

$$\Sigma_0^{-1} \bar{V}_i \Sigma_0^{-1} = \begin{vmatrix} 0 & \dots & 0 & \dots & 0 \\ \hdotsfor{5} \\ 0 & \dots & \bar{P}_i & \dots & 0 \\ \hdotsfor{5} \\ 0 & \dots & 0 & \dots & 0 \end{vmatrix} . \tag{5.96}$$

By substituting (5.93) and (5.95) in (4.15), (4.19) or (4.29) we find the estimate $\hat{\beta}$ of the vector β of unknown parameters by

$$\hat{\beta} = N^{-1}\left(\frac{1}{\sigma_1^2} X_1' P_1 y_1 + \dots + \frac{1}{\sigma_k^2} X_k' P_k y_k\right) \tag{5.97}$$

with N being the matrix of normal equations

$$N = \frac{1}{\sigma_1^2} X_1' P_1 X_1 + \dots + \frac{1}{\sigma_k^2} X_k' P_k X_k \ . \tag{5.98}$$

With (5.96) and (5.97) we get from (5.79)

$$\bar{q} = (\hat{e}_i' \bar{P}_i \hat{e}_i) , \quad i \in \{1, \ldots, k\} \tag{5.99}$$

with

$$\hat{e}_i = X_i \hat{\beta} - y_i . \tag{5.100}$$

Because of

$$\begin{aligned}
\mathrm{tr}(\bar{W}\Sigma_0) &= \mathrm{tr}(I - \Sigma_0^{-1} X (X'\Sigma_0^{-1} X)^{-1} X') \\
&= \mathrm{tr}(I) - \mathrm{tr}((X'\Sigma_0^{-1} X)^{-1} X'\Sigma_0^{-1} X) = n - u
\end{aligned}$$

the quantity r_i is interpreted with

$$r_i = \mathrm{tr}(\bar{W} \bar{V}_i) \quad \text{and} \quad \sum_{i=1}^{k} r_i = n - u \tag{5.101}$$

as *partial redundancy*, that is as contribution of the observation vector y_i to the overall redundancy $n - u$. The estimate $\hat{\sigma}_i^2$ of the variance component σ_i^2 is obtained from (5.82) with (5.79), (5.94), (5.99) and (5.101) by

$$\hat{\sigma}_i^2 = \hat{e}_i' P_i \hat{e}_i / r_i , \quad i \in \{1, \ldots, k\} . \tag{5.102}$$

As mentioned for (5.94) the estimates are iteratively computed.

The computation of the partial redundancy r_i in (5.101) can be simplified. The matrix $\bar{W}$ follows from (5.79) with (5.94), (5.95) and (5.98) by

$$\bar{W} = \begin{vmatrix} \frac{1}{\sigma_1^2}P_1 & \cdots & 0 \\ \hdotsfor{3} \\ 0 & \cdots & \frac{1}{\sigma_k^2}P_k \end{vmatrix} - \begin{vmatrix} \frac{1}{\sigma_1^2}P_1 & \cdots & 0 \\ \hdotsfor{3} \\ 0 & \cdots & \frac{1}{\sigma_k^2}P_k \end{vmatrix} \begin{vmatrix} X_1 \\ \cdots \\ X_k \end{vmatrix} N^{-1}$$

$$\begin{vmatrix} X_1', \ldots, X_k' \end{vmatrix} \begin{vmatrix} \frac{1}{\sigma_1^2}P_1 & \cdots & 0 \\ \hdotsfor{3} \\ 0 & \cdots & \frac{1}{\sigma_k^2}P_k \end{vmatrix} = \begin{vmatrix} \frac{1}{\sigma_1^2}P_1 & \cdots & 0 \\ \hdotsfor{3} \\ 0 & \cdots & \frac{1}{\sigma_k^2}P_k \end{vmatrix}$$

$$- \begin{vmatrix} \frac{1}{\sigma_1^2}P_1 X_1 N^{-1} X_1' \left(\frac{1}{\sigma_1^2}P_1\right) & \cdots & \frac{1}{\sigma_1^2}P_1 X_1 N^{-1} X_k' \left(\frac{1}{\sigma_k^2}P_k\right) \\ \hdotsfor{3} \\ \frac{1}{\sigma_k^2}P_k X_k N^{-1} X_1' \left(\frac{1}{\sigma_1^2}P_1\right) & \cdots & \frac{1}{\sigma_k^2}P_k X_k N^{-1} X_k' \left(\frac{1}{\sigma_k^2}P_k\right) \end{vmatrix} .$$

The products $\bar{\boldsymbol{W}}(\sigma_1^2\boldsymbol{V}_1),\ldots,\bar{\boldsymbol{W}}(\sigma_k^2\boldsymbol{V}_k)$ are obtained by

$$\bar{\boldsymbol{W}}(\sigma_1^2\boldsymbol{V}_1)=\begin{vmatrix}\boldsymbol{I}&\cdots&\boldsymbol{0}\\ \cdots\cdots\cdots\\ \boldsymbol{0}&\cdots&\boldsymbol{0}\end{vmatrix}-\begin{vmatrix}\frac{1}{\sigma_1^2}\boldsymbol{P}_1\boldsymbol{X}_1\boldsymbol{N}^{-1}\boldsymbol{X}_1'&\cdots&\boldsymbol{0}\\ \cdots\cdots\cdots\cdots\cdots\cdots\cdots\cdots\\ \frac{1}{\sigma_k^2}\boldsymbol{P}_k\boldsymbol{X}_k\boldsymbol{N}^{-1}\boldsymbol{X}_1'&\cdots&\boldsymbol{0}\end{vmatrix},\ldots,$$

$$\bar{\boldsymbol{W}}(\sigma_k^2\boldsymbol{V}_k)=\begin{vmatrix}\boldsymbol{0}&\cdots&\boldsymbol{0}\\ \cdots\cdots\cdots\\ \boldsymbol{0}&\cdots&\boldsymbol{I}\end{vmatrix}-\begin{vmatrix}\boldsymbol{0}&\cdots&\frac{1}{\sigma_1^2}\boldsymbol{P}_1\boldsymbol{X}_1\boldsymbol{N}^{-1}\boldsymbol{X}_k'\\ \cdots\cdots\cdots\cdots\cdots\cdots\cdots\cdots\\ \boldsymbol{0}&\cdots&\frac{1}{\sigma_k^2}\boldsymbol{P}_k\boldsymbol{X}_k\boldsymbol{N}^{-1}\boldsymbol{X}_k'\end{vmatrix}.\qquad(5.103)$$

The partial redundancy r_i therefore follows from (5.101) by

$$r_i=n_i-\operatorname{tr}\left(\frac{1}{\sigma_i^2}\boldsymbol{X}_i'\boldsymbol{P}_i\boldsymbol{X}_i\boldsymbol{N}^{-1}\right),\quad i\in\{1,\ldots,k\}\qquad(5.104)$$

where n_i denotes the number of observations $\boldsymbol{y}_i$ with $\sum_{i=1}^k n_i=n$. The value for the variance component σ_i^2 is taken from its estimate $\hat{\sigma}_i^2$ and is updated with each iteration.

For the computation of the partial redundancies $r_1,\ldots,r_k$ in (5.104) the inverse $\boldsymbol{N}^{-1}$ of the matrix of normal equations from (5.98) is needed. However, it might not be available for large systems of normal equations which are only solved and not inverted. In such a case, we can use the stochastic trace estimation, as proposed for computing variance components by KOCH and KUSCHE (2002), see also KUSCHE (2003). We apply the theorem by HUTCHINSON (1990), see also (2.165),

$$E(\boldsymbol{u}'\boldsymbol{B}\boldsymbol{u})=\operatorname{tr}\boldsymbol{B}\qquad(5.105)$$

where $\boldsymbol{B}$ denotes a symmetric $n\times n$ matrix and $\boldsymbol{u}$ an $n\times 1$ vector of n independent samples from a random variable U with $E(U)=0$ and $V(U)=1$. If U is a discrete random variable which takes with probability $1/2$ the values -1 and $+1$, then $\boldsymbol{u}'\boldsymbol{B}\boldsymbol{u}$ is an unbiased estimator of $\operatorname{tr}\boldsymbol{B}$ with minimum variance.

To apply (5.105) symmetric matrices have to be present in (5.104). They may be obtained by a Cholesky factorization of the matrices $\boldsymbol{X}_i'\boldsymbol{P}\boldsymbol{X}_i$ of normal equations. If these matrices are ill-conditioned, the Cholesky factorization will not work properly. It is therefore preferable, to apply the Cholesky factorization to the weight matrices $\boldsymbol{P}_i$ which will be already diagonal or have strong diagonal elements. With

$$\boldsymbol{P}_i=\boldsymbol{G}_i\boldsymbol{G}_i',\qquad(5.106)$$

where $\boldsymbol{G}_i$ denotes a regular lower triangular matrix, we obtain instead of (5.104)

$$r_i=n_i-\operatorname{tr}\left(\frac{1}{\sigma_i^2}\boldsymbol{G}_i'\boldsymbol{X}_i\boldsymbol{N}^{-1}\boldsymbol{X}_i'\boldsymbol{G}_i\right),\quad i\in\{1,\ldots,k\}.\qquad(5.107)$$

The Cholesky factorization may be approximately computed, for instance, by

$$G_i = \operatorname{diag}\left(\sqrt{p_{i11}}, \ldots, \sqrt{p_{in_in_i}}\right) \quad \text{with} \quad P_i = (p_{ijk}) , \tag{5.108}$$

because the results can be checked and improved with $\sum_{i=1}^{k} r_i = n - u$ from (5.101). By inserting the symmetric matrices of (5.107) into (5.105) the products

$$u'G_i'X_iN^{-1}X_i'G_iu \tag{5.109}$$

need to be determined. Since the computation of the inverse N^{-1} shall be avoided, the unknown parameter vectors δ_i are defined

$$\delta_i = N^{-1}X_i'G_iu \tag{5.110}$$

and the linear equations

$$N\delta_i = X_i'G_iu \tag{5.111}$$

are solved for δ_i so that (5.109) follows with

$$u'G_i'X_i\delta_i . \tag{5.112}$$

Different vectors u of independent samples of U give different values for the estimator of the trace so that the trace is obtained by the mean. GOLUB and VON MATT (1997) recommend just one sample vector u to compute the trace by (5.112) which was found to be sufficient by KOCH und KUSCHE (2002).

5.2.5 Distributions for Variance Components

The likelihood function (5.77) of the linear model (5.92) with the unknown variance components may be expanded into a series. One gets with the vector $\hat{e}_i$ of residuals from (5.100), which are computed by the iterative estimation of the variance components with (5.102), at the point of convergence (OU 1991; OU and KOCH 1994)

$$p(y|\sigma) \propto \prod_{i=1}^{k} \left(\frac{1}{\sigma_i^2}\right)^{\frac{r_i}{2}} \exp\left(-\frac{1}{2\sigma_i^2}\hat{e}_i'P_i\hat{e}_i\right) . \tag{5.113}$$

First, it will be assumed that no prior information is available for the variance components and afterwards that prior information is given.

a) Noninformative Priors
The noninformative prior density function (5.84) for the model (5.92) may

be also expanded into a series. One obtains approximately (OU and KOCH 1994)

$$p(\boldsymbol{\sigma}) \propto \prod_{i=1}^{k} \frac{1}{\sigma_i^2} \; . \tag{5.114}$$

This prior density function is a reasonable generalization of the noninformative prior for the variance factor σ^2 of the linear model (4.1) which is according to (2.218) proportional to $1/\sigma^2$.

Bayes' theorem (2.122) leads with (5.113) and (5.114) to the posterior density function for the vector $\boldsymbol{\sigma}$ of variance components

$$p(\boldsymbol{\sigma}|\boldsymbol{y}) \propto \prod_{i=1}^{k} \Big(\frac{1}{\sigma_i^2}\Big)^{\frac{r_i}{2}+1} \exp\Big(-\frac{1}{2\sigma_i^2}\hat{\boldsymbol{e}}_i'\boldsymbol{P}_i\hat{\boldsymbol{e}}_i\Big) \; . \tag{5.115}$$

This density function is formed by the product of the density functions of the k variance components σ_i^2 which according to (2.176) have the inverted gamma distribution. By adding the constants the posterior density function $p(\sigma_i^2|\boldsymbol{y})$ for the variance component σ_i^2 is obtained with

$$p(\sigma_i^2|\boldsymbol{y}) = \Big(\frac{1}{2}\hat{\boldsymbol{e}}_i'\boldsymbol{P}_i\hat{\boldsymbol{e}}_i\Big)^{\frac{r_i}{2}} \Gamma\Big(\frac{r_i}{2}\Big)^{-1} \Big(\frac{1}{\sigma_i^2}\Big)^{\frac{r_i}{2}+1} \exp\Big(-\frac{1}{2\sigma_i^2}\hat{\boldsymbol{e}}_i'\boldsymbol{P}_i\hat{\boldsymbol{e}}_i\Big) \; . \tag{5.116}$$

Since the joint posterior density function (5.115) follows from the product of the posterior density functions (5.116) for the variance components σ_i^2, the variance components are independent because of (2.110). We may therefore estimate the variance components σ_i^2 with (5.116), compute confidence intervals for them or test hypotheses. The Bayes estimate $\hat{\sigma}_{iB}^2$ of σ_i^2 is obtained with (2.177) and (3.9) by

$$\hat{\sigma}_{iB}^2 = \frac{\hat{\boldsymbol{e}}_i\boldsymbol{P}_i\hat{\boldsymbol{e}}_i}{r_i - 2} \tag{5.117}$$

or with substituting (5.102) by

$$\hat{\sigma}_{iB}^2 = \frac{r_i}{r_i - 2}\hat{\sigma}_i^2 \; . \tag{5.118}$$

This Bayes estimate differs for larger values of the partial redundancy r_i only slightly from the estmate $\hat{\sigma}_i^2$ in (5.102). Tables for computing the confidence interval (3.35) for σ_i^2 are found in OU (1991).

b) Informative Priors

The inverted gamma distributions shall again serve as informative priors for the variance compoments σ_i^2. The informative prior density function for $\boldsymbol{\sigma}$ is therefore obtained with (5.88) by

$$p(\boldsymbol{\sigma}) \propto \prod_{i=1}^{k} \Big(\frac{1}{\sigma_i^2}\Big)^{p_i+1} e^{-b_i/\sigma_i^2} \tag{5.119}$$

whose parameters b_i and p_i are determined by (5.89). Bayes' theorem (2.122) leads with the likelihood function (5.113) to the posterior density function for the vector $\boldsymbol{\sigma}$ of variance components

$$p(\boldsymbol{\sigma}|\boldsymbol{y}) \propto \prod_{i=1}^{k} \Big(\frac{1}{\sigma_i^2}\Big)^{\frac{r_i}{2}+p_i+1} \exp\Big[-\frac{1}{\sigma_i^2}\Big(\frac{1}{2}\hat{\boldsymbol{e}}_i'\boldsymbol{P}_i\hat{\boldsymbol{e}}_i+b_i\Big)\Big] . \tag{5.120}$$

This density function is formed again like (5.115) by the product of the density functions of k inverted gamma distributions for σ_i^2. With the appropriate constants the posterior density function for σ_i^2 is obtained according to (2.176) by

$$p(\sigma_i^2|\boldsymbol{y}) = \Big(\frac{1}{2}\hat{\boldsymbol{e}}_i'\boldsymbol{P}_i\hat{\boldsymbol{e}}_i+b_i\Big)^{\frac{r_i}{2}+p_i}\Gamma\Big(\frac{r_i}{2}+p_i\Big)^{-1}$$
$$\Big(\frac{1}{\sigma_i^2}\Big)^{\frac{r_i}{2}+p_i+1}\exp\Big[-\frac{1}{\sigma_i^2}\Big(\frac{1}{2}\hat{\boldsymbol{e}}_i'\boldsymbol{P}_i\hat{\boldsymbol{e}}_i+b_i\Big)\Big] . \tag{5.121}$$

The joint posterior density function (5.120) follows from the product of the k density functions (5.121). The variance components σ_i^2 are therefore independent because of (2.110). Thus, the estimate of σ_i^2, the computation of confidence intervals and the test of hypotheses is accomplished with (5.121). The Bayes estimate $\hat{\sigma}_{iB}^2$ of σ_i^2 follows with (2.177), (3.9) and after substituting (5.102) by

$$\hat{\sigma}_{iB}^2 = \frac{r_i\hat{\sigma}_i^2+2b_i}{r_i+2p_i-2} . \tag{5.122}$$

The interpretation of this result as weighted mean of the prior information and the estimate $\hat{\sigma}_i^2$ is found in OU and KOCH (1994). To compute the confidence interval (3.35) for σ_i^2, the tables of OU (1991) may be used.

5.2.6   Regularization

To extract information about a physical system from measurements, one has to solve an *inverse problem*. A typical example in geodesy and geophysics is the determination of the gravity field of the earth from satellite observations. The normal equations for estimating the unknown parameters of the gravity field, generally the coefficients of an expansion of the geopotential into spherical harmonics, tend to be ill-conditioned. For stabilizing and smoothing the solution a *regularization* is often applied, see for instance REIGBER et al. (2005), which generally is the Tikhonov-regularization (TIKHONOV and ARSENIN 1977, p.103). It means that one adds to the matrix of normal equations a positive definite matrix times a *regularization parameter* which generally is unknown. This regularization is also known as *ridge regression*, see for instance VINOD and ULLAH (1981, p.169). Its estimation of the unknown parameters can be obtained as Bayes estimate (4.87) with prior information, see also O'SULLIVAN (1986).

To estimate the regularization parameter, we recall that the Bayes estimate (4.87) with prior information can be derived with (4.90) by adding an additional vector $\boldsymbol{\mu}$ of observations with the covariance matrix $\sigma^2\boldsymbol{\Sigma}$ to the linear model. The additional observations will be defined such that they lead to the estimate of the Tikhonov-regularization. Inverse problems are often solved by different kinds of data, like satellite-to-satellite tracking beween two low flying satellites and between a low flying and a high flying satellite for computing the geopotential. The proper weighting of these different data needs to be determined which can be solved by estimating variance components, see the example to (5.69). To determine the regularization parameter in addition, the observations leading to the estimate of the Tikhonov-regularization are added to the linear model (5.92) with unknown variance components. The regularization parameter is then determined by means of the ratio of two variance components (KOCH and KUSCHE 2002).

The prior information on the unknown parameters $\boldsymbol{\beta}$ formulated as observation equation is introduced with (4.90) by

$$\boldsymbol{\beta} = \boldsymbol{\mu} + \boldsymbol{e}_\mu \quad \text{with} \quad D(\boldsymbol{\mu}|\sigma_\mu^2) = \sigma_\mu^2 \boldsymbol{P}_\mu^{-1} \tag{5.123}$$

with the $u \times 1$ vector $\boldsymbol{\mu}$ being the prior information on the expected values for $\boldsymbol{\beta}$ and with the variance component σ_μ^2 times the inverse of the $u \times u$ weight matrix $\boldsymbol{P}_\mu$ being the prior information on the covariance matrix of $\boldsymbol{\beta}$. By adding (5.123) to the model (5.92) we obtain instead of (5.93)

$$\boldsymbol{X} = \begin{vmatrix} \boldsymbol{X}_1 \\ \cdots \\ \boldsymbol{X}_k \\ \boldsymbol{I} \end{vmatrix}, \ \boldsymbol{y} = \begin{vmatrix} \boldsymbol{y}_1 \\ \cdots \\ \boldsymbol{y}_k \\ \boldsymbol{\mu} \end{vmatrix}, \ \boldsymbol{e} = \begin{vmatrix} \boldsymbol{e}_1 \\ \cdots \\ \boldsymbol{e}_k \\ \boldsymbol{e}_\mu \end{vmatrix},$$

$$D(\boldsymbol{y}|\boldsymbol{\sigma}) = \boldsymbol{\Sigma} = \sigma_1^2 \boldsymbol{V}_1 + \ldots + \sigma_k^2 \boldsymbol{V}_k + \sigma_\mu^2 \boldsymbol{V}_\mu,$$

$$\boldsymbol{\sigma} = |\sigma_1^2, \ldots, \sigma_k^2, \sigma_\mu^2|',$$

$$\boldsymbol{V}_i = \begin{vmatrix} \boldsymbol{0} & \cdots & \boldsymbol{0} & \cdots & \boldsymbol{0} \\ \cdots & & & & \cdots \\ \boldsymbol{0} & \cdots & \boldsymbol{P}_i^{-1} & \cdots & \boldsymbol{0} \\ \cdots & & & & \cdots \\ \boldsymbol{0} & \cdots & \boldsymbol{0} & \cdots & \boldsymbol{0} \end{vmatrix}, \ \boldsymbol{V}_\mu = \begin{vmatrix} \boldsymbol{0} & \cdots & \boldsymbol{0} \\ \cdots & & \cdots \\ \boldsymbol{0} & \cdots & \boldsymbol{0} \\ \cdots & & \cdots \\ \boldsymbol{0} & \cdots & \boldsymbol{P}_\mu^{-1} \end{vmatrix}. \tag{5.124}$$

The estimates $\hat{\boldsymbol{\beta}}$ of he unknown parameters $\boldsymbol{\beta}$ are obtained instead of (5.97) and (5.98) by

$$\hat{\boldsymbol{\beta}} = \boldsymbol{N}^{-1}\left(\frac{1}{\sigma_1^2}\boldsymbol{X}_1'\boldsymbol{P}_1\boldsymbol{y}_1 + \ldots + \frac{1}{\sigma_k^2}\boldsymbol{X}_k'\boldsymbol{P}_k\boldsymbol{y}_k + \frac{1}{\sigma_\mu^2}\boldsymbol{P}_\mu\boldsymbol{\mu}\right) \tag{5.125}$$

with $\boldsymbol{N}$ being the matrix of normal equations

$$\boldsymbol{N} = \frac{1}{\sigma_1^2}\boldsymbol{X}_1'\boldsymbol{P}_1\boldsymbol{X}_1 + \ldots + \frac{1}{\sigma_k^2}\boldsymbol{X}_k'\boldsymbol{P}_k\boldsymbol{X}_k + \frac{1}{\sigma_\mu^2}\boldsymbol{P}_\mu \ . \tag{5.126}$$

In case there is only one type of observations $\boldsymbol{y}_1$ together with the prior information $\boldsymbol{\mu}$, we find with $k=1$ in (5.125) and (5.126) the normal equations

$$\Big(\frac{1}{\sigma_1^2}\boldsymbol{X}_1'\boldsymbol{P}_1\boldsymbol{X}_1 + \frac{1}{\sigma_\mu^2}\boldsymbol{P}_\mu\Big)\hat{\boldsymbol{\beta}} = \frac{1}{\sigma_1^2}\boldsymbol{X}_1'\boldsymbol{P}_1\boldsymbol{y}_1 + \frac{1}{\sigma_\mu^2}\boldsymbol{P}_\mu\boldsymbol{\mu} \ . \tag{5.127}$$

By introducing the regularization parameter λ with

$$\lambda = \frac{\sigma_1^2}{\sigma_\mu^2} \tag{5.128}$$

we obtain

$$(\boldsymbol{X}_1'\boldsymbol{P}_1\boldsymbol{X}_1 + \lambda\boldsymbol{P}_\mu)\hat{\boldsymbol{\beta}} = \boldsymbol{X}_1'\boldsymbol{P}_1\boldsymbol{y}_1 + \lambda\boldsymbol{P}_\mu\boldsymbol{\mu} \ . \tag{5.129}$$

With $\boldsymbol{\mu} = \boldsymbol{0}$ the solution vector for the Tikhonov-regularization or the ridge regression is obtained. For two different types of observations we find with $k=2$ in (5.125) and (5.126) with

$$\lambda = \frac{\sigma_1^2}{\sigma_\mu^2} \quad \text{and} \quad \omega = \frac{\sigma_1^2}{\sigma_2^2} \tag{5.130}$$

the normal equations

$$(\boldsymbol{X}_1'\boldsymbol{P}_1\boldsymbol{X}_1 + \omega\boldsymbol{X}_2'\boldsymbol{P}_2\boldsymbol{X}_2 + \lambda\boldsymbol{P}_\mu)\hat{\boldsymbol{\beta}} = \boldsymbol{X}_1'\boldsymbol{P}_1\boldsymbol{y}_1 + \omega\boldsymbol{X}_2'\boldsymbol{P}_2\boldsymbol{y}_2 + \lambda\boldsymbol{P}_\mu\boldsymbol{\mu} \tag{5.131}$$

where λ again is the regularization parameter and ω expresses the relative weighting of the observations $\boldsymbol{y}_2$ with respect to $\boldsymbol{y}_1$.

The estimates of the variance components σ_i^2 and σ_μ^2 follow with (5.102) and $\hat{\boldsymbol{e}}_\mu = \hat{\boldsymbol{\beta}} - \boldsymbol{\mu}$ by

$$\begin{aligned}
\hat{\sigma}_i^2 &= \hat{\boldsymbol{e}}_i'\boldsymbol{P}_i\hat{\boldsymbol{e}}_i/r_i \ , \quad i \in \{1, \ldots, k\} \ , \\
\hat{\sigma}_\mu^2 &= \hat{\boldsymbol{e}}_\mu'\boldsymbol{P}_\mu\hat{\boldsymbol{e}}_\mu/r_\mu
\end{aligned} \tag{5.132}$$

with the partial redundancies r_i and r_μ from (5.104)

$$\begin{aligned}
r_i &= n_i - \text{tr}\Big(\frac{1}{\sigma_i^2}\boldsymbol{X}_i'\boldsymbol{P}_i\boldsymbol{X}_i\boldsymbol{N}^{-1}\Big) \ , \quad i \in \{1, \ldots, k\} \ , \\
r_\mu &= u - \text{tr}\Big(\frac{1}{\sigma_\mu^2}\boldsymbol{P}_\mu\boldsymbol{N}^{-1}\Big) \ .
\end{aligned} \tag{5.133}$$

If the inverse $\boldsymbol{N}^{-1}$ has not been computed, the partial redundancies $r_1,\ldots,$ r_k, r_μ can be obtained by the stochastic trace estimate following from (5.107)

$$r_i = n_i - \mathrm{tr}\Big(\frac{1}{\sigma_i^2}\boldsymbol{G}_i'\boldsymbol{X}_i\boldsymbol{N}^{-1}\boldsymbol{X}_i'\boldsymbol{G}_i\Big), \quad i \in \{1,\ldots,k\},$$

$$r_\mu = u - \mathrm{tr}\Big(\frac{1}{\sigma_\mu^2}\boldsymbol{G}_\mu'\boldsymbol{N}^{-1}\boldsymbol{G}_\mu\Big) \tag{5.134}$$

Noninformative priors are introduced for the variance components σ_i^2 so that they are according to (5.116) independently distributed like the inverted gamma distribution with the posterior density function

$$p(\sigma_i^2|\boldsymbol{y}_i) = \Big(\frac{1}{2}\hat{\boldsymbol{e}}_i'\boldsymbol{P}_i\hat{\boldsymbol{e}}_i\Big)^{\frac{r_i}{2}}\Gamma\Big(\frac{r_i}{2}\Big)^{-1}\Big(\frac{1}{\sigma_i^2}\Big)^{\frac{r_i}{2}+1}\exp\Big(-\frac{1}{2\sigma_i^2}\hat{\boldsymbol{e}}_i'\boldsymbol{P}_i\hat{\boldsymbol{e}}_i\Big). \tag{5.135}$$

By replacing the index i by μ the posterior density function $p(\sigma_\mu^2|\boldsymbol{\mu})$ follows. To obtain the posterior density function for the regularization parameter λ in (5.128) or for the parameter ω of the relative weighting in (5.130) the density function for the ratio v

$$v = \sigma_i^2/\sigma_j^2 \tag{5.136}$$

of two variance components σ_i^2 and σ_j^2 is needed. Since σ_i^2 and σ_j^2 are independently distributed, their joint posterior density function follows from (2.110) and (5.135) by

$$\begin{aligned}
p(\sigma_i^2,\sigma_j^2|\boldsymbol{y}_i,\boldsymbol{y}_j) &\propto \Big(\frac{1}{\sigma_i^2}\Big)^{\frac{r_i}{2}+1}\exp\Big(-\frac{1}{2\sigma_i^2}\hat{\boldsymbol{e}}_i'\boldsymbol{P}_i\hat{\boldsymbol{e}}_i\Big)\\
&\quad\Big(\frac{1}{\sigma_j^2}\Big)^{\frac{r_j}{2}+1}\exp\Big(-\frac{1}{2\sigma_j^2}\hat{\boldsymbol{e}}_j'\boldsymbol{P}_j\hat{\boldsymbol{e}}_j\Big)
\end{aligned} \tag{5.137}$$

where the constants need not be considered. The transformation $\sigma_i^2 = v\sigma_j^2$ of variables according to (5.136) with $\partial\sigma_i^2/\partial v = \sigma_j^2$ leads to, see for instance KOCH (1999, p.93),

$$p(v,\sigma_j^2|\boldsymbol{y}_i,\boldsymbol{y}_j) \propto \Big(\frac{1}{v}\Big)^{\frac{r_i}{2}+1}\Big(\frac{1}{\sigma_j^2}\Big)^{\frac{r_i+r_j}{2}+1}\exp\Big(-\frac{1}{2\sigma_j^2}\big(\frac{1}{v}\hat{\boldsymbol{e}}_i'\boldsymbol{P}_i\hat{\boldsymbol{e}}_i+\hat{\boldsymbol{e}}_j'\boldsymbol{P}_j\hat{\boldsymbol{e}}_j\big)\Big).$$

$$\tag{5.138}$$

To obtain the marginal density function for v, the variance component σ_j^2 is integrated out

$$\begin{aligned}
p(v|\boldsymbol{y}_i,\boldsymbol{y}_j) &\propto \Big(\frac{1}{v}\Big)^{\frac{r_i}{2}+1}\Big(\frac{1}{2v}\hat{\boldsymbol{e}}_i'\boldsymbol{P}_i\hat{\boldsymbol{e}}_i+\frac{1}{2}\hat{\boldsymbol{e}}_j'\boldsymbol{P}_j\hat{\boldsymbol{e}}_j\Big)^{-\frac{r_i+r_j}{2}}\\
&\quad \Gamma\Big(\frac{r_i+r_j}{2}\Big)\Big(\frac{1}{2v}\hat{\boldsymbol{e}}_i'\boldsymbol{P}_i\hat{\boldsymbol{e}}_i+\frac{1}{2}\hat{\boldsymbol{e}}_j'\boldsymbol{P}_j\hat{\boldsymbol{e}}_j\Big)^{\frac{r_i+r_j}{2}}\Gamma\Big(\frac{r_i+r_j}{2}\Big)^{-1}\\
&\quad \int_0^\infty \Big(\frac{1}{\sigma_j^2}\Big)^{\frac{r_i+r_j}{2}+1}\exp\Big(-\frac{1}{2\sigma_j^2}\big(\frac{1}{v}\hat{\boldsymbol{e}}_i'\boldsymbol{P}_i\hat{\boldsymbol{e}}_i+\hat{\boldsymbol{e}}_j'\boldsymbol{P}_j\hat{\boldsymbol{e}}_j\big)\Big)d\sigma_j^2.
\end{aligned} \tag{5.139}$$

As a comparison with (2.176) shows, integrating over σ_j^2 means integrating the density function of a special inverted gamma distribution so that we finally obtain the posterior density function for the ratio v of two variance components

$$p(v|\boldsymbol{y}_i,\boldsymbol{y}_j) \propto \left(\frac{1}{v}\right)^{\frac{r_i}{2}+1} \left(\frac{1}{2v}\hat{\boldsymbol{e}}_i\boldsymbol{P}_i\hat{\boldsymbol{e}}_i + \frac{1}{2}\hat{\boldsymbol{e}}_j'\boldsymbol{P}_j\hat{\boldsymbol{e}}_j\right)^{-\frac{r_i+r_j}{2}} . \tag{5.140}$$

An analytical integration of this density function could not be achieved so that the normalization constant is not derived. However, confidence intervals for the ratio v of two variance components or hypotheses tests for v are obtained by the numerical methods of Chapter 6.3.3, because random values can be generated which have the posterior density function (5.140). Random variates for the standard gamma distribution, for which $b = 1$ is valid in (2.172), may be drawn in case of large parameters, for instance for large partial redundancies, by the log-logistic method (DAGPUNAR 1988, p.110). The generated random values are transformed to random variates of the gamma distribution and then to the ones of the inverted gamma distribution (2.176) so that random values follow for the variance component σ_i^2 with the density function (5.135). By independently generating random values for σ_i^2 and σ_j^2 random variates for the ratio v from (5.136), hence for the regularization parameter λ from (5.128) or for the parameter ω of relative weighting from (5.130) are then obtained. Examples for computing confidence intervals of regularization parameters for determining the gravity field of the earth are given by KOCH and KUSCHE (2002).

MAYER-GÜRR et al. (2005), for instance, determined the gravity field of the earth from a kinematical orbit of the satellite CHAMP, which extended over one year, but which was broken up into short arcs. For each arc the variance factor of the contribution to the matrix of normal equations was iteratively estimated as a variance component and in addition the variance component for the regularization parameter by (5.132). XU et al. (2006) proposed improvements of estimated variance components by removing biases in the sense of traditional statistics caused by a regularization parameter. However, KOCH and KUSCHE (2007) pointed out that no biases occur when introducing the prior information (5.123), see the comments to (4.91).

5.3 Reconstructing and Smoothing of Three-dimensional Images

The reconstruction and smoothing of digital images from data of different sensors is a task for which Bayesian analysis is well suited, because prior information leading to smooth images can be introduced. However, the smoothing has to stop at the edges of the image where in case of three-dimensional images sudden changes of the intensities of the voxels, i.e. volume elements, occur which represent the image. The solution of such a task is given here for reconstructing and smoothing images of positron emission tomography.

5.3.1 Positron Emission Tomography

Positron emission tomography is applied to study metabolic activities like the distribution of a pharmaceutical in a part of a body of a human being or an animal. The pharmaceutical is combined with a radioactive isotope which produces a positron. The positron finds a nearby electron and annihilates with it to form a pair of photons. The two photons move in opposite directions along a straight line and collide at nearly the same time with a pair of detectors, thus establishing a coincidence line. The detectors are placed around the body on several rings forming a tube. The three-dimensional image of the positions of the photon emitters is reconstructed from the photon counts for all coincidence lines, see for instance LEAHY and QI (2000) and GUNDLICH et al. (2006). The image is represented by a three-dimensional array of voxels with intensities proportional to the number of photon emissions.

For a statistical analysis of the photon counts it is assumed that the counts are Poisson distributed. The maximum likelihood estimation is then solved by the expectation maximization (EM) algorithm independently proposed by SHEPP and VARDI (1982) and LANGE and CARSON (1984). This algorithm has two disadvantages, it is slow to converge and the reconstruction has high variance so that it needs smoothing to reduce the noise. For a faster convergence gamma distributed priors have been introduced by LANGE et al. (1987) and WANG and GINDI (1997). For the smoothing one should keep in mind that the intensities of the voxels of the image represent a random field for which the Markov property can be assumed, because the intensity of a voxel is mainly influenced by the ones of the voxels of the neighborhood, see for instance KOCH and SCHMIDT (1994, p.299). Because of the equivalence of Markov random fields and neighbor Gibbs fields the prior information can be expressed by the Gibbs distribution. It may be defined such that large density values of the posterior distribution follow for smooth images and small ones for rough images so that a smooth image is obtained from the prior information, see GEMAN and MCCLURE (1987). However, the smoothing has to stop at the edges where sudden changes of the intensities of the voxels occur.

A promising way of handling the edges has been obtained by modeling the Gibbs distribution by the density functions (4.56) and (4.57) of HUBER (1964) for the robust parameter estimation, see for instance FESSLER et al. (2000) and QI et al. (1998). Voxels beyond edges are considered outliers and are accordingly downweighted. A similar effect results from the use of the median root prior (ALENIUS and RUOTSALAINEN 1997) which gives good spatial details, as shown by BETTINARDI et al. (2002). For a better edge preserving property KOCH (2005A) modified Huber's density function such that pixels beyond edges of two-dimensional images do not contribute to the smoothing. The method was tested for photographic images and showed an excellent edge preserving quality. The same modification of Huber' s density function is also applied in the following chapter.

5.3.2 Image Reconstruction

Let Ω be the set of voxels forming a three-dimensional array with $L+1$ rows, $M+1$ columns and $O+1$ slices

$$\begin{aligned} \Omega \; &= \; \{j = (l, m, o),\ 0 \le l \le L,\ 0 \le m \le M,\ 0 \le o \le O\}\,, \\ u &= (L+1)(M+1)(O+1) \end{aligned} \tag{5.141}$$

and let β_j with $j \in \{1, \ldots, u\}$ be the unknown intensity of voxel j which is proportional to the number of photon emissions of voxel j. The vector $\boldsymbol{\beta}$ with $\boldsymbol{\beta} = (\beta_j)$ is therefore the vector of unknown parameters of the reconstruction.

As mentioned in the previous chapter, the emission of two photons in opposite directions establishes a coincidence line between a pair of detectors. Let y_{ij} be the number of photons emitted by voxel j along a coincidence line and counted at detector pair i. It cannot be observed, because more than one voxel will be cut by the coincidence line and will emit photons along that line which are counted. The expected value $\bar{y}_{ij}$ of y_{ij} is connected to the unknown intensity β_j of voxel j by

$$\bar{y}_{ij} = E(y_{ij}) = p_{ij}\beta_j \tag{5.142}$$

where p_{ij} gives the probability of detecting an emission from voxel j at detector pair i. It is a deterministic quantity and results from the geometry of the scanner. It is therefore known. By summing over the voxels j which are cut by the coincidence line between detector pair i the observation y_i, i.e. the photon count, and its expectation $\bar{y}_i$ follow with (5.142) from

$$y_i = \sum_j y_{ij} \quad \text{and} \quad \bar{y}_i = E(y_i) = \sum_j p_{ij}\beta_j \tag{5.143}$$

and the expectation $\bar{\boldsymbol{y}}$ of the vector $\boldsymbol{y}$ of observations from

$$\bar{\boldsymbol{y}} = E(\boldsymbol{y}|\boldsymbol{\beta}) = \boldsymbol{P}\boldsymbol{\beta} \tag{5.144}$$

with $\boldsymbol{P} = (p_{ij})$. The matrix $\boldsymbol{P}$ represents an approximation only, it has to be corrected for a number of effects to find the so-called system matrix for the linear relation between $\bar{\boldsymbol{y}}$ and $\boldsymbol{\beta}$, see for instance LEAHY and QI (2000).

The random number y_{ij} results from counting photons so that it is assumed as Poisson distributed, see for instance KOCH (1999, p.87). Since the y_{ij} are independent, their sum, which gives the measurement y_i with expectation $\bar{y}_i$, is also Poisson distributed with density function

$$\begin{aligned} p(y_i|\boldsymbol{\beta}) \; &= \; \frac{\bar{y}_i^{\,y_i}\exp\left(-\bar{y}_i\right)}{y_i!} \\ &= \; \frac{(\sum_j p_{ij}\beta_j)^{y_i}\exp\left(-\sum_j p_{ij}\beta_j\right)}{y_i!}\,. \end{aligned} \tag{5.145}$$

The measurements y_i are independent, too, the joint density function for $\boldsymbol{y}$ therefore follows with (2.110) from

$$p(\boldsymbol{y}|\boldsymbol{\beta}) = \prod_i \frac{(\sum_j p_{ij}\beta_j)^{y_i} \exp\left(-\sum_j p_{ij}\beta_j\right)}{y_i!} \ . \tag{5.146}$$

This is the likelihood function for the Bayesian reconstruction.

The intensity β_j of voxel j with $j \in \Omega$ represents a Markoff random field, as already mentioned in the previous chapter. A special Gibbs distribution defined for cliques with two sites of the three-dimensional neighborhood N_p of order p, i.e. for voxel j and for each voxel in the neighborhood N_p of voxel j, is therefore chosen as prior distribution (KOCH 2005A)

$$p(\boldsymbol{\beta}) \propto \exp\left\{ -\frac{c_\beta}{2} \sum_{j\in\Omega} \sum_{s\in N_p} (\beta_j - \beta_{j+s})^2 \right\} \ . \tag{5.147}$$

This is a normal distribution where $\propto$ means proportionality. The constant c_β acts as a weight and determines the contribution of the prior information. The index s defines the index of a voxel in half of the neighborhood of voxel j, because one has to sum in (5.147) over all cliques with two sites in the set Ω. This is accomplished by summing over the cliques of half the neighborhood N_p (KOCH and SCHMIDT 1994, p.277). A three-dimensional neighborhood, for instance, which extends over a distance of two voxels on each side of the central voxel j contains 32 voxels. The larger the intensity difference in (5.147) between voxel j and voxel $j + s$ the smaller is the density value. The reconstruction of a rough image is therefore less likely than the reconstruction of a smooth one.

If voxel j and voxel $j + s$ are seperated by an edge, a sudden change in the intensity, the voxel $j + s$ should not contribute to the smoothing of voxel j. We therefore assume that the density function resulting from a given index s in (5.147) is defined by the density function in (4.56) and (4.57) of HUBER (1964) for a robust parameter estimation. It is modified such that we use in (5.147) (KOCH 2005A)

$$\begin{aligned} p(\beta_j) &\propto \exp -\{(\beta_j - \beta_{j+s})^2/2\} &&\text{for} \quad |\beta_j - \beta_{j+s}| \le c \\ p(\beta_j) &= 0 &&\text{for} \quad |\beta_j - \beta_{j+s}| > c \end{aligned} \tag{5.148}$$

where the constant c is set according to the jumps in the intensities of the edges which one wants to preserve.

The prior (5.147) together with (5.148) and the likelihood function (5.146) gives by Bayes' theorem (2.122) the posterior density function for $\boldsymbol{\beta}$

$$p(\boldsymbol{\beta}|\boldsymbol{y}) \propto \exp\left\{ -\frac{c_\beta}{2} \sum_{j\in\Omega} \sum_{s\in N_p} (\beta_j - \beta_{j+s})^2 \right\} \prod_i (\sum_j p_{ij}\beta_j)^{y_i}$$

$$\exp\left(-\sum_j p_{ij}\beta_j\right) \ . \tag{5.149}$$

The conditional density function for β_j given the unknown intensities $\partial\beta_j$ in the neighborhood N_p of voxel j follows from (5.149) with t being now the summation index by

$$p(\beta_j|\partial\beta_j, \boldsymbol{y}) \;=\; \frac{1}{C}\exp\left\{-\frac{c_\beta}{2}\sum_{\pm s\in N_p}(\beta_j - \beta_{j+s})^2\right\}\prod_i(\sum_t p_{it}\beta_t)^{y_i}$$

$$\exp\left(-\sum_t p_{it}\beta_t\right) \tag{5.150}$$

where C denotes the normalization constant and where the sum has to be extended over the whole neighborhood N_p of voxel j so that the index s becomes positive and negative (KOCH and SCHMIDT 1994, p.262).

5.3.3 Iterated Conditional Modes Algorithm

Because of the large number of unknown intensities β_j we do not estimate $\boldsymbol{\beta}$ from (5.149), but derive the MAP estimate (3.30) for the unknown intensity β_j from (5.150) and apply it iteratively for $j \in \Omega$. Thus, the iterated conditional modes (ICM) algorithm of BESAG (1986) results. Taking the logarithm of (5.150)

$$\ln p(\beta_j|\partial\beta_j, \boldsymbol{y}) \;=\; -\frac{c_\beta}{2}\sum_{\pm s\in N_p}(\beta_j - \beta_{j+s})^2 + \sum_i(y_i\ln\sum_t p_{it}\beta_t$$

$$-\sum_t p_{it}\beta_t) - \ln C \tag{5.151}$$

and the derivative with respect to β_j

$$\frac{d\ln p(\beta_j|\partial\beta_j, \boldsymbol{y})}{d\beta_j} = -c_\beta\sum_{\pm s\in N_p}(\beta_j - \beta_{j+s}) + \sum_i\left(\frac{p_{ij}y_i}{\sum_t p_{it}\beta_t} - p_{ij}\right) \tag{5.152}$$

and setting the result equal to zero gives the condition the MAP estimate for β_j has to fulfill given in a form explained below

$$1 = \frac{1}{\sum_i p_{ij} + c_\beta\sum_{\pm s\in N_p}(\beta_j - \beta_{j+s})}\sum_i\frac{p_{ij}y_i}{\sum_t p_{it}\beta_t}\,. \tag{5.153}$$

It leads to the ICM algorithm given for the kth step of the iteration

$$\beta_j^{(k+1)} = \frac{\beta_j^{(k)}}{\sum_i p_{ij} + c_\beta\sum_{\pm s\in N_p}(\beta_j^{(k)} - \beta_{j+s}^{(k)})}\sum_i\frac{p_{ij}y_i}{\sum_t p_{it}\beta_t^{(k)}}\,. \tag{5.154}$$

The prior information of this algorithm is weighted by c_β with respect to the contribution of the observations y_i. If c_β is too large, the iterations will not converge anymore, because the intensities will continuously increase unconstrained by the observations. If c_β is very small or equal to zero, the expectation maximization (EM) algorithm of SHEPP and VARDI (1982) and LANGE and CARSON (1984) is obtained, see also VARDI et al. (1985). The second parameter, which controlls the prior information in the ICM algorithm, is according to (5.148) the constant c which determines the intensity difference of the edges one wants to preserve.

GREEN (1990) proposed the one step late (OSL) approximation in order to solve the Bayesian reconstruction also by the EM algorithm. The name was chosen, because the derivative of the prior density function with respect to the unknown intensity β_j of voxel j is evaluated at the current estimate of the unknown parameter during the iterations. It was already mentioned that one has to sum over all cliques with two sites in the set Ω of voxels for computing the prior density function (5.147). When taking the derivative of (5.147) with respect to the unknown intensity β_j of voxel j, the difference $\beta_j^{(k)} - \beta_{j+s}^{(k)}$ which results has therefore to be summed over the whole neighborhood N_p of voxel j. Thus, the ICM algorithm (5.154) is obtained so that the OSL approximation and the ICM algorithm are identical for the prior density function (5.147) chosen here. To show this, the condition for the MAP estimate in the form (5.153) had been chosen. It leads to the ICM algorithm in the shape of the OSL algorithm. Examples for estimating intensities of voxels of the positron emission tomography by the ICM algorithm (5.154) are given by KOCH (2006).

Instead of iteratively applying the ICM algorithm random samples for the intensities distributed like the posterior density function (5.149) may be generated by the Gibbs sampler covered in Chapter 6.3.2 together with the sampling-importance-resampling (SIR) algorithm explained in Chapter 6.2.1. The generated samples lead to estimates, confidence regions and hypothesis tests for the unknown parameters, as described in Chapter 6.3.3. However, this method takes considerably more computer time than the ICM algorithm (KOCH 2007).

5.4 Pattern Recognition

The human being is able to recognize objects in two-dimensional images by their texture, form, boundary or color. To transfer this capability to computers is called *pattern recognition*. Digital images are produced by sensors frequently in different spectral bands. The images shall be analyzed automatically to detect the shown objects. From these tasks of digital image analysis, which at present is the subject of intensive research, the *segmentation* only, i.e. the decomposition of a digital image based on the textures of the objects, shall be discussed here as an application of Bayesian statistics. For further reading of this topic see for instance NIEMANN (1990), RIPLEY (1996) and BISHOP (2006).

Information on the objects to be detected, so-called *characteristics* or *features*, are taken from the digital images, for instance, the grey values of a pixel of a digital image in different frequency bands. Based on these characteristics the objects have to be allocated to classes which represent the objects. This is called *classification* or *discriminant analysis*.

5.4.1 Classification by Bayes Rule

Let u different classes ω_i occur which in turn represent an object. Let the $p \times 1$ random vector $\boldsymbol{y}$ contain p characteristics which are observed. Let the density function for the vector $\boldsymbol{y}$ of characteristics under the condition that $\boldsymbol{y}$ originates from the class ω_i be $p(\boldsymbol{y}|\boldsymbol{y} \in \omega_i)$. Let the prior density function that $\boldsymbol{y} \in \omega_i$ holds true be p_i. If for the sake of simplifying the notation the condition C of the background information is omitted, Bayes' theorem (2.122) leads to the posterior density function $p(\boldsymbol{y} \in \omega_i|\boldsymbol{y})$ that the vector $\boldsymbol{y}$ of characteristics is assigned to the class ω_i under the condition that $\boldsymbol{y}$ is given

$$p(\boldsymbol{y} \in \omega_i|\boldsymbol{y}) \propto p_i p(\boldsymbol{y}|\boldsymbol{y} \in \omega_i) \quad \text{for} \quad i \in \{1,\ldots,u\} \,. \tag{5.155}$$

The classification is a decision problem for which it is advisable to work with the zero-one loss (3.28). Let $L(\omega_j, \boldsymbol{y} \in \omega_i)$ denote the loss function for classifying $\boldsymbol{y}$ into ω_i, while actually the class ω_j is present. It takes the values

$$L(\omega_j, \boldsymbol{y} \in \omega_i) = \begin{cases} 0 & \text{for} \quad i = j \\ a & \text{for} \quad i \neq j \,. \end{cases} \tag{5.156}$$

The loss is equal to zero, if the classification is correct, and equal to a, if it is not correct.

The posterior expected loss of the classification of $\boldsymbol{y}$ into ω_i is computed from (3.1) with (5.155) by

$$E[L(\boldsymbol{y} \in \omega_i)] = \sum_{\substack{j=1 \\ j \neq i}}^{u} L(\omega_j, \boldsymbol{y} \in \omega_i) p(\boldsymbol{y} \in \omega_j|\boldsymbol{y}) \,. \tag{5.157}$$

Bayes rule requires to minimize the posterior expected loss. This leads to the decision rule that the vector $\boldsymbol{y}$ of characteristics is assigned to the class ω_i, if

$$\sum_{\substack{j=1 \\ j \neq i}}^{u} L(\omega_j, \boldsymbol{y} \in \omega_i) p(\boldsymbol{y} \in \omega_j|\boldsymbol{y}) < \sum_{\substack{j=1 \\ j \neq k}}^{u} L(\omega_j, \boldsymbol{y} \in \omega_k) p(\boldsymbol{y} \in \omega_j|\boldsymbol{y})$$

$$\text{for all} \quad k \in \{1,\ldots,u\} \quad \text{with} \quad i \neq k \,. \tag{5.158}$$

If the misclassifications obtain equal losses,

$$L(\omega_i, \boldsymbol{y} \in \omega_j) = L(\omega_k, \boldsymbol{y} \in \omega_l)$$

follows and with (2.47) instead of (5.158)

$$(1 - p(\boldsymbol{y} \in \omega_i|\boldsymbol{y})) < (1 - p(\boldsymbol{y} \in \omega_k|\boldsymbol{y}))$$
$$\text{for all} \quad k \in \{1, \ldots, u\} \quad \text{with} \quad i \neq k \ .$$

The decision rule now says that the vector $\boldsymbol{y}$ of characteristics is assigned to the class ω_i, if

$$p(\boldsymbol{y} \in \omega_i|\boldsymbol{y}) > p(\boldsymbol{y} \in \omega_k|\boldsymbol{y}) \quad \text{for all} \quad k \in \{1, \ldots, u\} \quad \text{with} \quad i \neq k \ . \tag{5.159}$$

Thus, the vector $\boldsymbol{y}$ of characteristics is classified into the class ω_i which maximizes the posterior density function $p(\boldsymbol{y} \in \omega_i|\boldsymbol{y})$.

Example: Digital images of the surface of the earth are received by satellites with special cameras which decompose the images into pixels and measure the grey values of the pixels in p different frequency bands of the visible and invisible light. The grey values of one pixel form the $p \times 1$ vector $\boldsymbol{y}$ of characteristics. If u objects ω_i have to be identified in the digital image based on different textures, each pixel is assigned according to (5.159) by the vector $\boldsymbol{y}$ of characteristics into a class ω_i. By classifying in this manner each pixel is looked at isolated from its neigborhood. This disadvantage is overcome by the method of segmentation described in Chapter 5.4.3. $\triangle$

A real-valued *discriminant function* $d_i(\boldsymbol{y})$ for $i \in \{1, \ldots, u\}$ depending on the vector $\boldsymbol{y}$ of characteristics is defined for classifying $\boldsymbol{y}$ into ω_i, if

$$d_i(\boldsymbol{y}) > d_k(\boldsymbol{y}) \quad \text{for all} \quad k \in \{1, \ldots, u\} \quad \text{with} \quad i \neq k \ . \tag{5.160}$$

As will become obvious in the following chapter, the computations can be simplified, if the discriminant function is determined by

$$d_i(\boldsymbol{y}) = \ln p(\boldsymbol{y} \in \omega_i|\boldsymbol{y}) \tag{5.161}$$

or with (5.155), since constants need not be considered in (5.160),

$$d_i(\boldsymbol{y}) = \ln p(\boldsymbol{y}|\boldsymbol{y} \in \omega_i) + \ln p_i \ . \tag{5.162}$$

5.4.2 Normal Distribution with Known and Unknown Parameters

The distribution of the vector $\boldsymbol{y}$ of characteristics has not been specified yet. The simple case is being dealt with first that the vector $\boldsymbol{y}$ is according to (2.195) normally distributed with known parameters. Thus, we obtain in (5.155)

$$p(\boldsymbol{y}|\boldsymbol{y} \in \omega_i) \propto (\det \boldsymbol{\Sigma}_i)^{-1/2} \exp\left[-\frac{1}{2}(\boldsymbol{y} - \boldsymbol{\mu}_i)' \boldsymbol{\Sigma}_i^{-1} (\boldsymbol{y} - \boldsymbol{\mu}_i) \right] \tag{5.163}$$

with the $p \times 1$ vector $\boldsymbol{\mu}_i$ and the $p \times p$ positive definite matrix $\boldsymbol{\Sigma}_i$ as known parameters. We then find from (5.162) the discrimant function

$$d_i(\boldsymbol{y}) = -\frac{1}{2}(\boldsymbol{y} - \boldsymbol{\mu}_i)' \boldsymbol{\Sigma}_i^{-1} (\boldsymbol{y} - \boldsymbol{\mu}_i) - \frac{1}{2} \ln \det \boldsymbol{\Sigma}_i + \ln p_i \tag{5.164}$$

for the decision rule (5.160). The boundary between the classes ω_i and ω_j following from the decision is defined by the vector $\boldsymbol{y}$ of characteristics which fulfills for the indices $i, j \in \{1, \ldots, u\}$ with $i \neq j$ the relation $d_i(\boldsymbol{y}) = d_j(\boldsymbol{y})$. The boundary is represented because of the quadratic form in (5.164) by a surface of second order.

By assuming $\boldsymbol{\Sigma}_i = \boldsymbol{\Sigma}$ and $p_i = c$ for all $i \in \{1, \ldots, u\}$ the discriminant function (5.164) simplifies, since constants in the discriminant function need not be taken care of because of (5.160), as negative discriminant function to

$$-d_i(\boldsymbol{y}) = (\boldsymbol{y} - \boldsymbol{\mu}_i)' \boldsymbol{\Sigma}^{-1} (\boldsymbol{y} - \boldsymbol{\mu}_i) \qquad (5.165)$$

which is called *Mahalanobis distance*. The vector $\boldsymbol{y}$ of characteristics is therefore assigned to the class ω_i from which it has the shortest Mahalanobis distance.

Since constants may be neglected, we introduce instead of (5.165) the discriminant function

$$d_i(\boldsymbol{y}) = \boldsymbol{y}' \boldsymbol{\Sigma}^{-1} \boldsymbol{\mu}_i - \frac{1}{2} \boldsymbol{\mu}'_i \boldsymbol{\Sigma}^{-1} \boldsymbol{\mu}_i \ . \qquad (5.166)$$

It is like (5.165) a linear function of the vector $\boldsymbol{y}$ of characteristics. The decision boundary $d_i(\boldsymbol{y}) = d_j(\boldsymbol{y})$ with $i \neq j$ is formed by a hyperplane which cuts the plane spanned by two coordinate axes in straight lines, if the coordinate planes are not parallel to the hyperplane.

If finally $\boldsymbol{\Sigma} = \sigma^2 \boldsymbol{I}$ is valid, we obtain from (5.165) after omitting the constant σ^2 the negative discriminant function

$$-d_i(\boldsymbol{y}) = (\boldsymbol{y} - \boldsymbol{\mu}_i)' (\boldsymbol{y} - \boldsymbol{\mu}_i) \qquad (5.167)$$

which is called *minimum distance classifier*. Instead of (5.166) we get

$$d_i(\boldsymbol{y}) = \boldsymbol{y}' \boldsymbol{\mu}_i - \frac{1}{2} \boldsymbol{\mu}'_i \boldsymbol{\mu}_i \ . \qquad (5.168)$$

If ideal prototypes or templates are available for the classes ω_i which are represented by the vectors $\boldsymbol{\mu}_i$, the classification by (5.168) means a template matching for which in analogy to (2.150) the covariance or after a normalization the correlation of the vector $\boldsymbol{y}$ of characteristics and of the template $\boldsymbol{\mu}_i$ is computed. For instance, standardized numbers may be automatically read by first decomposing them into pixels and then comparing them with templates represented by pixels of equal size.

In general, the parameters $\boldsymbol{\mu}_i$ and $\boldsymbol{\Sigma}_i$ of the normal distribution (5.163) for the vector $\boldsymbol{y}$ of characteristics are not given, but have to be estimated because of (2.196) as expected values and covariances of the vectors of characteristics. For the estimation training samples, i.e. classified vectors of characteristics, are needed, thus

$$
\begin{aligned}
&\boldsymbol{y}_{11}, \boldsymbol{y}_{12}, \ldots, \boldsymbol{y}_{1n_1} \quad \text{from} \quad \omega_1 \\
&\boldsymbol{y}_{21}, \boldsymbol{y}_{22}, \ldots, \boldsymbol{y}_{2n_2} \quad \text{from} \quad \omega_2 \\
&\cdots\cdots\cdots\cdots\cdots\cdots\cdots\cdots\cdots\cdots \\
&\boldsymbol{y}_{u1}, \boldsymbol{y}_{u2}, \ldots, \boldsymbol{y}_{un_u} \quad \text{from} \quad \omega_u
\end{aligned}
\qquad (5.169)
$$

where the $p \times 1$ vectors $\boldsymbol{y}_{ij}$ of characteristics are independent and normally distributed like

$$\boldsymbol{y}_{ij} \sim N(\boldsymbol{\mu}_i, \boldsymbol{\Sigma}_i) \quad \text{for} \quad i \in \{1, \ldots, u\}, \; j \in \{1, \ldots, n_i\} \, . \tag{5.170}$$

The vector $\boldsymbol{y}_{ij}$ of characteristics of class ω_i for $i \in \{1, \ldots, u\}$ contains with $\boldsymbol{y}_{ij} = (y_{ijk})$, $j \in \{1, \ldots, n_i\}$ and $k \in \{1, \ldots, p\}$ as mentioned p characteristics. The estimates $\hat{\boldsymbol{\mu}}_i$ with $\hat{\boldsymbol{\mu}}_i = (\hat{\mu}_{ik})$ of the expected values $\boldsymbol{\mu}_i$ of these characteristics and the estimates $\hat{\boldsymbol{\Sigma}}_i$ with $\hat{\boldsymbol{\Sigma}}_i = (\hat{\sigma}_{ikl})$ and $k, l \in \{1, \ldots, p\}$ of the covariance matrices $\boldsymbol{\Sigma}_i$ of the characteristics are obtained with $p \leq n_i - 1$ in the multivariate linear model from the observations (5.169) arranged in the following table, see for instance KOCH (1999, p.251),

$$
\begin{aligned}
\boldsymbol{y}'_{i1} &= |y_{i11}, y_{i12}, \ldots, y_{i1p}| \\
\boldsymbol{y}'_{i2} &= |y_{i21}, y_{i22}, \ldots, y_{i2p}| \\
&\ldots\ldots\ldots\ldots\ldots\ldots\ldots\ldots\ldots\ldots \\
\boldsymbol{y}'_{in_i} &= |y_{in_i1}, y_{in_i2}, \ldots, y_{in_ip}| \, .
\end{aligned}
$$

We get

$$\hat{\mu}_{ik} = \frac{1}{n_i} \sum_{j=1}^{n_i} y_{ijk} \quad \text{for} \quad k \in \{1, \ldots, p\}$$

or

$$\hat{\boldsymbol{\mu}}_i = \frac{1}{n_i} \sum_{j=1}^{n_i} \boldsymbol{y}_{ij} \tag{5.171}$$

and

$$\hat{\sigma}_{ikl} = \frac{1}{n_i - 1} \sum_{j=1}^{n_i} (\hat{\mu}_{ik} - y_{ijk})(\hat{\mu}_{il} - y_{ijl}) \quad \text{for} \quad k, l \in \{1, \ldots, p\}$$

or

$$\hat{\boldsymbol{\Sigma}}_i = \frac{1}{n_i - 1} \sum_{j=1}^{n_i} (\hat{\boldsymbol{\mu}}_i - \boldsymbol{y}_{ij})(\hat{\boldsymbol{\mu}}_i - \boldsymbol{y}_{ij})' \, . \tag{5.172}$$

The parameters $\boldsymbol{\mu}_i$ and $\boldsymbol{\Sigma}_i$ have to be replaced for the classification by their estimates $\hat{\boldsymbol{\mu}}_i$ and $\hat{\boldsymbol{\Sigma}}_i$ in the discriminant functions (5.164) to (5.168). An improvement of this method, which takes care of different sizes n_i of the training samples, may be found by the predictive analysis of Bayesian statistics, see for instance KOCH (1990, p.142).

5.4.3 Parameters for Texture

So far, the vector $\boldsymbol{y}$ of characteristics was assumed to be normally distributed according to (2.195) with the parameters $\boldsymbol{\mu}_i$ and $\boldsymbol{\Sigma}_i$ determined by the vector of expected values and the covariance matrix of $\boldsymbol{y}$. If a texture is registered by the vector $\boldsymbol{y}$ of characteristics, for instance, such that $\boldsymbol{y}$ contains the grey values of a pixel of a digital image in several frequency bands, the parameters of the distribution should also characterize the texture. The digital image can then be segmented based on the textures. Distributions with parameters for textures may be derived from Gibbs distributions which were already mentioned in Chapter 5.3.1.

Let r with $r \in \Omega$ be a pixel of a digital image and Ω the set of pixels. Furthermore, let $\boldsymbol{y}_r$ with $\boldsymbol{y}_r = (y_{rb})$ and $b \in \{1, \ldots, B\}$ be the $B \times 1$ vector of characteristics which contains for the pixel r the B grey values of a digital image in B frequency bands. The textures of the image are numbered and let ϵ with $\epsilon \in \{1, \ldots, T\}$ be the number of one of the T textures and Ω_ϵ the set of pixels of the texture ϵ. Let the measurement y_{rb} and its error e_{rb} of the grey value of the frequency band b for the texture ϵ in pixel r with $r \in \Omega_\epsilon$ be represented in form of the observation equation (4.3) of the linear model by the observations y_{r+s} in the neighborhood N_p of pixel r. They are multiplied by the unknown parameters $\beta_{s\epsilon b}$ valid for the texture ϵ and summed over the neighborhood

$$\sum_{s \in N_p} \beta_{s\epsilon b}(y_{r+s,b} - \mu_{\epsilon b} + y_{r-s,b} - \mu_{\epsilon b}) = y_{rb} - \mu_{\epsilon b} + e_{rb} . \qquad (5.173)$$

Here, $\mu_{\epsilon b}$ denotes the expected value of the observations y_{rb} for the texture ϵ with $r \in \Omega_\epsilon$. The index s defines the index of a pixel in half of the neighborhood N_p as in (5.147) so that with $r + s$ and $r - s$ the whole neighborhood of pixel r is covered as in (5.150). Pixels which lie symmetrically with respect to r obtain identical parameters $\beta_{s\epsilon b}$ for the texture which therefore have to be multiplied by $y_{r+s,b}$ and $y_{r-s,b}$. Thus, the neighborhood of a pixel is taken care of and a texture is parameterized.

Let the observations y_{rb} in the different frequency bands be independent. The covariance matrix $D(\boldsymbol{y}_r)$ of $\boldsymbol{y}_r$ then follows because of (2.153) from the diagonal matrix $D(\boldsymbol{y}_r) = \mathrm{diag}(\sigma_{\epsilon 1}^2, \sigma_{\epsilon 2}^2, \ldots, \sigma_{\epsilon B}^2)$, where $\sigma_{\epsilon b}^2$ denotes the variance of y_{rb} with $r \in \Omega_\epsilon$ in the frequency band b. If, in addition, it is assumed that the vector $\boldsymbol{y}_r$ of characteristics is normally distributed, we get from (2.195) with (5.173) the density function for $\boldsymbol{y}_r$ by

$$p(\boldsymbol{y}_r | \partial \boldsymbol{y}_r, \epsilon_r, \partial \epsilon_r) \propto \exp \Big\{ - \sum_{b=1}^{B} \Big\{ \frac{1}{2\sigma_{\epsilon b}^2} [y_{rb} - \mu_{\epsilon b}$$

$$- \sum_{s \in N_p} \beta_{s\epsilon b}(y_{r+s,b} - \mu_{\epsilon b} + y_{r-s,b} - \mu_{\epsilon b})]^2 \Big\} \Big\} \qquad (5.174)$$

subject to the condition that the vectors $\boldsymbol{y}_r$ of characteristics in the neigborhood N_p of pixel r, denoted by $\partial \boldsymbol{y}_r$, and ϵ_r as well as $\partial \epsilon_r$ are given. The random variable ϵ_r with $\epsilon_r = \epsilon$ and $\epsilon \in \{1, \ldots, T\}$ denotes the label of the pixel r and expresses its affiliation to a texture by the number of the texture to which the pixel belongs. The quantity $\partial\epsilon_r$ denotes the textures of the pixels in the neighborhood of r. The random variable ϵ_r is an unknown parameter, which has to be estimated, in order to solve the problem of segmentation by attributing each pixel r to a texture. The density function (5.174) is the likelihood function for the estimation.

This density function is given by means of the Markov property which says that vectors of characteristics, which do not belong to the neigborhood N_p of the pixel r, are independent from the vector $\boldsymbol{y}_r$ of characteristics. According to (2.117) the components of these vectors of characteristics therefore do not appear in the density function (5.174). It can be shown that the density function (5.174) of the normal distribution is a special Gibbs distribution (KOCH and SCHMIDT 1994, p.308; KLONOWSKI 1999, p.30).

The parameters $\mu_{\epsilon b}, \sigma^2_{\epsilon b}$ and $\beta_{s\epsilon b}$ in (5.174) are unknown and have to be estimated. Because of the observation equations (5.173), where the expected value $\mu_{\epsilon b}$ is subtracted from the measurements y_{rb}, the parameters $\beta_{s\epsilon b}$ for texture are invariant with respect to $\mu_{\epsilon b}$. To show this, we sum the pixels r over the set Ω_ϵ of pixels belonging to the texture ϵ and obtain by assuming $\sum_{r\in\Omega_\epsilon} e_{rb} \approx 0$ in (5.173)

$$\sum_{r\in\Omega_\epsilon}\sum_{s\in N_p} \beta_{s\epsilon b}(y_{r+s,b} - \mu_{\epsilon b} + y_{r-s,b} - \mu_{\epsilon b}) = \sum_{r\in\Omega_\epsilon}(y_{rb} - \mu_{\epsilon b})\ .$$

Because of $r+s \in \Omega_\epsilon$ and $r-s \in \Omega_\epsilon$ we get in addition

$$\sum_{r\in\Omega_\epsilon}(y_{rb} - \mu_{\epsilon b}) = \sum_{r\in\Omega_\epsilon}(y_{r+s,b} - \mu_{\epsilon b}) = \sum_{r\in\Omega_\epsilon}(y_{r-s,b} - \mu_{\epsilon b}) \neq 0$$

so that by exchanging the two summations over r and s we have

$$\sum_{s\in N_p} \beta_{s\epsilon b} = 0,5\ . \tag{5.175}$$

Substituting this result in (5.173) leads to

$$\sum_{s\in N_p} \beta_{s\epsilon b}(y_{r+s,b} + y_{r-s,b}) = y_{rb} + e_{rb} \tag{5.176}$$

which shows the invariance of the parameters $\beta_{s\epsilon b}$ with respect to $\mu_{\epsilon b}$. This property of invariance is not desired, if identical textures, which have different expected values of grey values, shall be distinguished for instance based on different colors. For estimating the parameters the constraint may therefore be introduced

$$\sum_{s\in N_p} \beta_{s\epsilon b} = 0 \tag{5.177}$$

which leads with (5.173) to

$$\sum_{s \in N_p} \beta_{s\epsilon b}(y_{r+s,b} + y_{r-s,b}) = y_{rb} - \mu_{\epsilon b} + e_{rb} \tag{5.178}$$

so that the invariance is removed.

In a supervised classification the parameters $\mu_{\epsilon b}, \sigma^2_{\epsilon b}$ and $\beta_{s\epsilon b}$ in (5.174) are estimated from training sets and in an unsupervised classification from an approximate segmentation which is iteratively improved. The parameter $\mu_{\epsilon b}$ may be estimated in advance as the mean $\hat{\mu}_{\epsilon b}$ by

$$\hat{\mu}_{\epsilon b} = \frac{1}{n_\epsilon} \sum_{r \in \Omega_\epsilon} y_{rb} \tag{5.179}$$

if n_ϵ denotes the number of pixels in Ω_ϵ. Let the joint distribution of the vectors $\boldsymbol{y}_r$ of characteristics be approximately determined by the product of the density functions (5.174). The maximum likelihood estimate (3.33) of the parameters $\beta_{s\epsilon b}$ for texture in the frequency band b then leads to the method (4.28) of least squares, since the sum of squares in the exponent of the density function (5.174) has to be minimized. The observation equation of the measurement y_{rb} for this parameter estimation follows from (5.173) with

$$\sum_{s \in N_p} \beta_{s\epsilon b}(y_{r+s,b} - \hat{\mu}_{\epsilon b} + y_{r-s,b} - \hat{\mu}_{\epsilon b}) = y_{rb} - \hat{\mu}_{\epsilon b} + e_{rb} \; . \tag{5.180}$$

This observation equation has to be set up for all pixels $r \in \Omega_\epsilon$. The estimates of the parameters $\beta_{s\epsilon b}$ for texture and the estimate of the variance $\sigma^2_{\epsilon b}$ of the observation y_{rb} follow from (4.29) and (4.39) or when introducing the constraints (5.177) from (4.42) and (4.55), see also KOCH and SCHMIDT (1994, p.328) and KLONOWSKI (1999, p.36).

Textures contain in general not only a few pixels but extend over a larger part of the digital image. This prior information may be used to define the prior density function $p(\epsilon_r | \partial \epsilon_r)$ for the label ϵ_r, the affiliation of the pixel r to a texture, as Gibbs distribution (KOCH and SCHMIDT 1994, p.313)

$$p(\epsilon_r | \partial \epsilon_r) \propto \exp\left\{ -\left[\alpha_\epsilon + \sum_{s \in N_0} \beta_s(I(\epsilon_r, \epsilon_{r+s}) + I(\epsilon_r, \epsilon_{r-s}))\right]\right\} \tag{5.181}$$

by means of the indicator function

$$I(\epsilon_r, \epsilon_q) = \begin{cases} 1 & \text{for} \quad \epsilon_r \neq \epsilon_q \\ 0 & \text{for} \quad \epsilon_r = \epsilon_q \; . \end{cases} \tag{5.182}$$

It registers, whether pixels in the neighborhood N_0 of the pixel r belong to different textures. The unknown parameters β_s control with $\beta_s > 0$ the affiliation to the same texture, since for large values of β_s pairs of pixels from different textures obtain large negative exponents and therefore small values for the density function $p(\epsilon_r|\partial\epsilon_r)$. The unknown parameters α_ϵ control the number of pixels which are related to the texture ϵ with $\epsilon \in \{1,\ldots,T\}$. With $\alpha_\epsilon > 0$ large values for α_ϵ give rise to associating few pixels to the texture ϵ. The parameters α_ϵ and β_s may be estimated from given segmentations (KOCH and SCHMIDT 1994, p.326).

The posterior density function $p(\epsilon_r|\boldsymbol{y}_r,\partial\boldsymbol{y}_r,\partial\epsilon_r)$ for the unknown label ϵ_r, the affiliation of the pixel r to the texture ϵ with $\epsilon \in \{1,\ldots,T\}$, follows from Bayes' theorem (2.122) with the prior density function (5.181) for the label ϵ_r and the likelihood function (5.174) by

$$
p(\epsilon_r|\boldsymbol{y}_r,\partial\boldsymbol{y}_r,\partial\epsilon_r) \propto \exp\Big\{ -\big[\alpha_\epsilon + \sum_{s\in N_0}\beta_s(I(\epsilon_r,\epsilon_{r+s}) + I(\epsilon_r,\epsilon_{r-s}))\big]
$$

$$
- \sum_{b=1}^{B}\Big\{\frac{1}{2\sigma_{\epsilon b}^2}[y_{rb} - \mu_{\epsilon b} - \sum_{s\in N_p}\beta_{s\epsilon b}(y_{r+s,b} - \mu_{\epsilon b} + y_{r-s,b} - \mu_{\epsilon b})]^2\Big\}\Big\} \; .
$$

$$(5.183)$$

The parameters $\alpha_\epsilon, \beta_s, \mu_{\epsilon b}, \sigma_{\epsilon b}^2$ and $\beta_{s\epsilon b}$ are assumed as known by their estimates.

The conditional density function (5.183) can be used to estimate the labels ϵ_r of all pixels $r \in \Omega$ of a digital image by an iterative deterministic procedure like the ICM algorithm mentioned in Chapter 5.3.3 or by a stochastic procedure like the Gibbs sampler presented in Chapter 6.3.2, see also KOCH and SCHMIDT (1994, p.323).

For the extension of the described method of texture recognition to the automatic interpretation of digital images see for instance KLONOWSKI (1999), KOCH (1995B), KÖSTER (1995) and MODESTINO and ZANG (1992).

5.5 Bayesian Networks

5.5.1 Systems with Uncertainties

Bayesian networks are designed for making decisions in systems with uncertainties. The system is represented by an n-dimensional discrete random variable $X_1,\ldots,X_n$. Dependencies exist between the discrete values which the random variables X_1 to X_n may take on. They are expressed by the n-dimensional discrete density function defined in (2.65) for the random variables $X_1,\ldots,X_n$ and by their probability, respectively,

$$
p(x_{1j_1},\ldots,x_{nj_n}|C) = P(X_1 = x_{1j_1},\ldots,X_n = x_{nj_n}|C)
$$
$$
\text{with} \quad j_k \in \{1,\ldots,m_k\} \,, \; k \in \{1,\ldots,n\} \quad (5.184)
$$

or in a more compact notation with (2.77)

$$
p(x_1,\ldots,x_n|C) = P(X_1 = x_1,\ldots,X_n = x_n|C) \; . \tag{5.185}
$$

The dependency of the values of the random variables $X_1, \ldots, X_n$ is therefore not deterministically controlled but governed by uncertainty which is expressed by the density function (5.184) or (5.185), where C gives additional information on the system. The n-dimensional random variable $X_1, \ldots, X_n$ with its density function defines the system where the uncertainties appear. The values of the random variables $X_1, \ldots, X_n$ represent the state of the system and the density function of the random variables the probability and the uncertainty, respectively, of this state. The values of the random variables may be expressed by numbers which represent certain states. This corresponds to the definition of the random variables given in Chapter 2.2 by statements in form of real numbers. The values of the random variable, however, may also be given by propositions about certain states.

Example: It has to be decided, whether documents shall be delivered on foot, by a bicycle, a motorcycle or a car. The decision depends first on the weight of the documents which is expressed by the random variable X_1 in (5.184) with the values $x_{11} = 1$ (1 to 500 gr), $x_{12} = 2$ (500 to 1000 gr), $x_{13} = 3$ (1000 to 1500 gr), $x_{14} = 4$ (1500 to 2000 gr), $x_{15} = 5$ (above 2000 gr). Then, the weather X_2 with the values $x_{21} = 1$ (snow), $x_{22} = 2$ (ice), $x_{23} = 3$ (rain) and so on has to be considered for the decision. Furthermore, the persons who deliver the documents influence the decision. They are determined by the values of the random variable X_3 with $x_{31} = 1$ (person A), $x_{32} = 2$ (person B), $x_{33} = 3$ (person C), if three persons are available. The traffic conditions X_4 are considered with $x_{41} = 1$ (light traffic), $x_{42} = 2$ (heavy traffic) and so on. The random variable X_5 finally describes the possibilities to deliver the documents. It takes on the values $x_{51} = 1$ (on foot), $x_{52} = 2$ (bicycle), $x_{53} = 3$ (motorcycle), $x_{54} = 4$ (car). The probabilities of the values of the random variables express the uncertainty in the system for which one has to decide upon the way of delivery. $\triangle$

To reach a decision with respect to a random variable, for instance, X_i of the system, the probabilities of the values x_i of the random variable X_i have to be computed from (5.185). We obtain according to (2.83) as marginal density function $p(x_i|C)$ for X_i

$$p(x_i|C) = \sum_{x_1} \cdots \sum_{x_{i-1}} \sum_{x_{i+1}} \cdots \sum_{x_n} p(x_1, \ldots, x_{i-1}, x_i, x_{i+1}, \ldots, x_n|C) \ . \quad (5.186)$$

As will be explained in connection with (5.197), the marginal density function $p(x_i|C)$ may be interpreted as posterior density function in the sense of Bayes' theorem (2.122). Thus, we decide with Bayes rule and the zero-one loss according to (3.30) for the MAP estimate $\hat{x}_{Mi}$ of the random variable X_i, which maximizes the posterior density function and therefore the probability,

$$\hat{x}_{Mi} = \arg\max_{x_i} p(x_i|C) \ . \quad (5.187)$$

Decisions in a system with uncertainties, which is represented by the n-dimensional discrete random variable $X_1, \ldots, X_n$, can therefore be made, if the density function (5.185) of the random variable is known. It will be difficult for a large number of random variables, however, to determine their joint distribution. It is more advantageous, to replace the joint distribution with using the chain rule (2.116) by the conditional distributions and thus by the probabilities of the individual random variables so that the knowledge about the system with uncertainties is expressed by the probabilities. This was already mentioned in Chapter 2.1.3. Furthermore, independencies will exist between the random variables which simplify the conditional distributions.

5.5.2 Setup of a Bayesian Network

The density function (5.185) is expressed by the chain rule (2.116) to obtain

$$p(x_1, x_2, \ldots, x_n | C) = p(x_n | x_1, x_2, \ldots, x_{n-1}, C)$$
$$p(x_{n-1} | x_1, x_2, \ldots, x_{n-2}, C) \ldots p(x_2 | x_1, C) p(x_1 | C) \ . \quad (5.188)$$

The conditional density functions on the right-hand side of the chain rule may be graphically represented by a network whose nodes denote the random variables $X_1, \ldots, X_n$. The conditional density functions are shown by arrows. Thus, the joint density function is related to a network which is formally proved by NEAPOLITAN (1990, p.173). The head of an arrow points to the random variable, whose conditional density function is expressed, and the end of the arrow is connected with the random variable whose value appears as condition in the density function. By this manner a *directed acyclical graph* is developed, that is a graph with directed edges which do not produce cycles.

Example 1: A directed acyclical graph is shown in Figure 5.1 for the representation of (5.188) in case of $n = 6$. $\qquad\qquad$ Δ

The directed acyclical graph which represents (5.188) has one *root node*, from where arrows start only, and one *leaf node*, where arrows end only. The root node is represented by the random variable X_1 and the leaf node by the random variable X_n.

As mentioned independencies between random variables can be assumed for large systems. The chain rule (5.188) then simplifies, as was shown already with (2.117). If the random variable X_i is independent from X_k and if $i > k$ holds true, then the arrow directed from the random variable k to the random variable i drops out of the directed acyclical graph. This graph resulting from the independencies of the random variables is called a *Bayesian network*.

Example 2: Let in the Example 1 with Figure 5.1 the random variable X_2 be independent from X_1, the random variable X_4 independent from X_1 and X_3 and the random variable X_6 independent from X_1, X_2, X_3 and X_5. Figure 5.2 shows the directed acyclical graph, the Bayesian network, which results. It has in comparison to Figure 5.1 two root nodes and two leaf nodes. By means of the Bayesian network of Figure 5.2 the density function of the six dimensional random variable $X_1, \ldots, X_6$ may be immediately written down

$$p(x_1, \ldots, x_6|C) = p(x_6|x_4, C)p(x_5|x_1, x_2, x_3, x_4, C)p(x_4|x_2, C)$$
$$p(x_3|x_1, x_2, C)p(x_2|C)p(x_1|C) \ . \quad (5.189)$$

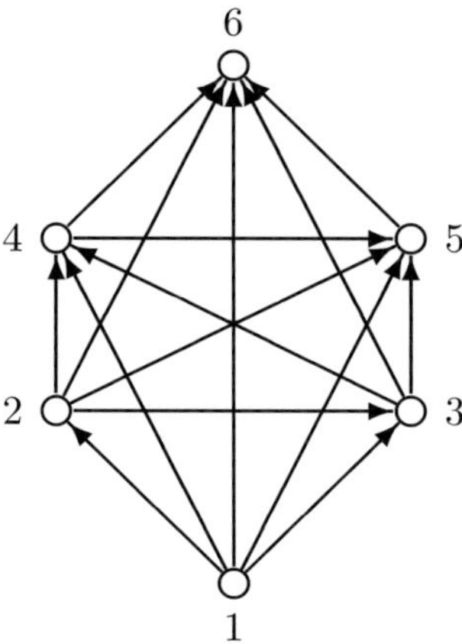

Figure 5.1: Directed Acyclical Graph

In the conditional density functions of the random variables X_1 to X_6 on the right-hand side of (5.189) only the values of the random variables enter as conditions from which arrows are pointing to random variables.

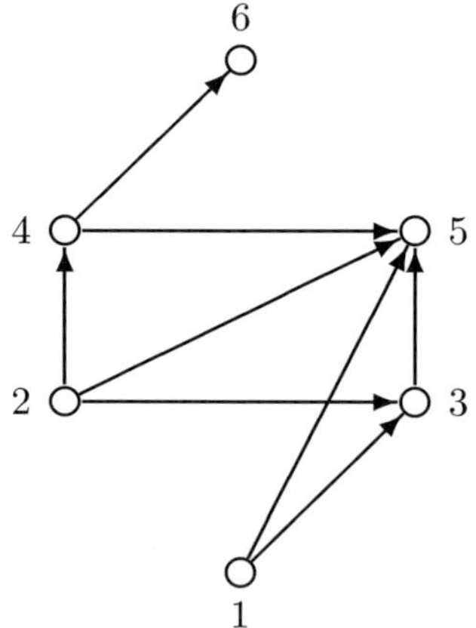

Figure 5.2: Bayesian Network

Of course, the density function (5.189) is also obtained, if the chain rule (5.188) is applied for $n = 6$ and the independencies between the random variables are then considered according to (2.117). Δ

This example shows how to proceed when setting up a Bayesian network. The system with uncertainties is represented by the n-dimensional random variable $X_1, \ldots, X_n$ with the random variables X_i for $i \in \{1, \ldots, n\}$ as nodes of the network. Dependencies in the Bayesian network are identified by arrows. The magnitude of the dependencies is expressed by the conditional density functions. The arrows point from the random variables whose values enter as conditions to the random variables whose probabilites have to be specified or generally speaking from the conditions to the consequences.

Information on the system with uncertainties, which is represented by the Bayesian network, is introduced as observations or measurements depending on the random variables of the Bayesian network. The density function for a node X_k takes on according to (5.188) with $i < j < k$ the form

$$p(x_k | x_{k-j}, \ldots, x_{k-i}, C) \ . \tag{5.190}$$

If the node X_k represents a leaf node of the Bayesian network, this density function may be interpreted as density function for an observation X_k depending on the random variables X_{k-j} to X_{k-i}. Leaf nodes may therefore express observations or data. If observations exist for the random variable X_i, which is not a leaf node, the leaf node X_k need only be appended to X_i, see Figure 5.3, in order to introduce the density function $p(x_k | x_i, C)$ for the data X_k depending on X_i. If the data X_k depend on additional random variables, arrows have to be inserted from these random variables to X_k and the conditional density function for X_k has to be specified correspondingly.

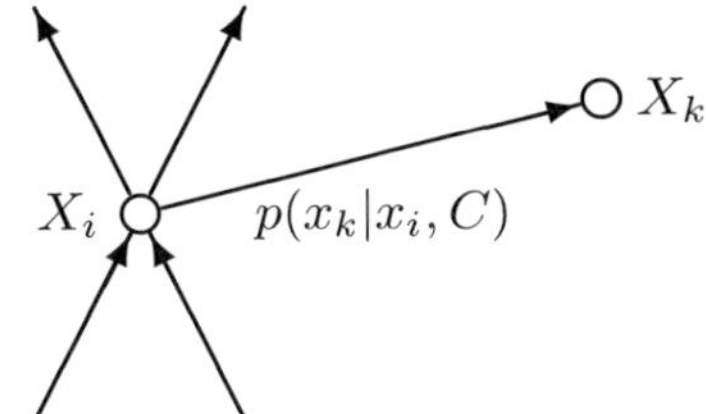

Figure 5.3: Leaf Node X_k for Data

As can be seen from (5.188) or (5.189), a root node of a Bayesian network, for instance X_i, has the density function $p(x_i | C)$. This density function may be interpreted in the sense of Bayes' theorem (2.122) as a prior density function stemming from prior information. Root nodes are therefore distinguished by relating prior information to them. But this does not mean that only root nodes can have prior density functions. If based on prior information a prior density function exists for the node X_i which is not a root node, the leaf node X_k is inserted, whose value is given by $X_k = x_{k0}$ as a constant, see Figure 5.4. The prior density function for X_i follows with the density function $p(x_{k0} | x_i, C)$ where x_{k0} is a constant and only x_i varies. The density function $p(x_{k0} | x_i, C)$ therefore depends only on the values for X_i and is identical with the density function $p(x_i | C)$, if X_i is a true root node.

The density function $p(x_i | C)$ of a root node X_i need not be interpreted as prior, it may be also understood as density function of an observation X_i

of the Bayesian network. Leaf nodes as well as root nodes may therefore represent data.

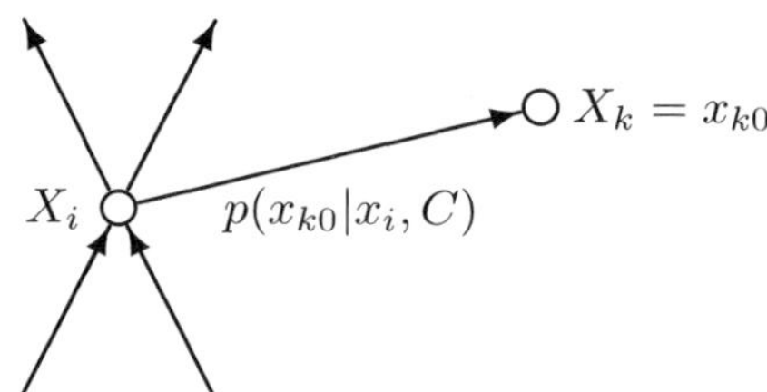

Figure 5.4: Leaf Node X_k for Prior Information

Example 3: To solve the decision problem of the example to (5.184), the Bayesian network of Figure 5.2 of the Example 2 is applied. The possibilities to deliver documents on foot, by a bicycle, a motorcycle or a car are represented by the random variable X_5, that is by the node 5. These possibilities of delivery are dependent on the weight X_1 of the documents, the weather X_2 at the time of the delivery, the person X_3, which takes care of the delivery, and of the traffic conditions X_4 to be encountered for the delivery. The random variables X_1 to X_4 are represented by the nodes 1 to 4. The traffic conditions X_4 also depend on the weather X_2. The employment of the person X_3 for the delivery is also influenced by the weather X_2 and in addition by the weight X_1. Based on experience prior information on the root node 1, that is on the weight X_1 of the documents, is available and the weather forecast provides prior information for the root node 2, that is for the weather X_2. The random variable X_6, i.e. the leaf node 6, represents prior information on the traffic conditions X_4. As will be explained in the following chapter, additional knowledge may be introduced into the Bayesian network, if values of the random variables are known, for instance, the weight of the documents to be delivered or the actual weather at the time of the delivery.   △

Deterministic variables instead of random variables are also allowed as nodes of a Bayesian network. The values x_k of the variable X_k shall depend on the values of X_{k-j} to X_{k-i} for $i < j < k$ deterministically with

$$x_k = f(x_{k-j}, \ldots, x_{k-i}) \tag{5.191}$$

where $f(x_{k-j}, \ldots, x_{k-i})$ denotes a function of $x_{k-j}, \ldots, x_{k-i}$. The node for X_k is then designated as leaf node, because the remaining random variables of the network cannot depend on X_k but only on X_{k-j} to X_{k-i}. The function (5.191) takes the place of the conditional density function $p(x_k|x_{k-j}, \ldots, x_{k-i}, C)$. Bayesian networks with deterministic variables are called *decision networks*.

Thus, Bayesian networks may be set up in a flexible manner to describe systems with uncertainties. The system is represented by the values of random variables. Prior information on the random variables is introduced by prior density functions and information on the random variables by observations. The magnitude of the dependencies between random variables result from the conditional density functions. The simplest way for obtaining them is to compute them from (2.24) as relative frequencies based on the knowledge about the system. The probabilities for the random variables of the system represent the uncertainties. The computation of these probabilities is the subject of the following chapter.

5.5.3 Computation of Probabilities

Based on the information, which enters the Bayesian network via the conditional density functions, decisions are made by computing the probability of a random variable of the Bayesian network and then applying (5.187). Additional information may be brought into the Bayesian network by *instantiating* random variables, that is by assigning values to them. Data, which have been taken, determine the values of random variables which represent the data, but also the remaining random variables, which do not denote data, may be instantiated. Based on this information the marginal density function $p(x_i|C)$ for any random variable X_i is computed from (5.186) with the chain rule (5.188). A decision is then made with (5.187).

Let the information first be determined by the conditional density functions of the network only. We then obtain the marginal density function $p(x_i|C)$ for X_i from (5.186) with (5.188) by

$$p(x_i|C) = \sum_{x_1} p(x_1|C) \sum_{x_2} p(x_2|x_1, C) \ldots \sum_{x_{i-1}} p(x_{i-1}|x_1, \ldots, x_{i-2}, C)$$

$$p(x_i|x_1, \ldots, x_{i-1}, C) \sum_{x_{i+1}} p(x_{i+1}|x_1, \ldots, x_i, C)$$

$$\ldots \sum_{x_n} p(x_n|x_1, \ldots, x_{n-1}, C) . \tag{5.192}$$

The summation over the values of X_n to X_{i+1} leads because of (2.103) to one, thus

$$p(x_i|C) = \sum_{x_1} p(x_1|C) \sum_{x_2} p(x_2|x_1, C) \ldots \sum_{x_{i-1}} p(x_{i-1}|x_1, \ldots, x_{i-2}, C)$$

$$p(x_i|x_1, \ldots, x_{i-1}, C) . \tag{5.193}$$

The remaining summations over the values of X_{i-1} to X_1 do not give one. This is true, because when summing over the values of X_{i-1} each value of the density function $p(x_{i-1}|x_1, \ldots, x_{i-2}, C)$ is multiplied by a different value of $p(x_i|x_1, \ldots, x_{i-1}, C)$. Thus,

$$\sum_{x_{i-1}} p(x_{i-1}|x_1, \ldots, x_{i-2}, C) p(x_i|x_1, \ldots, x_{i-1}, C) \neq 1 \tag{5.194}$$

follows. The same holds true for the summations over the values of X_{i-2} to X_1.

Let the information of the Bayesian network now be determined in addition by the instantiation of random variables. For example, let the value x_{k0} with $x_{k0} \in \{x_{k1}, \ldots, x_{km_k}\}$ from (5.184) be assigned to the random variable X_k with $i < k < n$. We then obtain instead of (5.192)

$$p(x_i|C) = \alpha \sum_{x_1} p(x_1|C) \sum_{x_2} p(x_2|x_1, C) \ldots \sum_{x_{i-1}} p(x_{i-1}|x_1, \ldots, x_{i-2}, C)$$

$$p(x_i|x_1, \ldots, x_{i-1}, C) \sum_{x_{i+1}} p(x_{i+1}|x_1, \ldots, x_i, C)$$

$$\ldots \sum_{x_{k-1}} p(x_{k-1}|x_1, \ldots, x_{k-2}, C)p(x_{k0}|x_1, \ldots, x_{k-1}, C) , \qquad (5.195)$$

since summing over the values of the random variables X_n to X_{k+1} which are not instantiated gives one because of (2.103). On the contrary, the summation over the values of X_{k-1} leads to a value unequal to one out of the same reason as explained by (5.194).

The quantity α in (5.195) denotes the normalization constant. It must be introduced so that (2.103) is fulfilled for $p(x_i|C)$. The reason is that the value x_{k0} is assigned to the random variable X_k and it is not summed over the remaining values of X_k. A normalization constant is therefore always mandatory, if a marginal density function is computed in the presence of instantiated random variables. If a random variable is instantiated which has a prior density function, the prior density gives a constant which can be absorbed in the normalization constant α.

We have first to sum in (5.195) over the values of X_{k-1}, then over the ones of X_{k-2} and so on until over the values of X_{i+1}. Then, the summation over X_{i-1} to X_1 follows. If we set

$$\lambda(x_1, \ldots, x_{k-1}) = p(x_{k0}|x_1, \ldots, x_{k-1}, C)$$

$$\lambda(x_1, \ldots, x_{k-2}) = \sum_{x_{k-1}} p(x_{k-1}|x_1, \ldots, x_{k-2}, C)\lambda(x_1, \ldots, x_{k-1})$$

$$\ldots\ldots\ldots\ldots\ldots\ldots\ldots\ldots\ldots\ldots\ldots\ldots\ldots\ldots\ldots\ldots\ldots\ldots\ldots\ldots\ldots\ldots\ldots\ldots \qquad (5.196)$$

$$\lambda(x_1, \ldots, x_i) = \sum_{x_{i+1}} p(x_{i+1}|x_1, \ldots, x_i, C)\lambda(x_1, \ldots, x_{i+1})$$

we obtain instead of (5.195)

$$p(x_i|C) = \alpha \sum_{x_1} p(x_1|C) \sum_{x_2} p(x_2|x_1, C) \ldots \sum_{x_{i-1}} p(x_{i-1}|x_1, \ldots, x_{i-2}, C)$$

$$p(x_i|x_1, \ldots, x_{i-1}, C)\lambda(x_1, \ldots, x_i) . \qquad (5.197)$$

This marginal density function may be interpreted in the sense of Bayes' theorem (2.122) as posterior density function, if $\lambda(x_1, \ldots, x_i)$ denotes the likelihood function which originates from the instantiation of a random variable, for instance of an observation. The summations over the remaining conditional density functions and their multiplications give the prior density function.

If the marginal density function $p(x_i|C)$ has to be computed for each node of the Bayesian network, it can be seen from (5.196) that the likelihood function may be obtained in advance for each random variable by a summation. However, the part which is interpreted as prior density function in (5.197) cannot be determined in advance independent from the likelihood function, since one has to sum over the values x_1 to x_{i-1} on which the likelihood function is also depending. Hence, for each node X_i of the Bayesian network the product

$$p(x_i|x_1,\ldots,x_{i-1},C)\lambda(x_1,\ldots,x_i)$$

has to be computed. The result is then multiplied by the conditional density function for X_{i-1} and the product has to be summed over the values of X_{i-1}. This procedure is repeated up to the random variable X_1. The computational effort, to get the marginal density function $p(x_i|C)$ of each node in a larger Bayesian network, is therefore quite considerably.

The discussion of computing the marginal density function was based on the chain rule (5.188) where independencies of random variables have not been considered. By independencies random variables disappear in the list of the variables whose values enter as conditions in the conditional density functions, as was shown with (5.189) for the example of the Bayesian network of Figure 5.2. The number of density values and therefore the number of summations decreases by the factor m_l, if the random variable X_l with m_l values disappears from the list of conditions. This can be seen from (2.115). Because of the independencies more than one leaf node or one root node might appear, as was already mentioned in the preceding chapter.

The computation of the marginal density function is especially simple, if not only the likelihood function as in (5.196) but also the prior density function can be separately determined from each other for every node of the network. A local computation is then possible by giving the contribution of each node of the Bayesian network to the marginal density function. Prerequisite of such a procedure, which goes back to PEARL (1986), is the *singly connected Bayesian network* for which only one path exists between any two nodes, the directions of the arrows not being considered. These networks are treated in the following three chapters where the computation of the probabilities is derived. Methods for transforming networks with multiple connections into singly connected ones are given in (PEARL 1988, p.195; DEAN and WELLMAN 1991, p.289).

Applications of the Bayesian networks, e.g. for the navigation of roboters, are described in (DEAN and WELLMAN 1991, p.297), for the interpretation of digital images in (KOCH 1995A; KULSCHEWSKI 1999), for decisions in connection with geographic information systems in (STASSOPOULOU et al. 1998) and for industrial applications in (OLIVER and SMITH 1990, p.177). Hints for setting up Bayesian networks and for computer programs to determine the probabilities are found in (JENSEN 1996, p.33).

Before concluding the chapter three simple examples shall be presented.

Example 1: It has to be decided which means of transportation shall be chosen depending on the weather. As means of transportation, which is represented by the random variable T, a bicycle and a car are available. Not integers with special meanings shall be assigned to the values of the random variables, as done in the example to (5.184), but the meanings are introduced immediately as values. The random variable T therefore gets the two values t

$$t = \left| \begin{array}{c} \text{bicycle} \\ \text{car} \end{array} \right| .$$

The weather represented by the random variable W is provided with the meanings w

$$w = \left| \begin{array}{c} \text{rain} \\ \text{overcast} \\ \text{sun} \end{array} \right| .$$

The values of the density function $p(t|w, C)$ for T given w and C, where C denotes the information about the habit of choosing a means of transportation, result, for instance, according to (2.24) as relative frequencies based on the habits from the following table. It is set up corresponding to (2.114) by

$$p(t|w, C) = \left| \begin{array}{cc} \text{bicycle} & \text{car} \\ 0.2 & 0.8 \\ 0.7 & 0.3 \\ 0.9 & 0.1 \end{array} \right| \begin{array}{c} \\ \text{rain} \\ \text{overcast} \\ \text{sun} \end{array} .$$

The density function $p(w|C)$ of the prior information on the weather W is determined by

$$p(w|C) = \left| \begin{array}{c} 0.3 \\ 0.5 \\ 0.2 \end{array} \right| \begin{array}{c} \text{rain} \\ \text{overcast} \\ \text{sun} \end{array} .$$

Prior information $V = v_0$ exists for the choice of the means of transportation T with

$$p(v_0|t, C) = \left| \begin{array}{c} 0.6 \\ 0.4 \end{array} \right| \begin{array}{c} \text{bicycle} \\ \text{car} \end{array} .$$

Summarizing we obtain the Bayesian network shown in Figure 5.5.

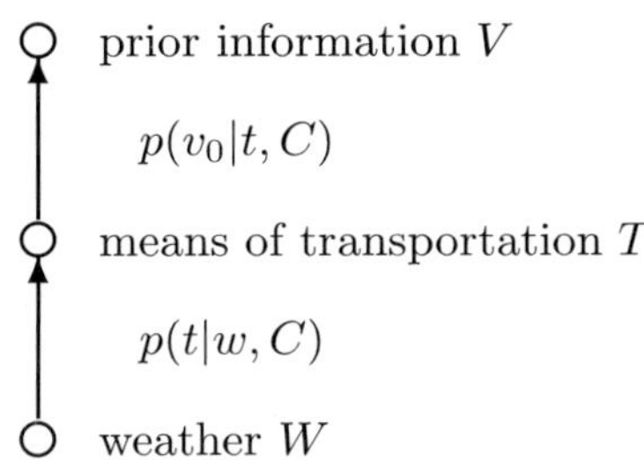

Figure 5.5: Bayesian Network of Example 1

The marginal density function $p(t|C)$ for the means of transportation T follows with (5.195) by

$$p(t|C) = \alpha \sum_w p(w|C)p(t|w, C)p(v_0|t, C)$$

$$= \alpha \left| \begin{array}{l} (0.3 \times 0.2 + 0.5 \times 0.7 + 0.2 \times 0.9) \times 0.6 \\ (0.3 \times 0.8 + 0.5 \times 0.3 + 0.2 \times 0.1) \times 0.4 \end{array} \right|$$

$$= \frac{1}{0.354 + 0.164} \left| \begin{array}{l} 0.354 \\ 0.164 \end{array} \right|$$

$$= \left| \begin{array}{l} 0.68 \\ 0.32 \end{array} \right. \left. \begin{array}{l} \text{bicycle} \\ \text{car} \end{array} \right. .$$

Based on the information, which is contained in the conditional density functions of the Bayesian network for the choice of a means of transportation, a probability of 68% results for the bicycle and a probability of 32% for the car. Because of (5.187) the bicycle is chosen for the transportation.

Information is now added such that the random variable W is instantiated by "w = overcast". The prior density function for W then gives a constant which can be absorbed in the normalization constant, as was already mentioned in connection with (5.195). However, this possibility is not chosen here in order to continue as before

$$p(t|w = \text{overcast}, C)$$

$$= \alpha\, p(w = \text{overcast}|C)p(t|w = \text{overcast}, C)p(v_0|t, C)$$

$$= \alpha \left| \begin{array}{l} 0.5 \times 0.7 \times 0.6 \\ 0.5 \times 0.3 \times 0.4 \end{array} \right|$$

$$= \left| \begin{array}{l} 0.78 \\ 0.22 \end{array} \right. \left. \begin{array}{l} \text{bicycle} \\ \text{car} \end{array} \right. .$$

For a clouded sky the probability of taking the bicycle is considerably greater than the one of choosing the car. Because of (5.187) the bicycle is therefore selected as means of transportation. With "$w = \text{rain}$" we obtain

$$p(t|w = \text{rain}, C)$$
$$= \alpha\, p(w = \text{rain}|C) p(t|w = \text{rain}, C) p(v_0|t, C)$$
$$= \alpha \begin{vmatrix} 0.3 \times 0.2 \times 0.6 \\ 0.3 \times 0.8 \times 0.4 \end{vmatrix}$$

$$= \begin{vmatrix} 0.27 \\ 0.73 \end{vmatrix} \begin{matrix} \text{bicycle} \\ \text{car} \end{matrix} .$$

The car will therefore be taken. $\triangle$

Example 2: Not only the weather W but also the feeling of the person choosing the means of transportation T influences the choice of transportation. Let the feeling, the random variable B, have the values b

$$b = \begin{vmatrix} \text{fresh} \\ \text{tired} \end{vmatrix} .$$

The values of the density function $p(t|w, b, C)$ for T given w, b and C result from the following two tables which are built up like (2.115)

$$p(t|w, b = \text{fresh}, C) = \begin{array}{cc|l} \text{bicycle} & \text{car} & \\ 0.2 & 0.8 & \text{rain} \\ 0.7 & 0.3 & \text{overcast} \\ 0.9 & 0.1 & \text{sun} \end{array}$$

$$p(t|w, b = \text{tired}, C) = \begin{array}{cc|l} \text{bicycle} & \text{car} & \\ 0.1 & 0.9 & \text{rain} \\ 0.6 & 0.4 & \text{overcast} \\ 0.8 & 0.2 & \text{sun} \end{array} .$$

The prior information on the feeling B is contained in the prior density function $p(b|C)$. It expresses the fact that the person, who selects the means of transportation T, is often tired

$$p(b|C) = \begin{vmatrix} 0.3 \\ 0.7 \end{vmatrix} \begin{matrix} \text{fresh} \\ \text{tired} \end{matrix} .$$

The prior information on the weather and the means of transportation is determined by the prior density functions of Example 1. The Bayesian network shown in Figure 5.6 is obtained.

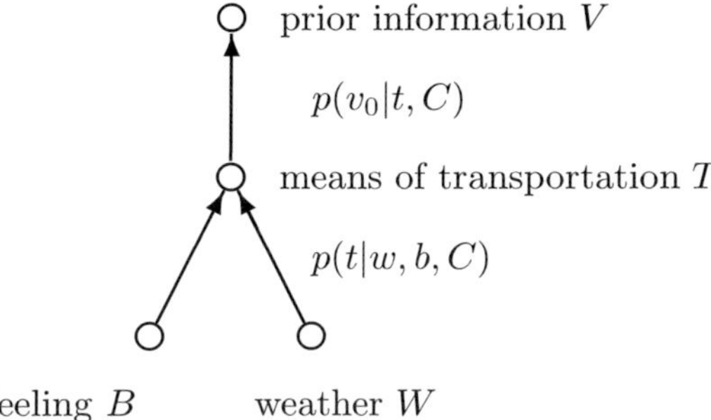

Figure 5.6: Bayesian Network of Example 2

The marginal density function $p(t|C)$ for the means of transportation T follows with (5.195) by

$$p(t|C) = \alpha \sum_b p(b|C) \sum_w p(w|C)p(t|w, b, C)p(v_0|t, C)$$

$$= \alpha \left(0.3 \left| \begin{array}{l} (0.3 \times 0.2 + 0.5 \times 0.7 + 0.2 \times 0.9) \times 0.6 \\ (0.3 \times 0.8 + 0.5 \times 0.3 + 0.2 \times 0.1) \times 0.4 \end{array} \right. \right.$$

$$\left. + 0.7 \left| \begin{array}{l} (0.3 \times 0.1 + 0.5 \times 0.6 + 0.2 \times 0.8) \times 0.6 \\ (0.3 \times 0.9 + 0.5 \times 0.4 + 0.2 \times 0.2) \times 0.4 \end{array} \right| \right)$$

$$= \left| \begin{array}{l} 0.62 \\ 0.38 \end{array} \right| \begin{array}{l} \text{bicycle} \\ \text{car} \end{array} .$$

Despite the fact that a tired feeling B prevails, the bicycle is chosen as means of transportation.

The weather W shall now be instantiated by "$w = $ overcast". We then obtain

$$p(t|w = \text{overcast}, C)$$

$$= \alpha \sum_b p(b|C)p(w = \text{overcast}|C)p(t|w = \text{overcast}, b, C)p(v_0|t, C)$$

$$= \alpha \left(0.3 \left| \begin{array}{l} 0.5 \times 0.7 \times 0.6 \\ 0.5 \times 0.3 \times 0.4 \end{array} \right. + 0.7 \left| \begin{array}{l} 0.5 \times 0.6 \times 0.6 \\ 0.5 \times 0.4 \times 0.4 \end{array} \right| \right)$$

$$= \left| \begin{array}{l} 0.72 \\ 0.28 \end{array} \right| \begin{array}{l} \text{bicycle} \\ \text{car} \end{array} .$$

With a clouded sky the probability increases like in Example 1 to use the bicycle.

If the random variables feeling B and weather W are instantiated by "$b = $

tired" and "$w = $ overcast" we get

$$p(t|b = \text{tired}, w = \text{overcast}, C) = \alpha\, p(b = \text{tired}|C)$$
$$p(w = \text{overcast}, C)p(t|w = \text{overcast}, b = \text{tired}, C)p(v_0|t, C)$$

$$= \alpha\left(0.7 \,\middle|\, \begin{array}{c} 0.5 \times 0.6 \times 0.6 \\ 0.5 \times 0.4 \times 0.4 \end{array} \,\middle|\,\right)$$

$$= \left|\begin{array}{c} 0.69 \\ 0.31 \end{array}\right| \begin{array}{l} \text{bicycle} \\ \text{car} \end{array}.$$

Despite a tired feeling the bicycle is chosen, when the sky is overcast. Δ

Example 3: As is well known, the weather influences the feelings of persons. Instead of introducing a prior density function for the feeling B, as done in Example 2, the feeling B shall now depend on the weather W. The density function $p(b|w, C)$ has the values shown in the following table which is set up like in (2.114) by

$$p(b|w, C) = \left|\begin{array}{cc} \text{fresh} & \text{tired} \\ 0.2 & 0.8 \\ 0.4 & 0.6 \\ 0.7 & 0.3 \end{array}\right| \begin{array}{l} \\ \text{rain} \\ \text{overcast} \\ \text{sun} \end{array}.$$

The Bayesian network of Figure 5.7 is therefore obtained for choosing the means of transportation T. This network by the way does not belong to the singly connected Bayesian network, because two paths exist from the random variable W to the random variable T. The probabilities for the random variables of this network therefore cannot be computed by the formulas given in the following three chapters.

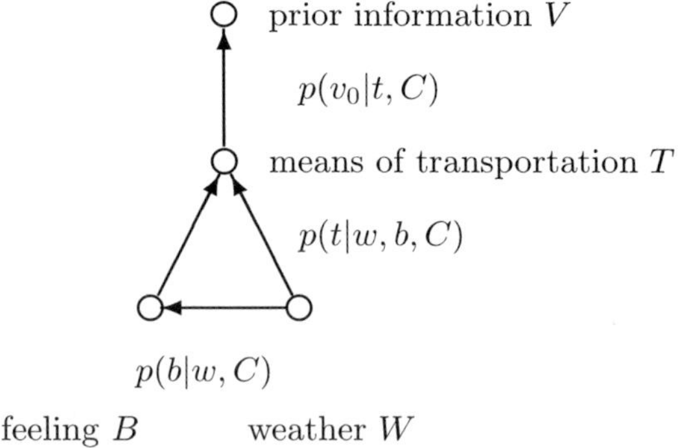

Figure 5.7: Bayesian Network of Example 3

The marginal density function $p(t|C)$ for the means of transportation T follows from (5.195) with the prior density functions for W and T of Example 1 and the conditional density function for T of the Example 2 by

$$p(t|C) = \alpha \sum_w p(w|C) \sum_b p(b|w,C)p(t|w,b,C)p(v_0|t,C)$$

$$= \alpha \Bigg(0.3 \left| \begin{array}{l} (0.2 \times 0.2 + 0.8 \times 0.1) \times 0.6 \\ (0.2 \times 0.8 + 0.8 \times 0.9) \times 0.4 \end{array} \right.$$

$$+\, 0.5 \left| \begin{array}{l} (0.4 \times 0.7 + 0.6 \times 0.6) \times 0.6 \\ (0.4 \times 0.3 + 0.6 \times 0.4) \times 0.4 \end{array} \right.$$

$$+\, 0.2 \left| \begin{array}{l} (0.7 \times 0.9 + 0.3 \times 0.8) \times 0.6 \\ (0.7 \times 0.1 + 0.3 \times 0.2) \times 0.4 \end{array} \right| \Bigg)$$

$$= \left| \begin{array}{l} 0.63 \\ 0.37 \end{array} \right| \begin{array}{l} \text{bicycle} \\ \text{car} \end{array} .$$

The dependency of the feeling B on the weather W changes in comparison to the Example 2 slightly the propabilities for the transportation by bicycle or by car.

Let the weather W be instantiated again by "w = overcast". Then the marginal density function for T follows with

$$p(t|w = \text{overcast}, C) = \alpha\, p(w = \text{overcast}|C)$$

$$\sum_b p(b|w = \text{overcast}, C)p(t|w = \text{overcast}, b, C)p(v_0|t, C)$$

$$= \alpha \Bigg(0.5 \left| \begin{array}{l} (0.4 \times 0.7 + 0.6 \times 0.6) \times 0.6 \\ (0.4 \times 0.3 + 0.6 \times 0.4) \times 0.4 \end{array} \right| \Bigg)$$

$$= \left| \begin{array}{l} 0.73 \\ 0.27 \end{array} \right| \begin{array}{l} \text{bicycle} \\ \text{car} \end{array} .$$

If we put "b = tired" and "w = overcast", we find

$$p(t|b = \text{tired}, w = \text{overcast}, C) = \alpha\, p(w = \text{overcast}, C)$$

$$p(b = \text{tired}|w = \text{overcast}, C)p(t|w = \text{overcast}, b = \text{tired}, C)p(v_0|t, C)$$

$$= \alpha \Bigg(\left| \begin{array}{l} 0.6 \times 0.6 \times 0.6 \\ 0.6 \times 0.4 \times 0.4 \end{array} \right| \Bigg)$$

$$= \left| \begin{array}{l} 0.69 \\ 0.31 \end{array} \right| \begin{array}{l} \text{bicycle} \\ \text{car} \end{array} .$$

For the first case a small change of the probabilities for the two means of transportation follows in comparison to Example 2, for the second case no change occurs. $\triangle$

5.5.4   Bayesian Network in Form of a Chain

A singly connected Bayesian network where only one path exists between any two nodes of the network is in its simplest form a chain which is depicted in

Figure 5.8. The random variable X_i depends only on the random variable

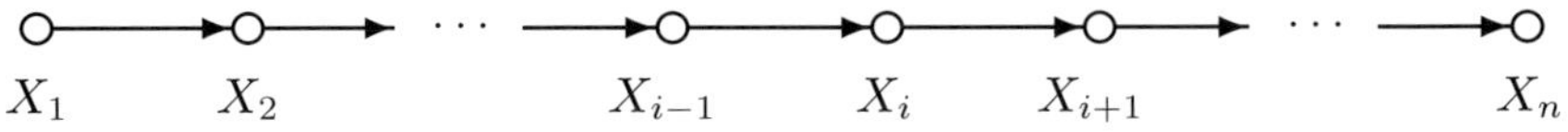

Figure 5.8: Bayesian Network as Chain

X_{i-1} for $i \in \{2, \ldots, n\}$. We therefore get with (2.117) and the chain rule (5.188)

$$p(x_1, \ldots, x_n|C) = p(x_1|C)p(x_2|x_1, C) \ldots p(x_{i-1}|x_{i-2}, C)$$
$$p(x_i|x_{i-1}, C)p(x_{i+1}|x_i, C) \ldots p(x_n|x_{n-1}, C) \ . \quad (5.198)$$

This is the density function of a Markov chain with transition probabilities $p(x_i|x_{i-1}, C)$, see for instance KOCH and SCHMIDT (1994, p.183).

The marginal density function $p(x_i|C)$ for the random variable X_i follows from (5.198) with (5.192)

$$p(x_i|C) = \sum_{x_1} p(x_1|C) \sum_{x_2} p(x_2|x_1, C) \ldots \sum_{x_{i-1}} p(x_{i-1}|x_{i-2}, C)$$
$$p(x_i|x_{i-1}, C) \sum_{x_{i+1}} p(x_{i+1}|x_i, C) \ldots \sum_{x_n} p(x_n|x_{n-1}, C) \ . \quad (5.199)$$

In contrast to (5.197) the prior density function, which will be called $\pi(x_i)$, in the marginal density function $p(x_i|C)$ can now be computed independently from the likelihood function $\lambda(x_i)$. We therefore obtain

$$p(x_i|C) = \alpha\pi(x_i)\lambda(x_i) \qquad (5.200)$$

with

$$\pi(x_1) \quad = p(x_1|C)$$
$$\pi(x_2) \quad = \sum_{x_1} \pi(x_1)p(x_2|x_1, C)$$
$$\pi(x_3) \quad = \sum_{x_2} \pi(x_2)p(x_3|x_2, C)$$
$$\cdots\cdots\cdots\cdots\cdots\cdots\cdots\cdots\cdots\cdots\cdots\cdots\cdots \qquad (5.201)$$
$$\pi(x_{i-1}) = \sum_{x_{i-2}} \pi(x_{i-2})p(x_{i-1}|x_{i-2}, C)$$
$$\pi(x_i) \quad = \sum_{x_{i-1}} \pi(x_{i-1})p(x_i|x_{i-1}, C)$$

and

$$\lambda(x_{n-1}) = \sum_{x_n} p(x_n|x_{n-1}, C)\lambda(x_n)$$

$$\cdots\cdots\cdots\cdots\cdots\cdots\cdots\cdots\cdots\cdots\cdots\cdots$$

$$
\begin{aligned}
\lambda(x_{i+1}) &= \sum_{x_{i+2}} p(x_{i+2}|x_{i+1}, C)\lambda(x_{i+2}) \\
\lambda(x_i) &= \sum_{x_{i+1}} p(x_{i+1}|x_i, C)\lambda(x_{i+1}) \ .
\end{aligned}
\tag{5.202}
$$

Again, α denotes the normalization constant in (5.200), since corresponding to (5.195) instantiated random variables shall be introduced. Beginning with the random variable X_1 the prior density function $\pi(x_i)$ can be computed for each random variable X_i with $i \in \{1, \ldots, n\}$. Starting from X_n the likelihood function $\lambda(x_i)$ is obtained for each random variable X_i with $i \in \{n-1, \ldots, 1\}$ after defining $\lambda(x_n)$ as follows.

If the leaf node X_n is not instantiated,

$$\lambda(x_n) = 1 \tag{5.203}$$

holds true for all values x_n, because the summation over the density function for X_n in (5.202) has to give the value one, as was already mentioned in connection with (5.192). If the leaf node is instantiated by

$$X_n = x_{n0} \ , \ x_{n0} \in \{x_{n1}, \ldots, x_{nm_n}\} \tag{5.204}$$

because of (5.184),

$$\lambda(x_{n0}) = 1 \quad \text{and} \quad \lambda(x_n) = 0 \quad \text{for the remaining values } x_n \tag{5.205}$$

must hold true so that from (5.202)

$$\lambda(x_{n-1}) = p(x_{n0}|x_{n-1}, C) \tag{5.206}$$

follows. If with $i < k < n$ the node X_k is instantiated by

$$X_k = x_{k0}, \ x_{k0} \in \{x_{k1}, \ldots, x_{km_k}\} \ , \tag{5.207}$$

we obtain corresponding to the derivation of (5.195)

$$\lambda(x_n) = 1$$

$$\cdots\cdots\cdots\cdots\cdots\cdots\cdots\cdots\cdots\cdots\cdots\cdots\cdots\cdots$$

$$
\begin{aligned}
\lambda(x_{k+1}) &= 1 \\
\lambda(x_{k0}) &= 1 \quad \text{and} \quad \lambda(x_k) = 0 \quad \text{for the remaining values } x_k \\
\lambda(x_{k-1}) &= p(x_{k0}|x_{k-1}, C) \ .
\end{aligned}
\tag{5.208}
$$

The computations for the node X_i shall be summarized (PEARL 1988, p.161). Let the node X_i be denoted by X, X_{i-1} as parent of X by U and X_{i+1} as child of X by Y. Instead of (5.200) we then obtain

$$p(x|C) = \alpha\pi(x)\lambda(x) \tag{5.209}$$

with the prior density function $\pi(x)$ from (5.201) which the parent U gives to its child X

$$\pi(x) = \sum_u \pi(u) p(x|u, C) \tag{5.210}$$

and with the likelihood function $\lambda(x)$ from (5.202) which the child Y reports to its parent X

$$\lambda(x) = \sum_y p(y|x, C) \lambda(y) \ . \tag{5.211}$$

To obtain from (5.209) the marginal density function $p(x|C)$ of each node X of the Bayesian network in form of a chain, the prior density function $\pi(y)$ which the node X gives to its child Y has to be computed from (5.210) and the likelihood function $\lambda(u)$ from (5.211) which the node X sends to parent U, thus

$$\begin{aligned} \pi(y) &= \sum_x p(y|x, C) \pi(x) \\ \lambda(u) &= \sum_x p(x|u, C) \lambda(x) \ . \end{aligned} \tag{5.212}$$

5.5.5 Bayesian Network in Form of a Tree

A further example of a singly connected Bayesian network, for which the prior density function can be given separately from the likelihood function for each node of the network, is the Bayesian network in form of a tree. Each node has several children but only one parent, as shown in Figure 5.9. The nodes X_1 to X_n possess the parent U_l which in turn has $(UU_l)_p$ as parent. The nodes Y_1 to Y_m are the children of X_i, $(YY_1)_1$ to $(YY_1)_{my_1}$ the children of Y_1 and $(YY_m)_1$ to $(YY_m)_{my_m}$ the children of Y_m. If the independencies are considered according to (2.117), the joint density function of the random variables of the Bayesian network follows with the chain rule (5.188) by

$$\begin{aligned} p(\ldots, &(uu_l)_p, \ldots, u_l, x_1, \ldots, x_i, \ldots, x_n, \ldots, y_1, \ldots, y_m, (yy_1)_1, \ldots, \\ &(yy_1)_{my_1}, \ldots, (yy_m)_1, \ldots, (yy_m)_{my_m}, \ldots | C) \\ &= \ldots p((uu_l)_p | \ldots) \ldots p(u_l|(uu_l)_p, C) p(x_1|u_l, C) \ldots \\ &\quad p(x_i|u_l, C) \ldots p(x_n|u_l, C) \ldots p(y_1|x_i, C) \ldots p(y_m|x_i, C) \\ &\quad p((yy_1)_1|y_1, C) \ldots p((yy_1)_{my_1}|y_1, C) \ldots \\ &\quad p((yy_m)_1|y_m, C) \ldots p((yy_m)_{my_m}|y_m, C) \ldots \end{aligned} \tag{5.213}$$

and with (5.195) the marginal density function $p(x_i|C)$ for the random variable X_i because of instantiated random variables by

$$p(x_i|C) = \alpha\Big[\ldots \sum_{(uu_l)_p} p((uu_l)_p|\ldots)\ldots \sum_{u_l} p(u_l|(uu_l)_p, C)$$

$$\sum_{x_1} p(x_1|u_l, C)\ldots p(x_i|u_l, C)\ldots \sum_{x_n} p(x_n|u_l, C)\ldots$$

$$\sum_{y_1} p(y_1|x_i, C)\ldots \sum_{y_m} p(y_m|x_i, C) \sum_{(yy_1)_1} p((yy_1)_1|y_1, C)\ldots$$

$$\sum_{(yy_1)_{my_1}} p((yy_1)_{my_1}|y_1, C)\ldots \sum_{(yy_m)_1} p((yy_m)_1|y_m, C)\ldots$$

$$\sum_{(yy_m)_{my_m}} p((yy_m)_{my_m}|y_m,, C)\ldots\Big]. \tag{5.214}$$

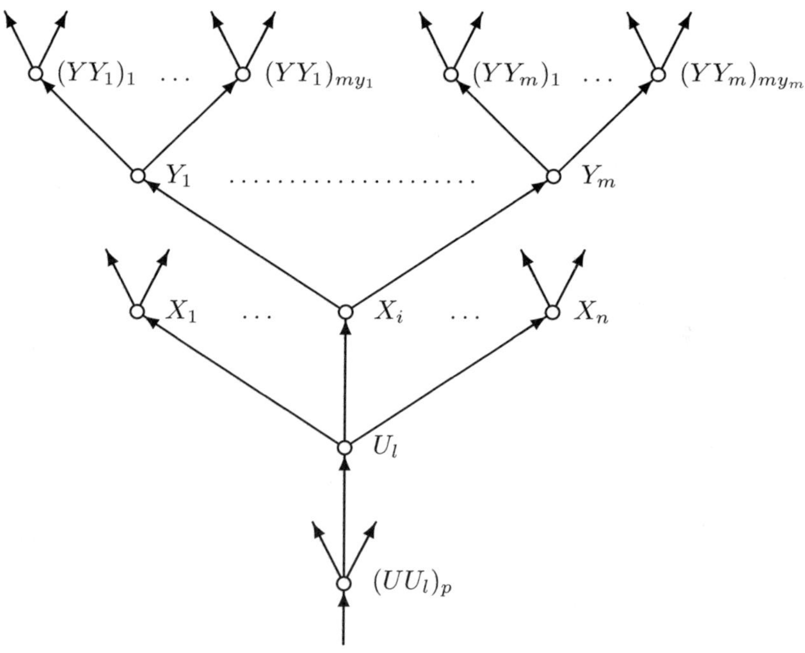

Figure 5.9: Bayesian Network as Tree

As in (5.200) the prior density function $\pi(x_i)$ in the marginal density function can be separately computed from the likelihood function $\lambda(x_i)$, thus

$$p(x_i|C) = \alpha\pi(x_i)\lambda(x_i) \tag{5.215}$$

with

$$\pi(x_i) = \sum_{u_l} p(x_i|u_l, C)\pi_{X_i}(u_l) \tag{5.216}$$

where $\pi_{X_i}(u_l)$ denotes the contribution to the prior density function which the parent U_l gives to the child X_i. We obtain with (5.214)

$$\pi_{X_i}(u_l) = \sum_{(uu_l)_p} \pi_{U_l}((uu_l)_p)p(u_l|(uu_l)_p, C) \sum_{x_1} p(x_1|u_l, C) \dots$$

$$\sum_{x_{i-1}} p(x_{i-1}|u_l, C) \sum_{x_{i+1}} p(x_{i+1}|u_l, C) \dots \sum_{x_n} p(x_n|u_l, C) \dots \tag{5.217}$$

and in addition

$$\lambda(x_i) = \prod_j \lambda_{Y_j}(x_i) \tag{5.218}$$

where $\lambda_{Y_j}(x_i)$ names the contribution to the likelihood function which the node Y_j delivers to the parent X_i. We obtain with (5.214) and (5.218)

$$\lambda_{Y_j}(x_i) = \sum_{y_j} p(y_j|x_i, C) \sum_{(yy_j)_1} p((yy_j)_1|y_j, C) \dots$$

$$\sum_{(yy_j)_{my_j}} p((yy_j)_{my_j}|y_j, C) \dots$$

$$= \sum_{y_j} p(y_j|x_i, C) \prod_k \lambda_{(YY_j)_k}(y_j)$$

$$= \sum_{y_j} p(y_j|x_i, C)\lambda(y_j) . \tag{5.219}$$

With this result and with (5.216) we finally get instead of (5.217)

$$\pi_{X_i}(u_l) = \sum_{(uu_l)_p} \pi_{U_l}((uu_l)_p)p(u_l|(uu_l)_p, C) \prod_{k \neq i} \lambda_{X_k}(u_l)$$

$$= \pi(u_l) \prod_{k \neq i} \lambda_{X_k}(u_l) . \tag{5.220}$$

The marginal density function $p(x_i|C)$ for the node X_i therefore follows from (5.215) together with (5.216), (5.218), (5.219) and (5.220). The contribution $\pi_{Y_j}(x_i)$ to the prior density function which the node X_i transmits to each of its children Y_j is computed from (5.220) by (PEARL 1988, p.169)

$$\pi_{Y_j}(x_i) = \pi(x_i) \prod_{k \neq j} \lambda_{Y_k}(x_i) \tag{5.221}$$

and the contribution $\lambda_{X_i}(u_l)$ from (5.219) to the likelihood function which X_i sends to the parent U_l by

$$\lambda_{X_i}(u_l) = \sum_{x_i} p(x_i|u_l, C)\lambda(x_i) . \tag{5.222}$$

If Y_k is a leaf node and not instantiated,

$$\lambda(y_k) = 1 \tag{5.223}$$

holds true for all values y_k as in (5.203). If Y_k is a leaf node or an arbitrary node and if it is instantiated by

$$Y_k = y_{k0} , \ y_{k0} \in \{y_{k1}, \ldots, y_{km_k}\} \tag{5.224}$$

from (5.184),

$$\lambda(y_{k0}) = 1 \quad \text{and} \quad \lambda(y_k) = 0 \quad \text{for the remaining values } y_k \tag{5.225}$$

result as in (5.205) or (5.208) and in case of an arbitrary node

$$\lambda((yy_k)_l) = 1 \quad \text{for all children } (YY_k)_l \text{ of } Y_k . \tag{5.226}$$

5.5.6 Bayesian Network in Form of a Polytreee

A last example for a singly connected Bayesian network is a Bayesian network as a polytree. Each node may have any number of parents and any number of children, but there is only one path from an arbitrary node to another one, see Figure 5.10. The node X possesses the parents U_1 to U_n. The node U_1 has besides X the children $(YU_1)_1$ to $(YU_1)_{ny_1}$ and U_n the children $(YU_n)_1$ to $(YU_n)_{ny_n}$ and so on, as shown in Figure 5.10. The joint density function of the random variables of the Bayesian network is given with

$$\begin{aligned}
p(\ldots, (uu_1)_1, &\ldots, (uu_1)_{nu_1}, \ldots, (uu_n)_1, \ldots, (uu_n)_{nu_n}, u_1, \ldots, u_n, \\
&(yu_1)_1, \ldots, (yu_1)_{ny_1}, \ldots, (yu_n)_1, \ldots, (yu_n)_{ny_n}, x, y_1, \ldots, y_m, \ldots, \\
&(uy_1)_1, \ldots, (uy_1)_{mu_1}, \ldots, (uy_m)_1, \ldots, (uy_m)_{mu_m}, \ldots, \\
&(yy_1)_1, \ldots, (yy_1)_{my_1}, \ldots, (yy_m)_1, \ldots, (yy_m)_{my_m}, \ldots | C) .
\end{aligned} \tag{5.227}$$

By applying the chain rule (5.188) with considering the independencies according to (2.117) and by computing the marginal density function $p(x|C)$

for the random variable X from (5.195) we obtain

$$p(x|C) = \alpha\Big[\cdots \sum_{(uu_1)_1} p((uu_1)_1|\ldots)\ldots \sum_{(uu_1)_{nu_1}} p((uu_1)_{nu_1}|\ldots)\ldots$$

$$\sum_{(uu_n)_1} p((uu_n)_1|\ldots)\ldots \sum_{(uu_n)_{nu_n}} p((uu_n)_{nu_n}|\ldots)$$

$$\sum_{u_1} p(u_1|(uu_1)_1,\ldots,(uu_1)_{nu_1},C)\ldots \sum_{u_n} p(u_n|(uu_n)_1,\ldots,(uu_n)_{nu_n},C)$$

$$\sum_{(yu_1)_1} p((yu_1)_1|u_1,\ldots,C)\ldots \sum_{(yu_1)_{ny_1}} p((yu_1)_{ny_1}|u_1,\ldots,C)\ldots$$

$$\sum_{(yu_n)_1} p((yu_n)_1|u_n,\ldots,C)\ldots \sum_{(yu_n)_{ny_n}} p((yu_n)_{ny_n}|u_n,\ldots,C)\ldots$$

$$p(x|u_1,\ldots,u_n,C) \sum_{y_1} p(y_1|x,(uy_1)_1,\ldots,(uy_1)_{mu_1},C)\ldots$$

$$\sum_{y_m} p(y_m|x,(uy_m)_1,\ldots,(uy_m)_{mu_m},C)\ldots$$

$$\sum_{(uy_1)_1} p((uy_1)_1|\ldots)\ldots \sum_{(uy_1)_{mu_1}} p((uy_1)_{mu_1}|\ldots)\ldots$$

$$\sum_{(uy_m)_1} p((uy_m)_1|\ldots)\ldots \sum_{(uy_m)_{mu_m}} p((uy_m)_{mu_m}|\ldots)\ldots$$

$$\sum_{(yy_1)_1} p((yy_1)_1|y_1,\ldots,C)\ldots \sum_{(yy_1)_{my_1}} p((yy_1)_{my_1}|y_1,\ldots,C)\ldots$$

$$\sum_{(yy_m)_1} p((yy_m)_1|y_m,\ldots,C)\ldots \sum_{(yy_m)_{my_m}} p((yy_m)_{my_m}|y_m,\ldots,C)\ldots\Big]\ .$$

$$(5.228)$$

The prior density function $\pi(x)$ and the likelihood function $\lambda(x)$ in the marginal density function may be separated again as in (5.215)

$$p(x|C) = \alpha\ \pi(x)\lambda(x) \tag{5.229}$$

with

$$\pi(x) = \sum_{u_1,\ldots,u_n} p(x|u_1,\ldots,u_n,C) \prod_{i=1}^{n} \pi_X(u_i) \tag{5.230}$$

where $\pi_X(u_i)$ denotes the contribution to the prior density function which the node U_i delivers to its child X. If one bears in mind that in (5.228) the density functions of the nodes are not given, which in Figure 5.10 are only indicated by arrows, one obtains

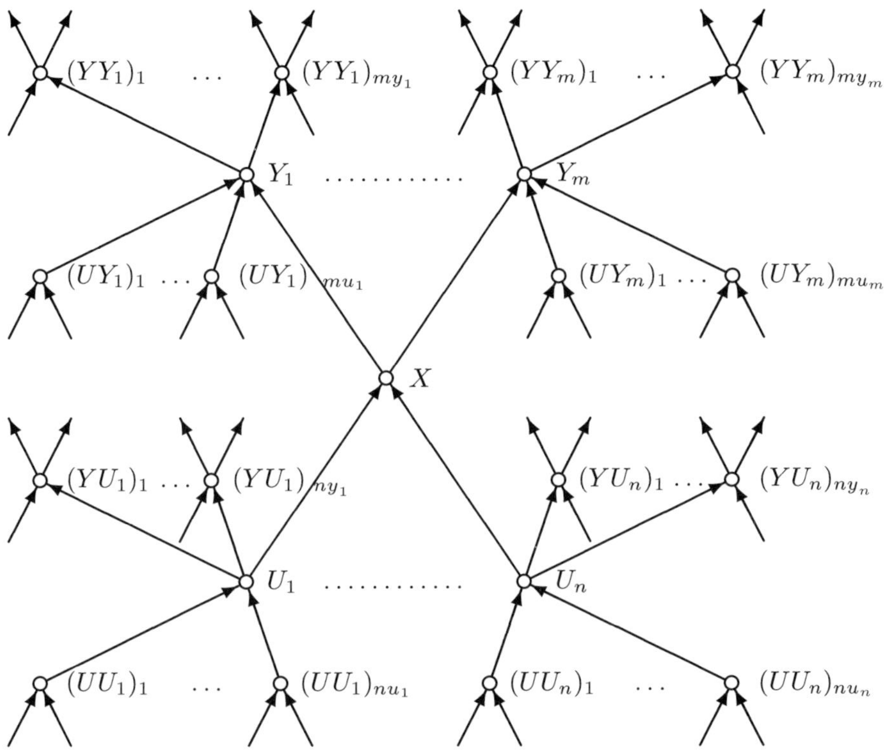

Figure 5.10: Bayesian Network as Polytree

$$\pi_X(u_i) = \prod_{l=1}^{ny_i} \Big[\sum_{(yu_i)_l} \prod_{\ldots} \ldots \sum_{\ldots} p((yu_i)_l|u_i,\ldots,C) \prod_{\ldots} \ldots \Big]$$

$$\sum_{(uu_i)_1,\ldots,(uu_i)_{nu_i}} p(u_i|(uu_i)_1,\ldots,(uu_i)_{nu_i},C)$$

$$\prod_{k=1}^{nu_i} \Big[\prod_{\ldots} \ldots \sum_{\ldots} p((uu_i)_k|\ldots) \prod_{\ldots} \ldots \Big] . \tag{5.231}$$

The likelihood function $\lambda(x)$ in (5.229) follows from

$$\lambda(x) = \prod_{j=1}^{m} \lambda_{Y_j}(x) \tag{5.232}$$

where $\lambda_{Y_j}(x)$ denotes the contribution which the node Y_j delivers to the parent X. We obtain with (5.228)

$$\lambda_{Y_j}(x) = \sum_{y_j} \prod_{l=1}^{my_j} \Big[\sum_{(yy_j)_l} \prod_{\cdots} \cdots \sum_{\cdots} p((yy_j)_l|y_j,\ldots,C) \prod_{\cdots} \cdots \Big]$$

$$\sum_{(uy_j)_1,\ldots,(uy_j)_{mu_j}} p(y_j|x,(uy_j)_1,\ldots,(uy_j)_{mu_j},C)$$

$$\prod_{k=1}^{mu_j} \Big[\prod_{\cdots} \cdots \sum_{\cdots} p((uy_j)_k|\cdots) \prod_{\cdots} \cdots \Big] \, . \tag{5.233}$$

The contribution $\pi_X(u_i)$ finally follows with (5.233) by

$$\pi_X(u_i) \;=\; \prod_{l=1}^{ny_i} \lambda_{(YU_{i)_l}}(u_i) \sum_{(uu_i)_1,\ldots,(uu_i)_{nu_i}}$$

$$p(u_i|(uu_i)_1,\ldots,(uu_i)_{nu_i},C) \prod_{k=1}^{nu_i} \pi_{U_i}(uu_i)_k \tag{5.234}$$

and $\lambda_{Y_j}(x)$ with (5.232) and (5.234) by

$$\lambda_{Y_j}(x) \;=\; \sum_{y_j} \lambda(y_j) \sum_{(uy_j)_1,\ldots,(uy_j)_{mu_j}}$$

$$p(y_j|x,(uy_j)_1,\ldots,(uy_j)_{mu_j},C) \prod_{k=1}^{mu_j} \pi_{Y_j}(uy_j)_k \, . \tag{5.235}$$

Thus, the marginal density function $p(x|C)$ for the random variable X is obtained from (5.229), (5.230), (5.232), (5.234) and (5.235). The contribution $\pi_{Y_j}(x)$ to the prior density function which the node X delivers to the child Y_j follows from (5.234) by

$$\pi_{Y_j}(x) = \prod_{l\neq j} \lambda_{Y_l}(x) \sum_{u_1,\ldots,u_n} p(x|u_1,\ldots,u_n,C) \prod_{k=1}^{n} \pi_X(u_k) \tag{5.236}$$

and the contribution $\lambda_X(u_i)$ which X sends to the parent U_i from (5.235) by

$$\lambda_X(u_i) = \sum_x \lambda(x) \sum_{u_k:k\neq i} p(x|u_1,\ldots,u_n,C) \prod_{k\neq i} \pi_X(u_k) \tag{5.237}$$

in agreement with PEARL (1988, p.183), who by the way normalizes the contributions $\pi_{Y_j}(x)$ and $\lambda_X(u_i)$. The values for $\lambda(y_k)$ of an instantiated or not instantiated leaf node Y_k or of an instantiated arbitrary node Y_k follow from (5.223) to (5.226).

Example: The marginal density function $p(x_2|C)$ for the random variable X_2 of the Bayesian network of Figure 5.11 shall be computed. We obtain from (5.229)

$$p(x_2|C) = \alpha \, \pi(x_2)\lambda(x_2)$$

and from (5.230)

$$\pi(x_2) = \sum_{u_2,u_3,u_4} p(x_2|u_2,u_3,u_4,C)\pi_{X_2}(u_2)\pi_{X_2}(u_3)\pi_{X_2}(u_4)$$

and from (5.232), (5.234) and (5.235)

$$\pi_{X_2}(u_2) = \lambda_{X_1}(u_2)p(u_2|C)$$

$$\lambda_{X_1}(u_2) = \sum_{x_1} \lambda(x_1) \sum_{u_1} p(x_1|u_1,u_2,C)\pi_{X_1}(u_1)$$

$$\pi_{X_1}(u_1) = p(u_1|C)$$

$$\pi_{X_2}(u_3) = \sum_{(uu_3)_1,(uu_3)_2} p(u_3|(uu_3)_1,(uu_3)_2,C)p((uu_3)_1|C)p((uu_3)_2|C)$$

$$\pi_{X_2}(u_4) = p(u_4|C) .$$

Thus, $\pi(x_2)$ finally follows from

$$\pi(x_2) = \sum_{u_2,u_3,u_4} p(u_2|C) \sum_{x_1} \lambda(x_1) \sum_{u_1} p(u_1|C)p(x_1|u_1,u_2,C)$$

$$\sum_{(uu_3)_1,(uu_3)_2} p((uu_3)_1|C)p((uu_3)_2|C)p(u_3|(uu_3)_1,(uu_3)_2,C)$$

$$p(u_4|C)p(x_2|u_2,u_3,u_4,C) .$$

The likelihood function $\lambda(x_2)$ is obtained with (5.232) by

$$\lambda(x_2) = \lambda_{Y_1}(x_2)\lambda_{Y_2}(x_2)\lambda_{Y_3}(x_2)$$

where with (5.235)

$$\lambda_{Y_1}(x_2) = \sum_{y_1} \prod_{l=1}^{2} \lambda_{(YY_1)_l}(y_1)p(y_1|x_2,C)$$

$$\lambda_{(YY_1)_1}(y_1) = \sum_{(yy_1)_1} \lambda((yy_1)_1)p((yy_1)_1|y_1,C)$$

$$\lambda_{(YY_1)_2}(y_1) = \sum_{(yy_1)_2} \lambda((yy_1)_2)p((yy_1)_2|y_1,C)$$

$$\lambda_{Y_2}(x_2) = \sum_{y_2} \lambda(y_2)p(y_2|x_2,C)$$

$$\lambda_{Y_3}(x_2) = \sum_{y_3} \lambda(y_3)p(y_3|x_2,C) ,$$

thus

$$\lambda(x_2) = \sum_{y_1,y_2,y_3} \sum_{(yy_1)_1,(yy_1)_2} \lambda((yy_1)_1)p((yy_1)_1|y_1,C)$$

$$\lambda((yy_1)_2)p((yy_1)_2|y_1,C)p(y_1|x_2,C)\lambda(y_2)p(y_2|x_2,C)\lambda(y_3)p(y_3|x_2,C) .$$

If the random variable $(YY_1)_1$ only is instantiated by

$$(YY_1)_1 = (yy_1)_{10} ,$$

summing the conditional density functions for the random variables $U_1, X_1,$ Y_2, Y_3 and $(YY_1)_2$ gives with (5.223) to (5.226) the values one, as was explained already for the derivation of (5.193). We obtain for $\pi(x_2)$ and $\lambda(x_2)$

$$\pi(x_2) = \sum_{u_2,u_3,u_4} p(u_2|C) \sum_{(uu_3)_1,(uu_3)_2} p((uu_3)_1|C)p((uu_3)_2|C)$$

$$p(u_3|(uu_3)_1, (uu_3)_2, C)p(u_4|C)p(x_2|u_2, u_3, u_4, C)$$

$$\lambda(x_2) = \sum_{y_1} p((yy_1)_{10}|y_1, C)p(y_1|x_2, C) .$$

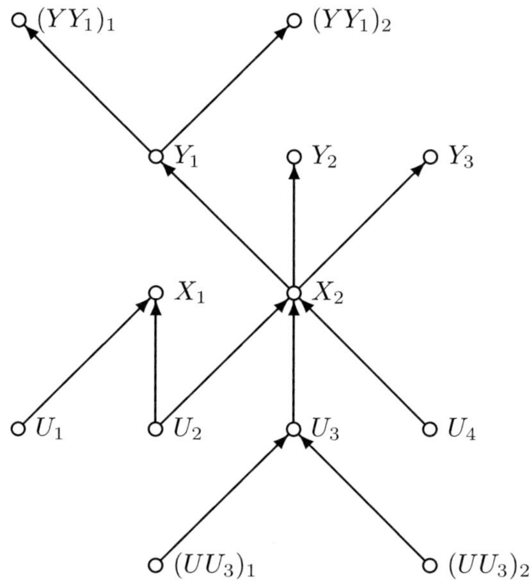

Figure 5.11: Example of a Bayesian Network as Polytree

To check this derivation, the joint density function for the random variables of the Bayesian network of Figure 5.11 is written down by applying the chain rule (5.188), the marginal density function for the random variable X_2 is then formed with (5.186) and the random variable $(YY_1)_1$ is instantiated. The marginal density function $p(x_2|C)$ obtained with (5.195) agrees with the result given above. $\qquad \triangle$

6 Numerical Methods

The posterior density function $p(\boldsymbol{x}|\boldsymbol{y}, C)$ for the continuous random vector $\boldsymbol{x}$ of the unknown parameters has to be integrated for the Bayes estimate from (3.10), for computing confidence regions with (3.35), for the test of hypotheses with (3.62), (3.71) or (3.74) and for determining marginal distributions with (3.5). Often, these integrals cannot be solved analytically so that numerical techniques have to be applied. Special methods of approximation exist, see for instance PRESS (1989, p.73), which however become inefficient, if the number of parameters is large. The stochastic method of Monte Carlo integration in Chapter 6.2 helps to overcome this deficiency. Random values have to be generated for the Monte Carlo integration from a distribution which approximates the posterior distribution for the unknown parameters.

Instead of integrals sums appear for discrete random variables, as already mentioned at the beginning of Chapter 3. The problem of integration therefore does not appear.

A different stochastic method is based on sampling from the posterior density function $p(\boldsymbol{x}|\boldsymbol{y}, C)$ itself. These random values contain all the information for the statistical inference on the unknown parameters. Bayes estimates, MAP estimates, covariance matrices, marginal distributions can be computed from these samples and hypotheses be tested. Markov Chain Monte Carlo methods of Chapter 6.3 provide the main tools to sample from the posterior distributions for the unknown parameters.

6.1 Generating Random Values

One distinguishes between generating *random numbers*, that are values of a random variable which possesses the uniform distribution (2.59) in the interval $[0, 1]$, and generating values of random variables with given distributions. Both mothods give *random values*, also called *random variates*.

6.1.1 Generating Random Numbers

To solve integrals by stochastic methods, very many random numbers are needed. They are generated by computers. Deterministic procedures are applied so that *pseudo random numbers* instead of true random numbers are obtained. However, one strives for generating pseudo random numbers with such properties that they can be used as true random numbers. One therefore talks of random numbers only.

Random numbers, i.e. values of a random variable X uniformly distributed in the interval $[0, 1]$, are often generated by the linear *congruential gen-*

erator

$$x_{i+1} = (ax_i + c)(\mathrm{mod}\, n) \quad \text{for} \quad i \in \{1, 2, \ldots\} \tag{6.1}$$

where the multiplicator a and the increment c denote nonnegative integers. The modulo notation $\mathrm{mod}\, n$ means that

$$x_{i+1} = ax_i + c - nl_i$$

holds true where $l_i = [(ax_i + c)/n]$ gives the largest positive integer in $(ax_i + c)/n$. Values in the interval $[0, 1]$ are obtained by x_i/n. Considerations for an appropriate choice of the constants in (6.1) and further methods are found, for example, in DAGPUNAR (1988, p.19), RIPLEY (1987, p.14) and RUBINSTEIN (1981, p.20).

If the random variable U with the values u has the uniform distribution in the interval $[0, 1]$, values x for the random variable X which according to (2.59) is uniformly distributed in the interval $[a, b]$ are obtained by the transformation

$$x = a + (b - a)u \,. \tag{6.2}$$

This result is obvious, but it is also proved in the Example 1 to the inversion method (6.5).

6.1.2   Inversion Method

An important procedure for generating random variates is the *inversion method*. It is based on generating random numbers.

Let X be a random variable with the distribution function $F(x)$ from (2.55). Since $F(x)$ is because of (2.58) a monotonically increasing function, its inverse function $F^{-1}(u)$ is defined for any value of u between 0 and 1 by the smallest value for x which fulfills $F(x) = u$, thus

$$F^{-1}(u) = \min\{x : F(x) = u \,, \ 0 \le u \le 1\} \,. \tag{6.3}$$

If the random variable U is uniformly distributed in the interval $[0, 1]$, the random variable

$$X = F^{-1}(U) \quad \text{has the distribution function} \quad F(x) \,. \tag{6.4}$$

This may be shown by the definition (2.52) of the distribution function, since for every $x \in \mathbb{R}$

$$P(X < x|C) = P(F^{-1}(U) < F^{-1}(u)|C) = P(U < u|C) = u = F(x)$$

holds true because of $F(u; 0, 1) = u$ from (2.60).

To generate a value x of the random variable X with the distribution function $F(x)$ by the inversion method, the following two steps are necessary:

1. a random number is generated, i.e. a value u of the random
   variable U uniformly distributed in the interval $[0, 1]$,   (6.5)
2. the value x follows from $x = F^{-1}(u)$.

Thus, the inversion method presents itself, if the distribution function $F(x)$ and its inverse function can be analytically given.

Example 1: Let values of the random variable X with the uniform distribution (2.59) be generated in the interval $[a, b]$. We obtain $(x - a)/(b - a) = u$ from the distribution function (2.60) and from (6.5) the relation

$$x = a + (b - a)u$$

which was already given in (6.2). $\triangle$

Example 2: Let values x of the random variable X having the Cauchy distribution (2.192) be generated. We obtain with the distribution function (2.193)

$$\frac{1}{\pi} \arctan\left(\frac{1}{\lambda}(x - \theta)\right) + \frac{1}{2} = u$$

and therefore

$$x - \theta = \lambda \tan\left(\pi(u - \frac{1}{2})\right)$$

or simplified, since with u also $u - \frac{1}{2}$ is uniformly distributed in the interval $[0, 1]$,

$$x - \theta = \lambda \tan(\pi u) .$$ (6.6)

$\triangle$

Additional examples like generating random values for the exponential distribution or the triangular distribution are found in DEVROYE (1986, p.29).

The inversion method is also well suited, to generate values of discrete random variables. Let $F(x_i)$ be the distribution function (2.50) of a discrete random variable X with the values x_i and $x_i \leq x_{i+1}$, $i \in \{1, \ldots, n\}$. The inverse function of $F(x_i)$ follows with (6.3) by

$$F^{-1}(u) = \min\{x_i : F(x_i) \geq u,\ 0 \leq u \leq 1\}$$ (6.7)

so that the following steps have to be taken when generating:

1. a random number u of the random variable U distributed
   uniformly in the interval $[0, 1]$ is generated and $i = 1$ is set,
2. as long as $F(x_i) < u$ is valid, i is replaced by $i + 1$,
3. else x_i is obtained .   (6.8)

To accelerate the steps of the computation, the random number x_i may not be searched sequentially, but the algorithm has to start at a better point than $i = 1$, see for instance RIPLEY (1987, p.71). The method of guide tables, for instance, accomplishes this (DEVROYE 1986, p.96).

The inversion method presented here for a discrete random variable with a univariate distribution is also valid for a discrete random vector with a multivariate distribution by applying a coding function which puts the vector into a one-to-one correspondance with the nonnegative integers of the real numbers (DEVROYE 1986, p.559; HÖRMANN et al. 2004, p.297).

6.1.3 Rejection Method

Random variates can be generated by the *rejection method* for a random variable with a density function which is not normalized. This happens, if the density function results from Bayes' theorem (2.122). Let X be the random variable and $\bar{p}(x)$ its density function which is not normalized and whose normalization constant follows from (2.131). For the application of the rejection method it is assumed that a density function $g(x)$ exists for which random numbers can be generated and for which

$$C \geq \bar{p}(x)/g(x) \quad \text{for every} \quad x \in \mathbb{R} \tag{6.9}$$

holds true where C means a constant with $C \geq 1$. Thus, $Cg(x)$ is the envelope of the density function $\bar{p}(x)$ which is not normalized. The rejection method is accomplished by the following steps, see for instance DAGPUNAR (1988, p.53), DEVROYE (1986, p.40), RUBINSTEIN (1981, p.45) and SMITH and GELFAND (1992):

1. a random value x for the random variable X with the density function $g(x)$ is generated,

2. a random number u for the random variable U uniformly distributed in the interval $[0, 1]$ is generated,

3. if $u < \bar{p}(x)/(Cg(x))$ holds true, x is a random value with the normalized density function $p(x)$, else the steps 1. to 3. have

   to be repeated. $\tag{6.10}$

The values x are predominantly accepted as random variates where $\bar{p}(x)$ approximates $Cg(x)$. The constant C should therefore be chosen subject to the constraint (6.9) such that it becomes minimal. A simple proof of the rejection method can be found in DAGPUNAR (1988, p.54). An example is given in Chapter 6.3.6.

The rejection method is not only valid for a random variable with a univariate distribution but also for a random vector with a multivariate distribution. If the bound C in (6.9) cannot be determined, there is an alternative to the rejection method, the sampling-importance-resampling (SIR) algorithm derived in Chapter 6.2.1.

6.1.4 Generating Values for Normally Distributed Random Variables

As explained in Chapter 2.4.1, the normal distribution is very often applied. It is therefore important to generate random variates for the normal distribution, and a number of techniques exist. A review is found, for instance, in DAGPUNAR (1988, p.93) and DEVROYE (1986, p.379). A method often applied goes back to BOX and MULLER (1958) with a modification of MARSAGLIA and BRAY (1964) for fast computations.

In general, values z for the random variable Z with the standard normal distribution $N(0,1)$, i.e. $Z \sim N(0,1)$, are generated. To obtain values x for the random variable X with the normal distribution $X \sim N(\mu, \sigma^2)$, the transformation (2.169) because of (2.202) is applied

$$x = \sigma z + \mu \,. \tag{6.11}$$

If random values for the $n \times 1$ random vector $\boldsymbol{x}$ with the multivariate normal distribution $\boldsymbol{x} \sim N(\boldsymbol{\mu}, \boldsymbol{\Sigma})$ shall be generated, where the covariance matrix $\boldsymbol{\Sigma}$ is assumed to be positive definite with (2.195) and (2.196), first n independent values for $Z \sim N(0,1)$ are generated and collected in the $n \times 1$ vector $\boldsymbol{z}$ which has the distribution $N(\boldsymbol{0}, \boldsymbol{I})$ because of (2.199). If the transformation is applied, which is based on the decomposition (3.38) of the covariance matrix $\boldsymbol{\Sigma}$ of the random vector $\boldsymbol{x}$ into its eigenvalues by $\boldsymbol{C}'\boldsymbol{\Sigma}\boldsymbol{C} = \boldsymbol{\Lambda}$, the $n \times 1$ vector $\boldsymbol{x}$ of random values for the random vector $\boldsymbol{x}$ follows from

$$\boldsymbol{x} = \boldsymbol{C}\boldsymbol{\Lambda}^{1/2}\boldsymbol{z} + \boldsymbol{\mu} \,. \tag{6.12}$$

The vector $\boldsymbol{x}$ has according to (2.202) because of $\boldsymbol{C}\boldsymbol{\Lambda}\boldsymbol{C}' = \boldsymbol{\Sigma}$ from (3.37) and (3.38) the distribution $\boldsymbol{x} \sim N(\boldsymbol{\mu}, \boldsymbol{\Sigma})$ we are looking for. The transformation (6.12) may be interpreted such that with $\boldsymbol{\Lambda}^{1/2}\boldsymbol{z}$ independent random variables with the variances λ_i are obtained and by the following transformation with $\boldsymbol{C}$ dependent random variables with the covariance matrix $\boldsymbol{\Sigma}$.

If a Cholesky factorization of the positive definite covariance matrix $\boldsymbol{\Sigma}$ by, say $\boldsymbol{\Sigma} = \boldsymbol{G}\boldsymbol{G}'$, is applied where $\boldsymbol{G}$ denotes a regular lower triangular matrix, see for instance KOCH (1999, p.30), the transformation is obtained

$$\boldsymbol{x} = \boldsymbol{G}\boldsymbol{z} + \boldsymbol{\mu} \,, \tag{6.13}$$

since the distribution $\boldsymbol{x} \sim N(\boldsymbol{\mu}, \boldsymbol{\Sigma})$, which we are looking for, follows with (2.202).

6.2 Monte Carlo Integration

The Monte Carlo integration is based on generating random variates with a density function which approximates the density function which needs to be integrated. The integral is then computed by a summation.

6.2.1   Importance Sampling and SIR Algorithm

Let $\boldsymbol{x}$ be a random vector of unknown parameters with $\boldsymbol{x} \in \mathcal{X}$, where $\mathcal{X}$ denotes the parameter space, and $p(\boldsymbol{x}|\boldsymbol{y},C)$ the posterior density function from Bayes' theorem (2.121). The integral I with

$$I = \int_{\mathcal{X}} g(\boldsymbol{x})p(\boldsymbol{x}|\boldsymbol{y},C)d\boldsymbol{x} = E_p(g(\boldsymbol{x}))_{\boldsymbol{x}\in\mathcal{X}} \tag{6.14}$$

needs to be computed where $g(\boldsymbol{x})$ denotes some function. The integral is equal to the expected value $E_p(g(\boldsymbol{x}))_{\boldsymbol{x}\in\mathcal{X}}$ of $g(\boldsymbol{x})$ computed by $p(\boldsymbol{x}|\boldsymbol{y},C)$ for $\boldsymbol{x} \in \mathcal{X}$. If $p(\boldsymbol{x}|\boldsymbol{y},C)$ is intractable, which means that random values cannot be generated from $p(\boldsymbol{x}|\boldsymbol{y},C)$, and if $u(\boldsymbol{x})$ is a tractable density function which approximates $p(\boldsymbol{x}|\boldsymbol{y},C)$, we rewrite the integral to obtain

$$\begin{aligned} I &= \int_{\mathcal{X}} g(\boldsymbol{x})(p(\boldsymbol{x}|\boldsymbol{y},C)/u(\boldsymbol{x}))u(\boldsymbol{x})d\boldsymbol{x} = E_u(g(\boldsymbol{x})(p(\boldsymbol{x}|\boldsymbol{y},C)/u(\boldsymbol{x})))_{\boldsymbol{x}\in\mathcal{X}} \\ &= E_p(g(\boldsymbol{x}))_{\boldsymbol{x}\in\mathcal{X}} \end{aligned} \tag{6.15}$$

where $E_u(g(\boldsymbol{x})(p(\boldsymbol{x}|\boldsymbol{y},C)/u(\boldsymbol{x})))_{\boldsymbol{x}\in\mathcal{X}}$ is the expected value of $g(\boldsymbol{x})(p(\boldsymbol{x}|\boldsymbol{y},C)/u(\boldsymbol{x}))$ computed by $u(\boldsymbol{x})$ for $\boldsymbol{x} \in \mathcal{X}$.

A sequence of m independent and identically with the density function $u(\boldsymbol{x})$ distributed random vectors $\boldsymbol{x}_i$, $i \in \{1,\ldots,m\}$ is now being generated. Because $u(\boldsymbol{x})$ approximates the density function $p(\boldsymbol{x}|\boldsymbol{y},C)$, the random values are generated at points which are important, that is at points where large values of $p(\boldsymbol{x}|\boldsymbol{y},C)$ are concentrated.

The estimate $\hat{I}$ of I from (6.15) follows as arithmetic mean, see for instance LEONARD and HSU (1999, p.275), by

$$\hat{I} = \frac{1}{m}\sum_{i=1}^{m} g(\boldsymbol{x}_i)(p(\boldsymbol{x}_i|\boldsymbol{y},C)/u(\boldsymbol{x}_i)) \; . \tag{6.16}$$

This is called the *importance sampling* of the Monte Carlo integration.

Let $g(\boldsymbol{x}_i)(p(\boldsymbol{x}_i|\boldsymbol{y},C)/u(\boldsymbol{x}_i))$ be independent and have equal variances σ^2

$$V(g(\boldsymbol{x}_i)(p(\boldsymbol{x}_i|\boldsymbol{y},C)/u(\boldsymbol{x}_i))) = \sigma^2 \quad \text{for} \quad i \in \{1,\ldots,m\} \; . \tag{6.17}$$

The variance $V(\hat{I})$ of the estimate $\hat{I}$ then follows with (2.158) by

$$V(\hat{I}) = \frac{1}{m}\sigma^2 \; . \tag{6.18}$$

If m goes to infinity, the variance goes to zero

$$V(\hat{I})_{m\to\infty} = 0 \tag{6.19}$$

and we obtain in the limit

$$\hat{I}_{m\to\infty} = I \; . \tag{6.20}$$

The number m of random variates depends on the variance $V(\hat{I})$ in (6.18), one wants to achieve. It also depends on, how well $p(\boldsymbol{x}|\boldsymbol{y}, C)$ is approximated by $u(\boldsymbol{x})$. If $u(\boldsymbol{x})$ is a poor choice, m needs to be large.

As shown in Chapter 4, one usually works because of Bayes' theorem (2.122) with a posterior density function which is not normalized. We will call it $\bar{p}(\boldsymbol{x}|\boldsymbol{y}, C)$, thus

$$p(\boldsymbol{x}|\boldsymbol{y}, C) = \bar{p}(\boldsymbol{x}|\boldsymbol{y}, C)/c \quad \text{with} \quad c = \int_{\mathcal{X}} \bar{p}(\boldsymbol{x}|\boldsymbol{y}, C)d\boldsymbol{x} \tag{6.21}$$

where c denotes the normalization constant (2.131). Its estimate $\hat{c}$ follows with $g(\boldsymbol{x}_i) = 1$ from (6.16) by

$$\hat{c} = \frac{1}{m} \sum_{i=1}^{m} \bar{p}(\boldsymbol{x}_i|\boldsymbol{y}, C)/u(\boldsymbol{x}_i) \tag{6.22}$$

and the estimate $\hat{I}$ of the integral I, if $\bar{p}(\boldsymbol{x}|\boldsymbol{y}, C)$ is applied, by

$$\hat{I} = \sum_{i=1}^{m} g(\boldsymbol{x}_i)(\bar{p}(\boldsymbol{x}_i|\boldsymbol{y}, C)/u(\boldsymbol{x}_i))/\sum_{i=1}^{m}(\bar{p}(\boldsymbol{x}_i|\boldsymbol{y}, C)/u(\boldsymbol{x}_i))$$

$$= \sum_{i=1}^{m} w_i\, g(\boldsymbol{x}_i) \tag{6.23}$$

with w_i

$$w_i = (\bar{p}(\boldsymbol{x}_i|\boldsymbol{y}, C)/u(\boldsymbol{x}_i))/\sum_{i=1}^{m}(\bar{p}(\boldsymbol{x}_i|\boldsymbol{y}, C)/u(\boldsymbol{x}_i)) \tag{6.24}$$

being the *importance weights*. Because of $\sum_{i=1}^{m} w_i = 1$ the estimate $\hat{I}$ can be interpreted as the weighted arithmetic mean (4.20) of $g(\boldsymbol{x}_i)$. For $m \to \infty$ we find in the limit with (6.14), (6.15) and (6.20) because of $g(\boldsymbol{x}) = 1$ in the intgral of (6.21)

$$\hat{I}_{m\to\infty} = E_u(g(\boldsymbol{x})(\bar{p}(\boldsymbol{x}|\boldsymbol{y}, C)/u(\boldsymbol{x})))_{\boldsymbol{x}\in\mathcal{X}}/E_u(1(\bar{p}(\boldsymbol{x}|\boldsymbol{y}, C)/u(\boldsymbol{x})))_{\boldsymbol{x}\in\mathcal{X}}$$

$$= E_{\bar{p}}(g(\boldsymbol{x}))_{\boldsymbol{x}\in\mathcal{X}}/E_{\bar{p}}(1)_{\boldsymbol{x}\in\mathcal{X}}$$

$$= \int_{\mathcal{X}} g(\boldsymbol{x})\bar{p}(\boldsymbol{x}|\boldsymbol{y}, C)d\boldsymbol{x}/\int_{\mathcal{X}} \bar{p}(\boldsymbol{x}|\boldsymbol{y}, C)d\boldsymbol{x} = \int_{\mathcal{X}} g(\boldsymbol{x})p(\boldsymbol{x}|\boldsymbol{y}, C)d\boldsymbol{x}$$

$$= I \tag{6.25}$$

If the prior density function $p(\boldsymbol{x}|C)$ in Bayes' theorem (2.122) gives a good approximation for the posterior density function $p(\boldsymbol{x}|\boldsymbol{y}, C)$ and if it is tractable, it can be used as approximate density function $u(\boldsymbol{x})$. The importance weights w_i then follow from (6.24) because of (2.122) by the likelihood function $p(\boldsymbol{y}|\boldsymbol{x},C)$ evaluated at the random variates $\boldsymbol{x}_i$ generated by the prior density function $p(\boldsymbol{x}|C)$. Thus,

$$w_i = p(\boldsymbol{y}|\boldsymbol{x}_i,C)/\sum_{i=1}^{m} p(\boldsymbol{y}|\boldsymbol{x}_i,C) \ . \tag{6.26}$$

To interpret the importance weights w_i in (6.24), we substitute $g(\boldsymbol{x})=1$ in (6.14) and introduce the restricted parameter space $\mathcal{X}_0 \subset \mathcal{X}$ with $\mathcal{X}_0 = \{x_1 \leq x_{10},\ldots,x_u \leq x_{u0}\}$ and its complement $\bar{\mathcal{X}}_0 = \mathcal{X}\backslash\mathcal{X}_0$. This splits the integral I into two parts, the first part becomes the distribution function $F(\boldsymbol{x} \in \mathcal{X}_0)$ for $p(\boldsymbol{x}|\boldsymbol{y},C)$ from (2.73) and the second part the rest, thus

$$I = F(\boldsymbol{x} \in \mathcal{X}_0) + \int_{\bar{\mathcal{X}}_0} p(\boldsymbol{x}|\boldsymbol{y},C)d\boldsymbol{x} \tag{6.27}$$

with

$$F(\boldsymbol{x} \in \mathcal{X}_0) \ = \ \int_{\mathcal{X}_0} p(\boldsymbol{x}|\boldsymbol{y},C)d\boldsymbol{x} = P(x_1 \leq x_{10},\ldots,x_u \leq x_{u0})$$

$$= \ E_p(1)_{\boldsymbol{x}\in\mathcal{X}_0} \tag{6.28}$$

The Monte Carlo estimate $\hat{F}(\boldsymbol{x} \in \mathcal{X}_0)$ follows with (6.23) by using again $\bar{p}(\boldsymbol{x}|\boldsymbol{y},C)$ instead of $p(\boldsymbol{x}|\boldsymbol{y},C)$ from

$$\hat{F}(\boldsymbol{x} \in \mathcal{X}_0) \ = \ \sum_{\boldsymbol{x}_i\in\mathcal{X}_0} (\bar{p}(\boldsymbol{x}_i|\boldsymbol{y},C)/u(\boldsymbol{x}_i))/\sum_{i=1}^{m}(\bar{p}(\boldsymbol{x}_i|\boldsymbol{y},C)/u(\boldsymbol{x}_i))$$

$$= \ \sum_{\boldsymbol{x}_i\in\mathcal{X}_0} w_i \tag{6.29}$$

where only the importance weights w_i for $\boldsymbol{x}_i \in \mathcal{X}_0$ are summed. For $m \to \infty$ we find with (6.25) in the limit

$$\hat{F}(\boldsymbol{x} \in \mathcal{X}_0)_{m\to\infty} \ = \ E_u(1(\bar{p}(\boldsymbol{x}|\boldsymbol{y},C)/u(\boldsymbol{x})))_{\boldsymbol{x}\in\mathcal{X}_0}/E_u(1(\bar{p}(\boldsymbol{x}|\boldsymbol{y},C)/u(\boldsymbol{x})))_{\boldsymbol{x}\in\mathcal{X}}$$

$$= \ E_{\bar{p}}(1)_{\boldsymbol{x}\in\mathcal{X}_0}/E_{\bar{p}}(1)_{\boldsymbol{x}\in\mathcal{X}}$$

$$= \ \int_{\mathcal{X}_0} \bar{p}(\boldsymbol{x}|\boldsymbol{y},C)d\boldsymbol{x}/\int_{\mathcal{X}} \bar{p}(\boldsymbol{x}|\boldsymbol{y},C)d\boldsymbol{x} = \int_{\mathcal{X}_0} p(\boldsymbol{x}|\boldsymbol{y},C)d\boldsymbol{x}$$

$$= \ F(\boldsymbol{x} \in \mathcal{X}_0) \tag{6.30}$$

The samples $\boldsymbol{x}_i$ for $\boldsymbol{x}_i \in \mathcal{X}$ can therefore be interpreted as values of a discrete distribution approximating $p(\boldsymbol{x}|\boldsymbol{y},C)$ with the importance weights w_i from (6.24) as probabilities, because the sum (6.29) of these probabilities for $\boldsymbol{x}_i \in \mathcal{X}_0$ gives according to (6.30) in the limit the distribution function for the posterior density function $p(\boldsymbol{x}|\boldsymbol{y},C)$.

If we sample from the posterior density function $\bar{p}(\boldsymbol{x}|\boldsymbol{y}, C)$ so that $u(\boldsymbol{x}) = \bar{p}(\boldsymbol{x}|\boldsymbol{y}, C)$, we obtain from (6.24) the constant weights $w_i = 1/m$ which are interpreted as constant probabilities for the samples $\boldsymbol{x}_i$. Summing over the probabilities $1/m$ for $g(\boldsymbol{x}_i) = 1$ in (6.23) gives $\hat{I} = 1$ in agreement with (2.74) that the integral over a density function is equal to one.

The interpretation of the samples $\boldsymbol{x}_i$ having probabilties w_i leads to the *sampling-importance-resampling* (SIR) algorithm of RUBIN (1988), see also SMITH and GELFAND (1992), as an alternative to the rejection method (6.10):

1. Draw let say M samples $\boldsymbol{x}_i$ for $\boldsymbol{x}_i \in \mathcal{X}$ from a tractable density function $u(\boldsymbol{x})$ approximating the target function $p(\boldsymbol{x}|\boldsymbol{y}, C)$.

2. Draw from these samples having probabilities equal to the importance weights w_i in (6.24) k samples $\boldsymbol{x}_i^*$ for $k < M$. They have the distribution $p(\boldsymbol{x}|\boldsymbol{y}, C)$ in case of $M \to \infty$. $\qquad$ (6.31)

Methods for sampling from a discrete distribution have been discussed in connection with (6.8). The samples $\boldsymbol{x}_i^*$ are drawn with replacemant, that is, they are not removed from the M samples. If there are only a few large weights and many small weights, GELMAN et al. (2004, p.316) suggest sampling without replacement to avoid that the same large weights will be picked up repeatedly, see also SKARE et al. (2003). If the posterior density function is well approximated by the prior density function, the importance weights w_i may be computed from (6.26), see SMITH and GELFAND (1992) and KOCH (2007).

The SIR algorithm has been applied for a recursive Bayesian filtering by GORDON and SALMOND (1995) who called their method bootstrap filter, see also DOUCET et al. (2000). The use of the Gibbs sampler together with the SIR algorithm is pointed out in Chapter 6.3.2.

6.2.2 Crude Monte Carlo Integration

The simplest approximation to the posterior density function $p(\boldsymbol{x}|\boldsymbol{y}, C)$ in (6.14) is given by the uniform distribution for $u(\boldsymbol{x})$ which in its univariate form is defined by (2.59). If the domain $\mathcal{X}$, where the values of the random vector $\boldsymbol{x}$ are defined, can be represented by parallels to the coordinate axes, the density function of the multivariate uniform distribution is obtained with $\boldsymbol{x} = (x_l)$ and $l \in \{1, \ldots, u\}$ by

$$u(\boldsymbol{x}) = \begin{cases} \prod_{l=1}^{u} [1/(b_l - a_l)] & \text{for} \quad a_l \leq x_l \leq b_l \\ 0 & \text{for} \quad x_l < a_l \quad \text{and} \quad x_l > b_l \,. \end{cases} \qquad (6.32)$$

The estimate $\hat{I}$ of the integral I of (6.14) then follows from (6.16) with

$$\hat{I} = \prod_{l=1}^{u} [(b_l - a_l)] \frac{1}{m} \sum_{i=1}^{m} g(\boldsymbol{x}_i) p(\boldsymbol{x}_i|\boldsymbol{y}, C) \,. \qquad (6.33)$$

This is the *crude Monte Carlo integration* for which the values $g(\boldsymbol{x}_i)p(\boldsymbol{x}_i|\boldsymbol{y},C)\prod_{l=1}^{u}(b_l-a_l)/m$ have to be summed. The number m of generated vectors must be large, because the uniform distribution is in general not a good approximation of any posterior density function $p(\boldsymbol{x}|\boldsymbol{y},C)$. The advantage of this technique is its simple application.

If the domain $\mathcal{X}$ cannot be expressed by parallels to the coordinate axes, the multivariate uniform distribution is obtained with the hypervolume $V_\mathcal{X}$ of the domain $\mathcal{X}$ by

$$u(\boldsymbol{x}) = 1/V_\mathcal{X} \tag{6.34}$$

and $\hat{I}$ by

$$\hat{I} = (V_\mathcal{X}/m)\sum_{i=1}^{m} g(\boldsymbol{x}_i)p(\boldsymbol{x}_i|\boldsymbol{y},C) \tag{6.35}$$

where the vectors $\boldsymbol{x}_i$ have to be generated in $\mathcal{X}$.

6.2.3   Computation of Estimates, Confidence Regions and Probabilities for Hypotheses

As already mentioned for (6.21), a posterior density function is because of Bayes' theorem (2.122) frequently applied which is not normalized and which follows from the right-hand side of (2.122). It will be denoted by $\bar{p}(\boldsymbol{x}|\boldsymbol{y},C)$, thus

$$\bar{p}(\boldsymbol{x}|\boldsymbol{y},C) = p(\boldsymbol{x}|C)p(\boldsymbol{y}|\boldsymbol{x},C) \ . \tag{6.36}$$

The normalization constant c for $\bar{p}(\boldsymbol{x}|\boldsymbol{y},C)$ is obtained with (2.129) and (2.131) by

$$p(\boldsymbol{x}|\boldsymbol{y},C) = \bar{p}(\boldsymbol{x}|\boldsymbol{y},C)/c \quad \text{from} \quad c = \int_\mathcal{X} \bar{p}(\boldsymbol{x}|\boldsymbol{y},C)d\boldsymbol{x} \ . \tag{6.37}$$

By importance sampling the estimate $\hat{c}$ is computed with (6.22) by

$$\hat{c} = \frac{1}{m}\sum_{i=1}^{m} \bar{p}(\boldsymbol{x}_i|\boldsymbol{y},C)/u(\boldsymbol{x}_i) \ . \tag{6.38}$$

The Bayes estimate $\hat{\boldsymbol{x}}_B$ of the random vector $\boldsymbol{x}$ of the unknown parameters is obtained from (3.10) with (6.37) by

$$\hat{\boldsymbol{x}}_B = E(\boldsymbol{x}|\boldsymbol{y}) = \frac{1}{c}\int_\mathcal{X} \boldsymbol{x}\bar{p}(\boldsymbol{x}|\boldsymbol{y},C)d\boldsymbol{x}$$

and the estimate by importance sampling with (6.23) and (6.38) by

$$\hat{\boldsymbol{x}}_B = \frac{1}{\hat{c}m}\sum_{i=1}^{m} \boldsymbol{x}_i\bar{p}(\boldsymbol{x}_i|\boldsymbol{y},C)/u(\boldsymbol{x}_i) = \sum_{i=1}^{m} w_i\boldsymbol{x}_i \ . \tag{6.39}$$

where w_i are the importance weights for $\boldsymbol{x}_i$ defined by (6.24). The MAP estimate $\hat{\boldsymbol{x}}_M$ of the unknown parameters $\boldsymbol{x}$ is computed from (3.30) by

$$\hat{\boldsymbol{x}}_M = \arg\max_{\boldsymbol{x}_i} \bar{p}(\boldsymbol{x}_i|\boldsymbol{y}, C) \; , \tag{6.40}$$

since the density functions $p(\boldsymbol{x}|\boldsymbol{y}, C)$ and $\bar{p}(\boldsymbol{x}|\boldsymbol{y}, C)$ become maximal at identical points.

The covariance matrix $D(\boldsymbol{x}|\boldsymbol{y})$ of the unknown parameters $\boldsymbol{x}$ is given with (6.37) by (3.11)

$$D(\boldsymbol{x}|\boldsymbol{y}) = \frac{1}{c} \int_{\mathcal{X}} (\boldsymbol{x} - E(\boldsymbol{x}|\boldsymbol{y}))(\boldsymbol{x} - E(\boldsymbol{x}|\boldsymbol{y}))' \bar{p}(\boldsymbol{x}|\boldsymbol{y}, C) d\boldsymbol{x} \; .$$

Its estimate by importance sampling follows from (6.23) with (6.38) and (6.39) by

$$\hat{D}(\boldsymbol{x}|\boldsymbol{y}) = \sum_{i=1}^{m} w_i(\boldsymbol{x}_i - \hat{\boldsymbol{x}}_B)(\boldsymbol{x}_i - \hat{\boldsymbol{x}}_B)' \; . \tag{6.41}$$

To determine the $1 - \alpha$ confidence region for the vector $\boldsymbol{x}$ of unknown parameters, the integral has to be solved according to (3.35) and (6.37)

$$\frac{1}{c} \int_{\mathcal{X}_B} \bar{p}(\boldsymbol{x}|\boldsymbol{y}, C) d\boldsymbol{x} = 1 - \alpha$$

with

$$\bar{p}(\boldsymbol{x}_1|\boldsymbol{y}, C) \geq \bar{p}(\boldsymbol{x}_2|\boldsymbol{y}, C) \quad \text{for} \quad \boldsymbol{x}_1 \in \mathcal{X}_B \; , \; \boldsymbol{x}_2 \notin \mathcal{X}_B \; .$$

To apply (6.23) with (6.38), the density values $\bar{p}(\boldsymbol{x}_i|\boldsymbol{y}, C)$ are sorted in decreasing order and the sequence $\bar{p}(\boldsymbol{x}_j|\boldsymbol{y}, C)$ for $j \in \{1, \ldots, m\}$ is formed. It leads to the sequence w_j for $j \in \{1, \ldots, m\}$ of the importance weights. By summing up to the index B for which

$$\sum_{j=1}^{B} w_j = 1 - \alpha \tag{6.42}$$

holds true, the point $\boldsymbol{x}_B$ at the boundary of the confidence region and its density value p_B from (3.41) is obtained with

$$p_B = \bar{p}(\boldsymbol{x}_B|\boldsymbol{y}, C)/\hat{c} \; . \tag{6.43}$$

The vectors $\boldsymbol{x}_j$ contain generated random values so that (6.42) can only be approximately fulfilled. If the vectors $\boldsymbol{x}_j$ of the space $\mathcal{X}$ are graphically depicted as points and if neighboring points are selected with smaller density values than p_B in (6.43) and with larger ones, the boundary of the confidence region is obtained by interpolating between these points.

To test the composite hypothesis (3.45), the ratio V of the integrals

$$V = \frac{\int_{\mathcal{X}_0} \bar{p}(\boldsymbol{x}|\boldsymbol{y}, C)d\boldsymbol{x}/\hat{c}}{\int_{\mathcal{X}_1} \bar{p}(\boldsymbol{x}|\boldsymbol{y}, C)d\boldsymbol{x}/\hat{c}}$$

has to be computed because of (3.62). Hence, m random vectors $\boldsymbol{x}_i$ are generated and as in (6.29) the importance weights w_i for $\boldsymbol{x}_i \in \mathcal{X}_0$ and $\boldsymbol{x}_i \in \mathcal{X}_1$, respectively, are summed

$$\hat{V} = \sum_{\boldsymbol{x}_i \in \mathcal{X}_0} w_i / \sum_{\boldsymbol{x}_i \in \mathcal{X}_1} w_i \; . \tag{6.44}$$

Correspondingly, the ratio of the integrals in (3.71) for the test of the composite hypothesis (3.45) and the integral in (3.74) for the test of the point null hypothesis have to be solved. To test a point null hypothesis by means of a confidence region according to (3.82), the density value p_B at the boundary of the confidence region is determined by (6.43).

6.2.4   Computation of Marginal Distributions

If estimates have to be computed, confidence regions to be established or hypotheses to be tested only for a subset of the unknown parameters $\boldsymbol{x}$, which is collected in the vector $\boldsymbol{x}_1$ with $\boldsymbol{x} = |\boldsymbol{x}_1', \boldsymbol{x}_2'|'$, the posterior marginal density function $p(\boldsymbol{x}_1|\boldsymbol{y}, C)$ is determined from (3.5) with

$$p(\boldsymbol{x}_1|\boldsymbol{y}, C) = \int_{\mathcal{X}_2} p(\boldsymbol{x}_1, \boldsymbol{x}_2|\boldsymbol{y}, C)d\boldsymbol{x}_2 \tag{6.45}$$

by a Monte Carlo integration. With random variates for $\boldsymbol{x}_1$, which have the marginal density function $p(\boldsymbol{x}_1|\boldsymbol{y}, C)$, the vector $\boldsymbol{x}_1$ is then estimated, confidence regions for $\boldsymbol{x}_1$ are computed or hypotheses for $\boldsymbol{x}_1$ are tested by the methods described in Chapter 6.2.3. Random variates for $\boldsymbol{x}_1$ have to be generated from a density function which approximates the marginal density function for $\boldsymbol{x}_1$. In addition, random variates for $\boldsymbol{x}_2$ with the density function $p(\boldsymbol{x}_1, \boldsymbol{x}_2|\boldsymbol{y}, C)$ are generated given the value for $\boldsymbol{x}_1$. This is the first method for determining marginal distributions which is presented in the following. If it is not possible to generate random variables for $\boldsymbol{x}_2$, a second method is being dealt with which is based on generating random variates for the entire vector $\boldsymbol{x} = |\boldsymbol{x}_1', \boldsymbol{x}_2'|'$. An approximate method which determines the marginal distribution for $\boldsymbol{x}_1$ by substituting estimates for $\boldsymbol{x}_2$ instead of generating random values for $\boldsymbol{x}_2$ has been proposed by KOCH (1990, p.58).

The first method for solving the integral (6.45) starts from the vector $\boldsymbol{x}_{1i}$ of random variates generated for the random vector $\boldsymbol{x}_1$. Since the density function $\bar{p}(\boldsymbol{x}_1, \boldsymbol{x}_2|\boldsymbol{y}, C)$ from (6.36) and the marginal density function $\bar{p}(\boldsymbol{x}_1|\boldsymbol{y}, C)$ are used which both are not normalized, the integral has to be solved

$$\bar{p}(\boldsymbol{x}_{1i}|\boldsymbol{y}, C) = \int_{\mathcal{X}_2} \bar{p}(\boldsymbol{x}_{1i}, \boldsymbol{x}_2|\boldsymbol{y}, C)d\boldsymbol{x}_2 \; . \tag{6.46}$$

To apply (6.16), l random values $\boldsymbol{x}_{2j}$ for the random vector $\boldsymbol{x}_2$ are generated by the density function $u(\boldsymbol{x}_2)$ which approximates the density function $\bar{p}(\boldsymbol{x}_{1i}, \boldsymbol{x}_2)$ for $\boldsymbol{x}_2$ given $\boldsymbol{x}_{1i}$. We obtain

$$\bar{p}(\boldsymbol{x}_{1i}|\boldsymbol{y}, C) = \frac{1}{l} \sum_{j=1}^{l} \bar{p}(\boldsymbol{x}_{1i}, \boldsymbol{x}_{2j}|\boldsymbol{y}, C)/u(\boldsymbol{x}_{2j}) \ . \tag{6.47}$$

For the random vector $\boldsymbol{x}_1$ altogether m values $\boldsymbol{x}_{1i}$ with the density function $u(\boldsymbol{x}_1)$ are generated which approximates the marginal density function $\bar{p}(\boldsymbol{x}_1|\boldsymbol{y}, C)$ not being normalized. The normalization constant c_1 for $\bar{p}(\boldsymbol{x}_1|\boldsymbol{y}, C)$ is corresponding to (6.37) defined with

$$p(\boldsymbol{x}_1|\boldsymbol{y}, C) = \bar{p}(\boldsymbol{x}_1|\boldsymbol{y}, C)/c_1 \quad \text{by} \quad c_1 = \int_{X_1} \bar{p}(\boldsymbol{x}_1|\boldsymbol{y}, C)d\boldsymbol{x}_1 \ . \tag{6.48}$$

Its estimate $\hat{c}_1$ is obtained as in (6.38) by

$$\hat{c}_1 = \frac{1}{m} \sum_{i=1}^{m} \bar{p}(\boldsymbol{x}_{1i}|\boldsymbol{y}, C)/u(\boldsymbol{x}_{1i}) \tag{6.49}$$

with $\bar{p}(\boldsymbol{x}_{1i}|\boldsymbol{y}, C)$ from (6.47). Thus, l values $\boldsymbol{x}_{2j}$ are generated for each of the m generated values $\boldsymbol{x}_{1i}$.

The Bayes estimate $\hat{\boldsymbol{x}}_{1B}$ of the vector $\boldsymbol{x}_1$ of the unknown parameters follows corresponding to (6.39) from

$$\hat{\boldsymbol{x}}_{1B} = \frac{1}{\hat{c}_1 m} \sum_{i=1}^{m} \boldsymbol{x}_{1i}\bar{p}(\boldsymbol{x}_{1i}|\boldsymbol{y}, C)/u(\boldsymbol{x}_{1i}) = \sum_{i=1}^{m} w_{1i}\boldsymbol{x}_{1i} \tag{6.50}$$

where w_{1i} denote the importance weights fom (6.24) for $\boldsymbol{x}_{1i}$. The MAP estimate $\hat{\boldsymbol{x}}_{1M}$ is obtained as in (6.40) from

$$\hat{\boldsymbol{x}}_{1M} = \arg\max_{\boldsymbol{x}_{1i}} \bar{p}(\boldsymbol{x}_{1i}|\boldsymbol{y}, C) \ . \tag{6.51}$$

The $1 - \alpha$ confidence region for the random vector $\boldsymbol{x}_1$ is determined by a summation according to (6.42) up to the index B, for which

$$\sum_{j=1}^{B} w_{1j} = 1 - \alpha \tag{6.52}$$

holds true. The sequence of the importance weights w_{1j} is formed by sorting the density values $\bar{p}(\boldsymbol{x}_{1i}|\boldsymbol{y}, C)$ in decreasing order. The density value p_B of a point at the boundary of the confidence region for $\boldsymbol{x}_1$ is obtained from (6.43) by

$$p_B = \bar{p}(\boldsymbol{x}_{1B}|\boldsymbol{y}, C)/\hat{c}_1 \ . \tag{6.53}$$

The ratio (3.62) of the two integrals for the test of the composite hypothesis (3.45) follows from (6.44) by

$$\sum_{\boldsymbol{x}_{1i}\in\mathcal{X}_{10}} w_{1i} \Big/ \sum_{\boldsymbol{x}_{1i}\in\mathcal{X}_{11}} w_{1i} \tag{6.54}$$

where $\mathcal{X}_{10}$ and $\mathcal{X}_{11}$ denote the domains over which $\boldsymbol{x}_1$ is integrated. The remaining integrals for testing hypotheses are correspondingly obtained. To test a point null hypothesis by means of a confidence region according to (3.82), the density value p_B at the boundary of the confidence region is determined by (6.53).

An example for computing a confidence region for the random vector $\boldsymbol{x}_1$ by means of (6.53) is given in the following Chapter 6.2.5.

For the second technique to determine the marginal distribution for $\boldsymbol{x}_1$, let again the posterior density function $\bar{p}(\boldsymbol{x}_1,\boldsymbol{x}_2|\boldsymbol{y},C)$ which is not normalized be available from (6.36). Its normalization constant c is determined by (6.37). Thus, the integral needs to be solved

$$p(\boldsymbol{x}_1|\boldsymbol{y},C) = \frac{1}{c}\int_{\mathcal{X}_2} \bar{p}(\boldsymbol{x}_1,\boldsymbol{x}_2|\boldsymbol{y},C)d\boldsymbol{x}_2 . \tag{6.55}$$

By means of the density function $u(\boldsymbol{x})$, which approximates $\bar{p}(\boldsymbol{x}_1,\boldsymbol{x}_2|\boldsymbol{y},C)$, m vectors $\boldsymbol{x}_i = |\boldsymbol{x}'_{1i},\boldsymbol{x}'_{2i}|'$ of random values are generated. As explained for the SIR algorithm (6.31), the samples $\boldsymbol{x}_i$ are interpreted as values of a discrete distribution having the importance weights w_i from (6.24) as probabilities. To obtain the marginal distribution according to (2.82), one has to sum the probabilities w_i over the values $\boldsymbol{x}_2$. However, one has to keep in mind that in the discrete density function $p(x_{1j_1},x_{2j_2},\ldots,x_{ij_i}|C)$ for each value x_{1j_1} also each value x_{2j_2}, each value x_{3j_3} and so on are given. This is not the case for the density values $w_i = \bar{p}(\boldsymbol{x}_{1i},\boldsymbol{x}_{2i}|\boldsymbol{y},C)/(\hat{c}mu(\boldsymbol{x}_i))$, since the vectors $\boldsymbol{x}_i = |\boldsymbol{x}'_{1i},\boldsymbol{x}'_{2i}|'$ originate from generating random variates. To sum over $\boldsymbol{x}_{2i}$, the space $\mathcal{X}_1$ with $\boldsymbol{x}_1 \in \mathcal{X}_1$ and $\boldsymbol{x}_2 \in \mathcal{X}_2$ is therefore devided by intervals on the coordinate axes into small subspaces $\Delta\mathcal{X}_{1j}$ with $j \in \{1,\ldots,J\}$, for instance, the plane into small squares by parallels to the coordinate axes. The density values w_i are then summed for all values $\boldsymbol{x}_{1i} \in \Delta\mathcal{X}_{1j}$ and $\boldsymbol{x}_{2i} \in \mathcal{X}_2$, in order to obtain values $p_d(\boldsymbol{x}_{1j}|\boldsymbol{y},C)$ for the discrete marginal density function

$$p_d(\boldsymbol{x}_{1j}|\boldsymbol{y},C) = \sum_{\boldsymbol{x}_{1i}\in\Delta\mathcal{X}_{1j},\boldsymbol{x}_{2i}\in\mathcal{X}_2} w_i \ , \ \ j \in \{1,\ldots,J\} \ . \tag{6.56}$$

Here, $\boldsymbol{x}_{1j}$ denotes a value which represents the space $\Delta\mathcal{X}_{1j}$. If I_j vectors $\boldsymbol{x}_{1i}$ are located in the space $\Delta\mathcal{X}_{1j}$, the value $\boldsymbol{x}_{1j}$ may be introduced as mean

$$\boldsymbol{x}_{1j} = \frac{1}{I_j}\sum_{i=1}^{I_j} \boldsymbol{x}_{1i} \ \ \text{with} \ \ \boldsymbol{x}_{1i} \in \Delta\mathcal{X}_{1j} \tag{6.57}$$

or $\boldsymbol{x}_{1j}$ defines the midpoint of $\Delta\mathcal{X}_{1j}$. The summation of the density values w_i gives according to (2.69) the probability $P(\boldsymbol{x}_{1j} \in \Delta\mathcal{X}_{1j}|\boldsymbol{y}, C)$ that $\boldsymbol{x}_{1j}$ lies in the space $\Delta\mathcal{X}_{1j}$ and therefore the discrete density function. The continuous marginal density function $p(\boldsymbol{x}_{1j}|\boldsymbol{y}, C)$ for $\boldsymbol{x}_{1j}$ follows because of (2.72) approximately by dividing the probability by the hypervolume $V_{\Delta\mathcal{X}_{1j}}$ of the space $\Delta\mathcal{X}_{1j}$

$$p(\boldsymbol{x}_{1j}|\boldsymbol{y}, C) = p_d(\boldsymbol{x}_{1j}|\boldsymbol{y}, C)/V_{\Delta\mathcal{X}_{1j}} \ . \tag{6.58}$$

The Bayes estimate $\hat{\boldsymbol{x}}_{1B}$ of $\boldsymbol{x}_1$ is obtained corresponding to (2.140) and (3.10) with (6.56) by

$$\hat{\boldsymbol{x}}_{1B} = \sum_{j=1}^{J} \boldsymbol{x}_{1j}p_d(\boldsymbol{x}_{1j}|\boldsymbol{y}, C) \ . \tag{6.59}$$

If the density values $p_d(\boldsymbol{x}_{1j}|\boldsymbol{y}, C)$ are arranged in decreasing order such that the series $p_d(\boldsymbol{x}_{1l}|\boldsymbol{y}, C)$ with $l \in \{1, \ldots, J\}$ is obtained, the index B for the point $\boldsymbol{x}_{1B}$ at the boundary of the confidence region for $\boldsymbol{x}_1$ follows from (6.42) with

$$\sum_{l=1}^{B} p_d(\boldsymbol{x}_{1l}|\boldsymbol{y}, C) = 1 - \alpha \ . \tag{6.60}$$

The density value p_B is found with the hypervolume $V_{\Delta\mathcal{X}_{1B}}$ of the subspace $\Delta\mathcal{X}_{1B}$ with the point $\boldsymbol{x}_{1B}$ from (6.58) by

$$p_B = p_d(\boldsymbol{x}_{1B}|\boldsymbol{y}, C)/V_{\Delta\mathcal{X}_{1B}} \ . \tag{6.61}$$

The ratio (3.62) of the two integrals for testing the composite hypothesis (3.45) follows from (6.44) with

$$\sum_{\boldsymbol{x}_{1j}\in\mathcal{X}_{10}} p_d(\boldsymbol{x}_{1j}|\boldsymbol{y}, C) \ / \sum_{\boldsymbol{x}_{1j}\in\mathcal{X}_{11}} p_d(\boldsymbol{x}_{1j}|\boldsymbol{y}, C) \tag{6.62}$$

where $\mathcal{X}_{10}$ and $\mathcal{X}_{11}$ denote the domains over which to integrate $\boldsymbol{x}_1$. Correspondingly, the remaining integrals to test hypotheses are obtained. To test a point null hypothesis by means of a confidence region according to (3.82), the density value p_B at the boundary of the confidence region is obtained by (6.61).

An example for determining with (6.61) the confidence region for the random vector $\boldsymbol{x}_1$ is given in the following chapter.

6.2.5 Confidence Region for Robust Estimation of Parameters as Example

Confidence regions will be determined from (3.35) for the unknown parameters $\boldsymbol{\beta}$ which are computed by the robust estimation of parameters presented in Chapter 4.2.5. Since the integration over the posterior density function $p(\boldsymbol{\beta}|\boldsymbol{y})$ for $\boldsymbol{\beta}$ from (4.62) could not be solved analytically, importance sampling (6.16) is applied. Thus, a distribution needs to be given which approximates the posterior density function $p(\boldsymbol{\beta}|\boldsymbol{y})$ from (4.62) and for which random values can be generated.

The density function $\bar{p}(\bar{e}_v|\boldsymbol{\beta})$ with $v \in \{1,\ldots,n\}$ in (4.62), which is not normalized, is determined by the right-hand sides of (4.56) and (4.57). If it is approximately assumed that the observations contain no outliers, the normal distribution (4.56) is valid, and one obtains instead of (4.62) with $\bar{e}_v$ from (4.60) and $\boldsymbol{P}$ from (4.58)

$$\prod_{v=1}^{n} \bar{p}(\bar{e}_v|\boldsymbol{\beta}) = \prod_{v=1}^{n} \exp(-\bar{e}_v^2/2)$$

$$= \exp\Big(-\frac{1}{2\sigma^2}\sum_{v=1}^{n} p_v(\boldsymbol{x}_v'\boldsymbol{\beta} - y_v)^2\Big)$$

$$= \exp\Big(-\frac{1}{2\sigma^2}(\boldsymbol{y} - \boldsymbol{X}\boldsymbol{\beta})'\boldsymbol{P}(\boldsymbol{y} - \boldsymbol{X}\boldsymbol{\beta})\Big)\,.$$

This posterior density function is identical with (4.11) so that the distribution for $\boldsymbol{\beta}$ follows from (4.14) and (4.29) with

$$\boldsymbol{\beta}|\boldsymbol{y} \sim N(\hat{\boldsymbol{\beta}}, \boldsymbol{\Sigma}) \tag{6.63}$$

and

$$\hat{\boldsymbol{\beta}} = (\boldsymbol{X}'\boldsymbol{P}\boldsymbol{X})^{-1}\boldsymbol{X}'\boldsymbol{P}\boldsymbol{y} \quad \text{and} \quad \boldsymbol{\Sigma} = \sigma^2(\boldsymbol{X}'\boldsymbol{P}\boldsymbol{X})^{-1}\,. \tag{6.64}$$

This normal distribution is an approximate distribution for the posterior distribution for $\boldsymbol{\beta}$ from (4.62).

Confidence regions for $\boldsymbol{\beta}$ are determined with respect to the estimate $\hat{\boldsymbol{\beta}}$. The transformation into the vector $\boldsymbol{\beta}_T$

$$\boldsymbol{\beta}_T = \boldsymbol{\beta} - \hat{\boldsymbol{\beta}} \tag{6.65}$$

is therefore applied. We then obtain with (2.202) instead of (6.63) the distribution

$$\boldsymbol{\beta}_T|\boldsymbol{y} \sim N(\boldsymbol{0}, \boldsymbol{\Sigma})\,. \tag{6.66}$$

A confidence region will not be established for all u parameters $\boldsymbol{\beta}_T$ but only for a subset of r parameters so that $\boldsymbol{\beta}_T$ is partitioned into the $r \times 1$ vector $\boldsymbol{\beta}_t$ and the $(u - r) \times 1$ vector $\boldsymbol{\beta}_q$ together with a corresponding partitioning of the $u \times u$ covariance matrix $\boldsymbol{\Sigma}$

$$\boldsymbol{\beta}_T = \left|\begin{array}{c} \boldsymbol{\beta}_t \\ \boldsymbol{\beta}_q \end{array}\right| \quad \text{and} \quad \boldsymbol{\Sigma} = \left|\begin{array}{cc} \boldsymbol{\Sigma}_{tt} & \boldsymbol{\Sigma}_{tq} \\ \boldsymbol{\Sigma}_{qt} & \boldsymbol{\Sigma}_{qq} \end{array}\right|\,. \tag{6.67}$$

A confidence region will be determined for the vector $\boldsymbol{\beta}_t$ of unknown parameters by the two methods presented in Chapter 6.2.4.

For the first method an approximate distribution for $\boldsymbol{\beta}_t$ is needed on the one hand. It follows as marginal distribution from (6.66) with (2.197) and (6.67) by

$$\boldsymbol{\beta}_t|\boldsymbol{y} \sim N(\boldsymbol{0}, \boldsymbol{\Sigma}_{tt}) \ . \tag{6.68}$$

On the other hand, an approximate distribution for $\boldsymbol{\beta}_q$ has to be specified for given values of $\boldsymbol{\beta}_t$. It is obtained as conditional normal distribution from (6.66) with (2.198). The rejection method, to be applied in Chapter 6.3.6, needs for generating random values for $\boldsymbol{\beta}_q$ the constants in the exponent of the conditional distribution, which is therefore derived in the following.

The coefficient matrix $\boldsymbol{X}$ in (6.64) is split up corresponding to the partitioning of $\boldsymbol{\beta}_t$ and $\boldsymbol{\beta}_q$ into

$$\boldsymbol{X} = |\boldsymbol{X}_t, \boldsymbol{X}_q| \ . \tag{6.69}$$

We then obtain for $\boldsymbol{X}'\boldsymbol{P}\boldsymbol{X}$ in (6.64)

$$\boldsymbol{X}'\boldsymbol{P}\boldsymbol{X} = \left| \begin{array}{cc} \boldsymbol{X}'_t\boldsymbol{P}\boldsymbol{X}_t & \boldsymbol{X}'_t\boldsymbol{P}\boldsymbol{X}_q \\ \boldsymbol{X}'_q\boldsymbol{P}\boldsymbol{X}_t & \boldsymbol{X}'_q\boldsymbol{P}\boldsymbol{X}_q \end{array} \right| \ . \tag{6.70}$$

The joint density function for $\boldsymbol{\beta}_t$ and $\boldsymbol{\beta}_q$ therefore follows from (6.66) with (2.195) by

$$p\left(\left| \begin{array}{c} \boldsymbol{\beta}_t \\ \boldsymbol{\beta}_q \end{array} \right| \Big| \boldsymbol{y} \right) \propto \exp\left(-\frac{1}{2\sigma^2}|\boldsymbol{\beta}'_t, \boldsymbol{\beta}'_q| \left| \begin{array}{cc} \boldsymbol{X}'_t\boldsymbol{P}\boldsymbol{X}_t & \boldsymbol{X}'_t\boldsymbol{P}\boldsymbol{X}_q \\ \boldsymbol{X}'_q\boldsymbol{P}\boldsymbol{X}_t & \boldsymbol{X}'_q\boldsymbol{P}\boldsymbol{X}_q \end{array} \right| \left| \begin{array}{c} \boldsymbol{\beta}_t \\ \boldsymbol{\beta}_q \end{array} \right| \right) \ . \tag{6.71}$$

We are looking for the density function $p(\boldsymbol{\beta}_q|\boldsymbol{\beta}_t, \boldsymbol{y})$ of $\boldsymbol{\beta}_q$ given the values for $\boldsymbol{\beta}_t$.

The exponent in (6.71) is rewritten

$$\boldsymbol{\beta}'_t\boldsymbol{X}'_t\boldsymbol{P}\boldsymbol{X}_t\boldsymbol{\beta}_t + 2\boldsymbol{\beta}'_q\boldsymbol{X}'_q\boldsymbol{P}\boldsymbol{X}_t\boldsymbol{\beta}_t + \boldsymbol{\beta}'_q\boldsymbol{X}'_q\boldsymbol{P}\boldsymbol{X}_q\boldsymbol{\beta}_q$$
$$= c_1 + (\boldsymbol{\beta}_q + \hat{\boldsymbol{\beta}}_q)'\boldsymbol{X}'_q\boldsymbol{P}\boldsymbol{X}_q(\boldsymbol{\beta}_q + \hat{\boldsymbol{\beta}}_q) + c_2 \tag{6.72}$$

with

$$\hat{\boldsymbol{\beta}}_q = (\boldsymbol{X}'_q\boldsymbol{P}\boldsymbol{X}_q)^{-1}\boldsymbol{X}'_q\boldsymbol{P}\boldsymbol{X}_t\boldsymbol{\beta}_t$$
$$c_1 = \boldsymbol{\beta}'_t\boldsymbol{X}'_t\boldsymbol{P}\boldsymbol{X}_t\boldsymbol{\beta}_t$$
$$c_2 = -\hat{\boldsymbol{\beta}}'_q\boldsymbol{X}'_q\boldsymbol{P}\boldsymbol{X}_q\hat{\boldsymbol{\beta}}_q \ . \tag{6.73}$$

If $\boldsymbol{\beta}_t$ is given, then $\hat{\boldsymbol{\beta}}_q$ is a constant vector and c_1 and c_2 are constants. Hence, the conditional density function $p(\boldsymbol{\beta}_q|\boldsymbol{\beta}_t, \boldsymbol{y})$ being looked for is determined by

$$p(\boldsymbol{\beta}_q|\boldsymbol{\beta}_t, \boldsymbol{y}) \propto \exp\left(-\frac{1}{2\sigma^2}(\boldsymbol{\beta}_q + \hat{\boldsymbol{\beta}}_q)'\boldsymbol{X}'_q\boldsymbol{P}\boldsymbol{X}_q(\boldsymbol{\beta}_q + \hat{\boldsymbol{\beta}}_q) \right) \tag{6.74}$$

so that the associated distribution follows by a comparison with (2.195) from

$$\boldsymbol{\beta}_q|\boldsymbol{\beta}_t, \boldsymbol{y} \sim N(-\hat{\boldsymbol{\beta}}_q, \sigma^2(\boldsymbol{X}_q'\boldsymbol{P}\boldsymbol{X}_q)^{-1}) \ . \tag{6.75}$$

This distribution is identical with the conditional distribution from (2.198), since the matrix identity (4.46) gives with (6.64), (6.67) and (6.70)

$$\frac{1}{\sigma^2}\left|\begin{array}{cc} \boldsymbol{X}_t'\boldsymbol{P}\boldsymbol{X}_t & \boldsymbol{X}_t'\boldsymbol{P}\boldsymbol{X}_q \\ \boldsymbol{X}_q'\boldsymbol{P}\boldsymbol{X}_t & \boldsymbol{X}_q'\boldsymbol{P}\boldsymbol{X}_q \end{array}\right| = \left|\begin{array}{cc} \boldsymbol{\Sigma}_{tt} & \boldsymbol{\Sigma}_{tq} \\ \boldsymbol{\Sigma}_{qt} & \boldsymbol{\Sigma}_{qq} \end{array}\right|^{-1}$$

$$= \left|\begin{array}{cc} \cdots & \cdots \\ -(\boldsymbol{\Sigma}_{qq} - \boldsymbol{\Sigma}_{qt}\boldsymbol{\Sigma}_{tt}^{-1}\boldsymbol{\Sigma}_{tq})^{-1}\boldsymbol{\Sigma}_{qt}\boldsymbol{\Sigma}_{tt}^{-1} & (\boldsymbol{\Sigma}_{qq} - \boldsymbol{\Sigma}_{qt}\boldsymbol{\Sigma}_{tt}^{-1}\boldsymbol{\Sigma}_{tq})^{-1} \end{array}\right| \ . \tag{6.76}$$

This result leads from (6.75) to

$$\boldsymbol{\beta}_q|\boldsymbol{\beta}_t, \boldsymbol{y} \sim N(\boldsymbol{\Sigma}_{qt}\boldsymbol{\Sigma}_{tt}^{-1}\boldsymbol{\beta}_t, \boldsymbol{\Sigma}_{qq} - \boldsymbol{\Sigma}_{qt}\boldsymbol{\Sigma}_{tt}^{-1}\boldsymbol{\Sigma}_{tq})$$

in agreement with (2.198).

To establish the confidence region for $\boldsymbol{\beta}_t$ by the first method of Chapter 6.2.4 with (6.53), m vectors $\boldsymbol{\beta}_{ti}$ of random values have to be generated for the $r \times 1$ random vector $\boldsymbol{\beta}_t$ with the approximate distribution (6.68). Hence, r independent random variates with the normal distribution $N(0,1)$ are generated by the technique mentioned in Chapter 6.1.4 and put into the $r \times 1$ vector $\boldsymbol{z}_i$. By the decomposition (3.38) of the covariance matrix $\boldsymbol{\Sigma}_{tt}$ from (6.68) into eigenvalues with

$$\boldsymbol{C}_t'\boldsymbol{\Sigma}_{tt}\boldsymbol{C}_t = \boldsymbol{\Lambda}_t \tag{6.77}$$

we obtain by the transformation (6.12)

$$\boldsymbol{\beta}_{ti} = \boldsymbol{C}_t\boldsymbol{\Lambda}_t^{1/2}\boldsymbol{z}_i \quad \text{for} \quad i \in \{1, \ldots, m\} \tag{6.78}$$

the m vectors $\boldsymbol{\beta}_{ti}$ of random values with the distribution (6.68) whose density function follows from (2.195) by

$$p(\boldsymbol{\beta}_{ti}|\boldsymbol{y}) \propto \exp\left(-\frac{1}{2}\boldsymbol{\beta}_{ti}'\boldsymbol{\Sigma}_{tt}^{-1}\boldsymbol{\beta}_{ti}\right) \tag{6.79}$$

with

$$\boldsymbol{\Sigma}_{tt}^{-1} = \boldsymbol{C}_t\boldsymbol{\Lambda}_t^{-1}\boldsymbol{C}_t' \tag{6.80}$$

from (3.37) and (6.77). The exponent in (6.79) is therefore obtained with (3.37) and (6.78) by

$$\boldsymbol{\beta}_{ti}'\boldsymbol{\Sigma}_{tt}^{-1}\boldsymbol{\beta}_{ti} = \boldsymbol{z}_i'\boldsymbol{\Lambda}_t^{1/2}\boldsymbol{C}_t'(\boldsymbol{C}_t\boldsymbol{\Lambda}_t^{-1}\boldsymbol{C}_t')\boldsymbol{C}_t\boldsymbol{\Lambda}_t^{1/2}\boldsymbol{z}_i$$
$$= \boldsymbol{z}_i'\boldsymbol{z}_i \ , \tag{6.81}$$

thus

$$p(\boldsymbol{\beta}_{ti}|\boldsymbol{y}) \propto \exp\left(-\frac{1}{2}\boldsymbol{z}_i'\boldsymbol{z}_i\right). \qquad (6.82)$$

For each value $\boldsymbol{\beta}_{ti}$ we have to generate l random variates $\boldsymbol{\beta}_{qj}$ for the $(u - r) \times 1$ random vector $\boldsymbol{\beta}_q$ with the distribution (6.75). Thus, $u - r$ random values with the normal distribution $N(0,1)$ are generated and collected in the $(u - r) \times 1$ vector $\boldsymbol{z}_j$. The decomposition (3.38) of the covariance matrix in (6.75) into its eigenvalues by

$$\sigma^2 \boldsymbol{C}_q'(\boldsymbol{X}_q'\boldsymbol{P}\boldsymbol{X}_q)^{-1}\boldsymbol{C}_q = \boldsymbol{\Lambda}_q \qquad (6.83)$$

leads with the transformation (6.12)

$$\boldsymbol{\beta}_{qj} = \boldsymbol{C}_q\boldsymbol{\Lambda}_q^{1/2}\boldsymbol{z}_j - \hat{\boldsymbol{\beta}}_q \quad \text{for} \quad j \in \{1,\ldots,l\} \qquad (6.84)$$

to the l vectors $\boldsymbol{\beta}_{qj}$ of random values with the distribution (6.75) whose density function follows from (2.195) with

$$p(\boldsymbol{\beta}_{qj}|\boldsymbol{y}) \propto \exp\left(-\frac{1}{2\sigma^2}(\boldsymbol{\beta}_{qj} + \hat{\boldsymbol{\beta}}_q)'\boldsymbol{X}_q'\boldsymbol{P}\boldsymbol{X}_q(\boldsymbol{\beta}_{qj} + \hat{\boldsymbol{\beta}}_q)\right). \qquad (6.85)$$

As in (6.80) to (6.82) we obtain

$$p(\boldsymbol{\beta}_{qj}|\boldsymbol{y}) \propto \exp\left(-\frac{1}{2}\boldsymbol{z}_j'\boldsymbol{z}_j\right). \qquad (6.86)$$

For each vector of random values for the vector $\boldsymbol{\beta}_T$ of unknown parameters, which is obtained with $\boldsymbol{\beta}_{ti}$ and $\boldsymbol{\beta}_{qj}$, the standardized errors $\bar{e}_v$ with $v \in \{1,\ldots,n\}$ are computed from (4.60) which lead to the posterior density function $p(\boldsymbol{\beta}|\boldsymbol{y})$ in (4.62). The transformation $\boldsymbol{\beta}_T = \boldsymbol{\beta} - \hat{\boldsymbol{\beta}}$ according to (6.65) corresponds to the transformation of the standardized error $\bar{e}_v$ into the transformed standardized error $\bar{e}_{Tv} = \bar{e}_v - \hat{\bar{e}}_v$. It is obtained because of (4.60), (4.70) and because of $\boldsymbol{\beta} = \boldsymbol{\beta}_T + \hat{\boldsymbol{\beta}}$ and $\hat{\boldsymbol{\beta}}_M = \hat{\boldsymbol{\beta}}$ by

$$\begin{aligned}
\bar{e}_{Tv} &= \bar{e}_v - \hat{\bar{e}}_v \\
&= \sqrt{p_v}(\boldsymbol{x}_v'|\boldsymbol{\beta}_{ti}',\boldsymbol{\beta}_{qj}'|' + \boldsymbol{x}_v'\hat{\boldsymbol{\beta}} - y_v)/\sigma - \sqrt{p_v}(\boldsymbol{x}_v'\hat{\boldsymbol{\beta}} - y_v)/\sigma
\end{aligned}$$

and finally by

$$\bar{e}_{Tv} = \sqrt{p_v}\boldsymbol{x}_v'|\boldsymbol{\beta}_{ti}',\boldsymbol{\beta}_{qj}'|'/\sigma \quad \text{for} \quad v \in \{1,\ldots,n\}. \qquad (6.87)$$

The transformation causes special observations y_v to be assumed for which $\boldsymbol{x}_v'\hat{\boldsymbol{\beta}} - y_v = 0$ is valid. The observations therefore do not enter into the computation of the posterior density function, they only serve to determine the midpoint of the confidence region by means of $\hat{\boldsymbol{\beta}}$.

The posterior density function $\bar{p}(\boldsymbol{\beta}_T|\boldsymbol{y})$ for the transformed vector $\boldsymbol{\beta}_T$ is obtained with the density functions $\bar{p}(\bar{e}_{Tk}|\boldsymbol{\beta})$ of the right-hand sides of (4.56) and (4.57) from (4.62) by

$$\bar{p}(\boldsymbol{\beta}_T|\boldsymbol{y}) = \prod_{v=1}^{n} \bar{p}(\bar{e}_{Tv}|\boldsymbol{\beta}_{ti}, \boldsymbol{\beta}_{qj}) \tag{6.88}$$

where the density functions are not normalized. The marginal density function $\bar{p}(\boldsymbol{\beta}_{ti}|\boldsymbol{y})$ for a vector $\boldsymbol{\beta}_{ti}$ of random values for $\boldsymbol{\beta}_t$, which is also not normalized, follows from (6.47) and (6.86) by

$$\bar{p}(\boldsymbol{\beta}_{ti}|\boldsymbol{y}) = \frac{1}{l} \sum_{j=1}^{l} \prod_{v=1}^{n} \bar{p}(\bar{e}_{Tv}|\boldsymbol{\beta}_{ti}, \boldsymbol{\beta}_{qj}) / \exp\left(-\frac{1}{2}\boldsymbol{z}_j'\boldsymbol{z}_j\right) \tag{6.89}$$

and the normalization constant $\hat{c}_1$ for $\bar{p}(\boldsymbol{\beta}_t|\boldsymbol{y})$ from (6.49) and (6.82) by

$$\hat{c}_1 = \frac{g}{m} \sum_{i=1}^{m} \bar{p}(\boldsymbol{\beta}_{ti}|\boldsymbol{y}) / \exp\left(-\frac{1}{2}\boldsymbol{z}_i'\boldsymbol{z}_i\right) \tag{6.90}$$

where g denotes the normalization constant of the normal distribution (6.68). It is computed with (2.195), (3.37), (6.77) and $\boldsymbol{\Lambda}_t = \mathrm{diag}(\lambda_1, \ldots, \lambda_r)$ by

$$g = (2\pi)^{r/2}(\det \boldsymbol{\Sigma}_{tt})^{1/2} = (2\pi)^{r/2} \det(\boldsymbol{\Lambda}_t \boldsymbol{C}_t' \boldsymbol{C}_t)^{1/2}$$

$$= (2\pi)^{r/2} \prod_{i=1}^{r} \lambda_i^{1/2} \ . \tag{6.91}$$

The density values $\bar{p}(\boldsymbol{\beta}_{ti}|\boldsymbol{y})$ are sorted in decreasing order and the values with the index p are obtained. They are then summed up to the index B for which according to (6.50) and (6.52)

$$\frac{g}{\hat{c}_1 m} \sum_{p=1}^{B} \bar{p}(\boldsymbol{\beta}_{tp}|\boldsymbol{y}) / \exp\left(-\frac{1}{2}\boldsymbol{z}_p'\boldsymbol{z}_p\right) = 1 - \alpha \tag{6.92}$$

is valid. The density value p_B of a point at the boundary of the $1-\alpha$ confidence region for the random vector $\boldsymbol{\beta}_t$ is then obtained from (6.53) with (6.90) by

$$p_B = \bar{p}(\boldsymbol{\beta}_{tB}|\boldsymbol{y})/\hat{c}_1 \ . \tag{6.93}$$

The vectors $\boldsymbol{\beta}_{ti}$ are graphically depicted as points. Neighboring points are selected with smaller density values than p_B from (6.93) and larger ones. The boundary of the confidence region for the vector $\boldsymbol{\beta}_t$ of the unknown parameters then follows by an interpolation.

Example 1: Let the polynomial model used by KOCH and YANG (1998A) be given. Examples of confidence regions for points of a simple network of distances are found in GUNDLICH (1998). The observation equations of the polynomial model corresponding to (4.59) are determined by

$$\beta_0 + x_v\beta_1 + x_v^2\beta_2 = y_v + e_v, \ p_v = 1, \ \sigma^2 = 1, \ v \in \{1,\ldots,6\} \qquad (6.94)$$

where β_0, β_1 and β_2 denote the three unknown parameters. The six values for the abszissa are

$$x_1 = 0.0, \ x_2 = 0.5, \ x_3 = 1.0, \ x_4 = 1.5, \ x_5 = 2.0, \ x_6 = 2.5 . \qquad (6.95)$$

By means of the density value p_B from (6.93) the confidence region with content $1 - \alpha = 95\%$ shall be determined for the two unknown parameters β_0 and β_1. To obtain a smooth boundary for the confidence region by the interpolation, very many random values were generated, that is $m = 20\,000$ for $\boldsymbol{\beta}_t$ with $\boldsymbol{\beta}_t = |\beta_0 - \hat{\beta}_0, \beta_1 - \hat{\beta}_1|'$ and $l = 10\,000$ for $\boldsymbol{\beta}_q$ with $\boldsymbol{\beta}_q = |\beta_3 - \hat{\beta}_3|$. The density value

$$p_B = 0.0066$$

was obtained and the confidence region with content 95% for β_0 and β_1 shown in Figure 6.1.

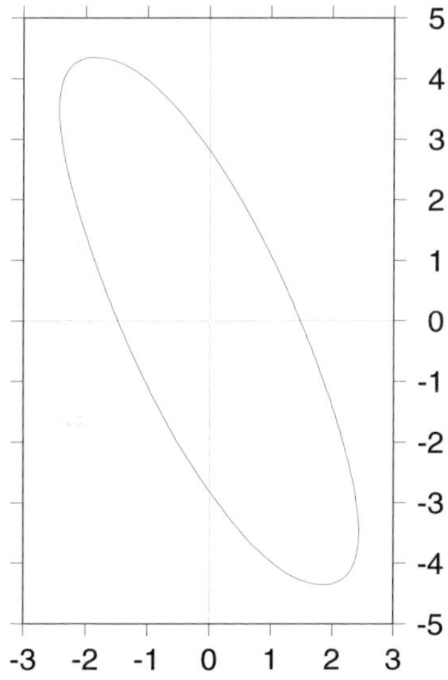

Figure 6.1: Confidence Region for β_0 and β_1 by the First Method of
Chapter 6.2.4

$\triangle$

The confidence region for $\boldsymbol{\beta}_t$ will be also determined with the second technique of Chapter 6.2.4 by (6.61). We generate m random values $\boldsymbol{\beta}_{Ti}$

with

$$\boldsymbol{\beta}_{Ti} = |\boldsymbol{\beta}'_{ti}, \boldsymbol{\beta}'_{qi}|' \tag{6.96}$$

for the $u \times 1$ random vector $\boldsymbol{\beta}_T$ transformed by (6.65) with the distribution (6.66). This means that u independent random variates with the normal distribution $N(0,1)$ are generated and collected in the $u \times 1$ vector $\boldsymbol{z}_i$. By the decomposition (3.38) of the covariance matrix $\boldsymbol{\Sigma}$ from (6.66) into its eigenvalues by

$$\boldsymbol{C}'\boldsymbol{\Sigma}\boldsymbol{C} = \boldsymbol{\Lambda} \tag{6.97}$$

the m vectors $\boldsymbol{\beta}_{Ti}$ of random values

$$\boldsymbol{\beta}_{Ti} = \boldsymbol{C}\boldsymbol{\Lambda}^{1/2}\boldsymbol{z}_i \quad \text{for} \quad i \in \{1,\ldots,m\} \tag{6.98}$$

are obtained by the transformation (6.12) with the distribution (6.66). As in (6.79) to (6.82)

$$p(\boldsymbol{\beta}_{Ti}|\boldsymbol{y}) \propto \exp\left(-\frac{1}{2}\boldsymbol{z}'_i\boldsymbol{z}_i\right) \tag{6.99}$$

follows. The transformed standardized errors $\bar{e}_{Tv}$ for $v \in \{1,\ldots,n\}$ are computed from (6.87) with $\boldsymbol{\beta}_{Ti}$ from (6.96) and (6.98). The density function $\bar{p}(\boldsymbol{\beta}_T|\boldsymbol{y})$, which is not normalized, for the transformed vector $\boldsymbol{\beta}_T$ is then obtained as in (6.88) by

$$\bar{p}(\boldsymbol{\beta}_T|\boldsymbol{y}) = \prod_{v=1}^{n} \bar{p}(\bar{e}_{Tv}|\boldsymbol{\beta}_{ti},\boldsymbol{\beta}_{qi}) \ . \tag{6.100}$$

The discrete marginal density function $p_d(\boldsymbol{\beta}_{tj}|\boldsymbol{y})$ for $\boldsymbol{\beta}_{tj}$ follows from (6.56) with (6.99) by

$$
\begin{aligned}
p_d(\boldsymbol{\beta}_{tj}|\boldsymbol{y}) \quad = \quad & \sum_{\boldsymbol{\beta}_{ti}\in\Delta\mathcal{X}_{1j},\boldsymbol{\beta}_{qi}\in\mathcal{X}_2} \prod_{v=1}^{n} \left(\bar{p}(\bar{e}_{Tv}|\boldsymbol{\beta}_{ti},\boldsymbol{\beta}_{qi})/\exp(-\tfrac{1}{2}\boldsymbol{z}'_i\boldsymbol{z}_i)\right)/ \\
& \sum_{i=1}^{m}\prod_{v=1}^{n} \left(\bar{p}(\bar{e}_{Tk}|\boldsymbol{\beta}_{ti},\boldsymbol{\beta}_{qi})/\exp(-\tfrac{1}{2}\boldsymbol{z}'_i\boldsymbol{z}_i)\right) \ , \\
& j \in \{1,\ldots,J\} \ .
\end{aligned}
\tag{6.101}
$$

Here, $\mathcal{X}_1$ and $\mathcal{X}_2$ denote with $\boldsymbol{\beta}_t \in \mathcal{X}_1$ and $\boldsymbol{\beta}_q \in \mathcal{X}_2$ the spaces where $\boldsymbol{\beta}_t$ and $\boldsymbol{\beta}_q$ are defined, $\Delta\mathcal{X}_{1j}$ with $j \in \{1,\ldots,J\}$ small subspaces into which $\mathcal{X}_1$ is divided by intervals on the coordinate axes and $\boldsymbol{\beta}_{tj}$ the midpoint of $\Delta\mathcal{X}_{1j}$.

The density values $p_d(\boldsymbol{\beta}_{tj}|\boldsymbol{y})$ from (6.101) are arranged in decreasing order so that the values $p_d(\boldsymbol{\beta}_{tp}|\boldsymbol{y})$ for $p \in \{1,\ldots,J\}$ are obtained. The index B for which according to (6.60)

$$\sum_{p=1}^{B} p_d(\boldsymbol{\beta}_{tp}|\boldsymbol{y}) = 1 - \alpha \tag{6.102}$$

is valid determines because of (6.61) the density value p_B for the point $\boldsymbol{\beta}_{tB}$ at the boundary of the $1 - \alpha$ confidence region for $\boldsymbol{\beta}_t$ by

$$p_B = p_d(\boldsymbol{\beta}_{tB}|\boldsymbol{y})/V_{\Delta\mathcal{X}_{1B}} \tag{6.103}$$

where $V_{\Delta\mathcal{X}_{1B}}$ denotes the hypervolume of the subspace $\Delta\mathcal{X}_{1B}$ with the point $\boldsymbol{\beta}_{tB}$. The boundary of the confidence region B follows again by interpolation like for the first method with p_B from (6.93).

Example 2: The confidence region of content $1 - \alpha = 95\%$ shall be determined again for the unknown parameters β_0 and β_1 in (6.94) of the Example 1 to (6.93). The large number $m = 20\,000\,000$ of random values was generated for $\boldsymbol{\beta}_T = |\beta_0 - \hat{\beta}_0, \beta_1 - \hat{\beta}_1, \beta_2 - \hat{\beta}_2|'$, in order to obtain a smooth boundary for the confidence region by the interpolation. With dividing the plane, where β_0 and β_1 are defined, into $20\,000$ surface elements the density value p_B from (6.103) was obtained by

$$p_B = 0.0064$$

in good agreement with the value for Example 1 to (6.93). Figure 6.2 shows the confidence region with content 95% for β_0 and β_1. It is almost identical with the confidence region of Figure 6.1.

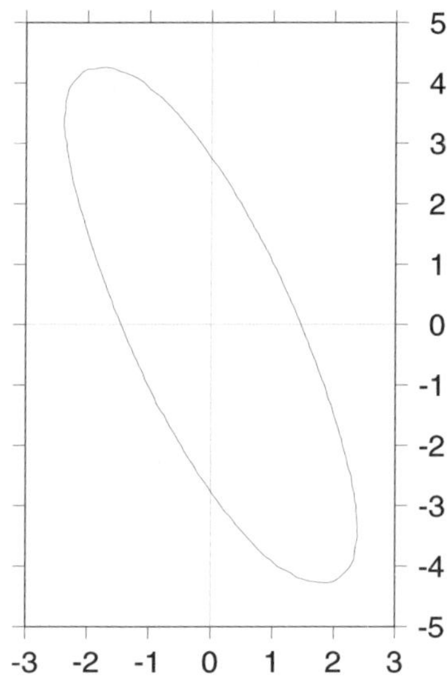

Figure 6.2: Confidence Region for β_0 and β_1 by the Second Method of Chapter 6.2.4

$\triangle$

6.3   Markov Chain Monte Carlo Methods

The numerical methods of the Monte Carlo integration rely on generating random samples from distributions which approximate the posterior density function $p(\boldsymbol{x}|\boldsymbol{y}, C)$ for the random vector $\boldsymbol{x}$ of unknown parameters from Bayes' theorem (2.122). Numerical methods are now covered which are based on generating random variates from the posterior density function $p(\boldsymbol{x}|\boldsymbol{y}, C)$ itself. For special cases $p(\boldsymbol{x}|\boldsymbol{y}, C)$ might be tractable so that random samples can be drawn. There are also cases, when $p(\boldsymbol{x}|\boldsymbol{y}, C)$ is intractable but results from the ratio of tractable density functions. Random variates can then be computed by the ratios of the random samples for the tractable density functions, as explained for the posterior density function (5.140). Furthermore, it is possible to generate samples from normally distributed observations and transform them by the estimates of the unknown parameters to obtain random variates for the parameters, see KOCH (2002) and ALKHATIB und SCHUH (2007).

General methods for generating random variates for the posterior density function $p(\boldsymbol{x}|\boldsymbol{y}, C)$ are given by the *Markov Chain Monte Carlo methods*. They simulate a Markov chain in the parameter space $\mathcal{X}$ for the unknown parameters $\boldsymbol{x}$ so that the limiting distribution of the chain is the target distribution, the distribution with the posterior density function $p(\boldsymbol{x}|\boldsymbol{y}, C)$. Random variates for $\boldsymbol{x}$ are generated from an approximate distribution and then moved towards a better approximation of the target distribution. The samples are drawn sequentially, and the distribution of one sample depends on the previous draw so that a Markov chain is formed. At each step of the simulation the approximate distribution is improved, until it converges to the target distribution. Two Markov Chain Monte Carlo methods will be presented, the Metropolis algorithm and the Gibbs sampler.

6.3.1   Metropolis Algorithm

The *Metropolis algorithm* was the first Markov Chain Monte Carlo method to be developed and goes back to METROPOLIS et al. (1953). It does not need a special distribution to sample from and can be applied for a posterior density function $\bar{p}(\boldsymbol{x}|\boldsymbol{y}, C)$ from (6.37) which is not normalized because ratios of density functions are computed, as will be seen in (6.104).

One samples a proposal $\boldsymbol{x}^*$ from a jumping or proposal distribution $p_t(\boldsymbol{x}^*|$ $\boldsymbol{x}^{t-1})$ for $t \in \{1, 2, \ldots\}$ with $\boldsymbol{x}^{t-1}$ being the previous generated vector. The jumping distribution has to be symmetric. This restriction is overcome by the Metropolis-Hastings procedure, see for instance GELMAN et al. (2004, p.291), which will not be considered here. Symmetry means that the probability of drawing $\bar{\boldsymbol{x}}$ from $\boldsymbol{x}$ is equal to the one of obtaining $\boldsymbol{x}$ from $\bar{\boldsymbol{x}}$. The ratio r of the density functions

$$r = \frac{p(\boldsymbol{x}^*|\boldsymbol{y})}{p(\boldsymbol{x}^{t-1}|\boldsymbol{y})} \tag{6.104}$$

is computed. One sets

$$
\boldsymbol{x}^t = \begin{cases} \boldsymbol{x}^* & \text{with probability } \min(r,1) \\ \boldsymbol{x}^{t-1} & \text{otherwise.} \end{cases} \tag{6.105}
$$

This means, if $r \geq 1$, the proposal $\boldsymbol{x}^*$ is accepted. If $r < 1$, a random number u for the random variable U is generated with (6.1) uniformly distributed in the interval $[0,1]$ and $\boldsymbol{x}^*$ is accepted, if $u < r$ since $P(U < u) = u$ from (2.60).

A simple algorithm which is frequently applied is the random-walk Metropolis, see for instance LIU (2001, p.114). The last generated vector $\boldsymbol{x}^{t-1}$ is perturbed by a random vector $\boldsymbol{\epsilon}^{t-1}$ to obtain the proposal $\boldsymbol{x}^* = \boldsymbol{x}^{t-1} + \boldsymbol{\epsilon}^{t-1}$. The components of $\boldsymbol{\epsilon}^{t-1}$ are assumed as being independent and identically distributed usually like the normal distribution which is symmetric. The vector $\boldsymbol{\epsilon}^{t-1}$ is then readily generated by (6.11).

The Metropolis algorithm is easy to apply. However, if the proposal distribution is far off the target distribution, the Metropolis algorithm becomes inefficient because of its slow convergence especially in higher dimensions. For the proof of convergence of the Metropolis algorithm to the target distribution see for instance GELMAN et al. (2004, p.290). A variant of the Metropolis algorithm is the technique of simulated annealing by KIRKPATRICK et al. (1983), although it is an optimization method instead of a simulation method. A scale parameter, called temperature, is introduced into the target distribution and gradually decreased to avoid that the annealing is trapped in local minima.

6.3.2   Gibbs Sampler

The *Gibbs sampler* was introduced by GEMAN and GEMAN (1984) for the Bayesian restoration of digital images and was then frequently used for different tasks of digital image analysis, see for instance GEMAN and MCCLURE (1987), GEMAN et al. (1987), KOCH and SCHMIDT (1994, p.310). After GELFAND and SMITH (1990) showed that the Gibbs sampler can be used for a variety of problems of Bayesian inference, see also SMITH and ROBERTS (1993), O'HAGAN (1994, p.225), GILKS (1996) and GELMAN et al. (2004, p.287), it became a frequently applied Markov Chain Monte Carlo method.

The Gibbs sampler decomposes the parameter space by sampling from the conditional distributions of the posterior distribution for each parameter x_k of the vector $\boldsymbol{x} = (x_k)$ of unknown parameters, thus diminishing the problem of high dimensions. The conditional density functions of the posterior density function $p(\boldsymbol{x}|\boldsymbol{y},C)$ are obtained with (2.102) by

$$
\begin{aligned}
& p(x_k|x_1,\ldots,x_{k-1},x_{k+1},\ldots,x_u,\boldsymbol{y},C) \\
& = \frac{p(x_1,\ldots,x_n|\boldsymbol{y},C)}{p(x_1,\ldots,x_{k-1},x_{k+1},\ldots,x_n|\boldsymbol{y},C)} \quad \text{for} \quad k \in \{1,\ldots,u\}\,.
\end{aligned} \tag{6.106}
$$

The posterior density function $p(\boldsymbol{x}|\boldsymbol{y}, C)$ is uniquely determined by these conditional density functions for x_k (BESAG 1974). Since only x_k is variable, while fixed values are assigned to the remaining components, the denominator on the right-hand side of (6.106) corresponds to a normalization constant, as a comparison with (6.37) shows. The conditional density function for x_k at the left-hand side of (6.106) is therefore found such that in the posterior density function $p(\boldsymbol{x}|\boldsymbol{y}, C)$ only the component x_k is considered being variable and that the appropriate normalization constant is introduced. The latter is not necessary, if the rejection method (6.10) is applied or the SIR algorithm (6.31) in case, the constant C in (6.9) cannot be determined. We therefore asssume that random values for x_k can be generated from the conditional density functions. An example of using the SIR algorithm for the Gibbs sampler is given by KOCH (2007), as already mentioned in Chapter 5.3.3.

The Gibbs sampler begins with arbitrary starting values

$$x_1^{(0)}, \ldots, x_u^{(0)} . \tag{6.107}$$

Then random values are sequentially drawn for x_k with $k \in \{1, \ldots, u\}$ from (6.106) to complete one iteration. For the qth iteration we generate

$$
\begin{aligned}
x_1^{(q)} \quad &\text{from} \quad p(x_1|x_2^{(q-1)}, \ldots, x_u^{(q-1)}, \boldsymbol{y}, C) \\
x_2^{(q)} \quad &\text{from} \quad p(x_2|x_1^{(q)}, x_3^{(q-1)}, \ldots, x_u^{(q-1)}, \boldsymbol{y}, C) \\
x_3^{(q)} \quad &\text{from} \quad p(x_3|x_1^{(q)}, x_2^{(q)}, x_4^{(q-1)}, \ldots, x_u^{(q-1)}, \boldsymbol{y}, C) \\
&\quad \cdots\cdots\cdots\cdots\cdots\cdots\cdots\cdots\cdots\cdots\cdots\cdots\cdots\cdots\cdots \\
x_u^{(q)} \quad &\text{from} \quad p(x_u|x_1^{(q)}, \ldots, x_{u-1}^{(q)}, \boldsymbol{y}, C) .
\end{aligned}
\tag{6.108}
$$

The sequence of random variates generated by the iterations forms a Markov chain. It will assumed that after o iterations convergence is reached so that the distribution of the generated random vector with values $x_1^{(o)}, \ldots, x_u^{(o)}$ is close enough to the target distribution. In the limit $o \to \infty$ it is the target distribution with density function $p(\boldsymbol{x}|\boldsymbol{y}, C)$ (GEMAN and GEMAN 1984). Conditions for the convergence are found, for instance, in ROBERTS and SMITH (1994). The process with o iterations is repeated with new starting values m times so that m random variates $\boldsymbol{x}_i$ generated for $\boldsymbol{x}$ are obtained which will be denoted by

$$\boldsymbol{x}_i = |x_{1i}, \ldots, x_{ui}|' \quad \text{with} \quad i \in \{1, \ldots, m\} . \tag{6.109}$$

One frequently generates only one Markov chain and discards during a burn-in phase of o iterations, until convergence is reached, all random samples. Afterwards, the random variates of only each sth iteration are collected to obtain the sample $\boldsymbol{x}_i$ in (6.109), because subsequent samples are correlated. The value of s, of course, depends on the correlation between the samples which in turn depends on the correlation of the unknown parameters.

In case of strong correlations it is helpful to use a *grouping*, also called *blocking* technique (LIU 2001, p.130). The vector of unknown parameters is grouped into subvectors, for instance $\boldsymbol{x} = |\boldsymbol{x}_1', \ldots, \boldsymbol{x}_r'|'$ where strongly correlated components of $\boldsymbol{x}$ are clustred in the same subvector. Sampling from the conditional density functions for the subvectors $\boldsymbol{x}_k$ for $k \in \{1, \ldots, r\}$ leads instead of (6.108) to the following algorithm: draw for the qth iteration

$$
\begin{aligned}
\boldsymbol{x}_1^{(q)} \quad &\text{from} \quad p(\boldsymbol{x}_1 | \boldsymbol{x}_2^{(q-1)}, \ldots, \boldsymbol{x}_r^{(q-1)}, \boldsymbol{y}, C) \\
\boldsymbol{x}_2^{(q)} \quad &\text{from} \quad p(\boldsymbol{x}_2 | \boldsymbol{x}_1^{(q)}, \boldsymbol{x}_3^{(q-1)}, \ldots, \boldsymbol{x}_r^{(q-1)}, \boldsymbol{y}, C) \\
\boldsymbol{x}_3^{(q)} \quad &\text{from} \quad p(\boldsymbol{x}_3 | \boldsymbol{x}_1^{(q)}, \boldsymbol{x}_2^{(q)}, \boldsymbol{x}_4^{(q-1)}, \ldots, \boldsymbol{x}_r^{(q-1)}, \boldsymbol{y}, C) \\
&\qquad\qquad \ldots\ldots\ldots\ldots\ldots\ldots\ldots\ldots\ldots\ldots\ldots\ldots\ldots\ldots \\
\boldsymbol{x}_r^{(q)} \quad &\text{from} \quad p(\boldsymbol{x}_r | \boldsymbol{x}_1^{(q)}, \ldots, \boldsymbol{x}_{r-1}^{(q)}, \boldsymbol{y}, C) \ .
\end{aligned}
\tag{6.110}
$$

After discarding the random variates of the first o iterations either in m parallel runs or after collecting the random variates of each sth iteration in a single run, the m random variates $\boldsymbol{x}_i$ are obtained

$$
\boldsymbol{x}_i = |\boldsymbol{x}_{1i}', \ldots, \boldsymbol{x}_{ri}'|' \quad \text{with} \quad i \in \{1, \ldots, m\} \ .
\tag{6.111}
$$

An example for the Gibbs sampler (6.108) is given in Chapter 6.3.6 and for the Gibbs sampler (6.110) in Chapter 6.3.5.

6.3.3 Computation of Estimates, Confidence Regions and Probabilities for Hypotheses

The random variates x_{ki} with $i \in \{1, \ldots, m\}$, $k \in \{1, \ldots, u\}$ generated for the component x_k of the random vector $\boldsymbol{x}$ of unknown parameters have the marginal density function $p(x_k | \boldsymbol{y}, C)$ of the joint density function $p(\boldsymbol{x} | \boldsymbol{y}, C)$. The mean of these random values therefore gives an estimate of the expected value $E(x_k | \boldsymbol{y})$ of x_k. Thus, the Bayes estimate $\hat{x}_{kB}$ of x_k follows with (3.9) from

$$
\hat{x}_{kB} = \frac{1}{m} \sum_{i=1}^{m} x_{ki}
\tag{6.112}
$$

and the Bayes estimate $\hat{\boldsymbol{x}}_B$ of the entire vector $\boldsymbol{x}$ of unknown parameters with (6.109) from

$$
\hat{\boldsymbol{x}}_B = \frac{1}{m} \sum_{i=1}^{m} \boldsymbol{x}_i \ .
\tag{6.113}
$$

The MAP estimate $\hat{\boldsymbol{x}}_M$ of $\boldsymbol{x}$ is obtained from (3.30) by

$$
\hat{\boldsymbol{x}}_M = \arg\max_{\boldsymbol{x}_i} p(\boldsymbol{x}_i | \boldsymbol{y}, C) \ .
\tag{6.114}
$$

The result (6.113) may be also derived as Bayes estimate (6.39) of a Monte Carlo integration. If the random values $\boldsymbol{x}_i$ in (6.39) which have the density function $u(\boldsymbol{x}_i)$ are generated by the Gibbs sampler, the importance weights w_i in (6.24) take on the values

$$w_i = 1/m \tag{6.115}$$

so that (6.113) follows immediately from (6.39).

The estimate of the covariance matrix $D(\boldsymbol{x}|\boldsymbol{y})$ is obtained from (6.41) with (6.115) by

$$\hat{D}(\boldsymbol{x}|\boldsymbol{y}) = \frac{1}{m} \sum_{i=1}^{m} (\boldsymbol{x}_i - \hat{\boldsymbol{x}}_B)(\boldsymbol{x}_i - \hat{\boldsymbol{x}}_B)' \ . \tag{6.116}$$

The index B for a point at the boundary of the $1 - \alpha$ confidence region for $\boldsymbol{x}$ is determined by (6.42) with (6.115). If the density values $p(\boldsymbol{x}_i|\boldsymbol{y}, C)$ with $\boldsymbol{x}_i$ from (6.109) are sorted in decreasing order so that the sequence $p(\boldsymbol{x}_j|\boldsymbol{y}, C)$ with $j \in \{1, \ldots, m\}$ is obtained, the index B follows from (6.42) with (6.115) by

$$B = m(1 - \alpha) \tag{6.117}$$

and the density value p_B for the point $\boldsymbol{x}_B$ at the boundary of the confidence region from (3.41) by

$$p_B = p(\boldsymbol{x}_B|\boldsymbol{y}, C) \ . \tag{6.118}$$

The estimate $\hat{V}$ of the ratio of the integrals for testing the composite hypothesis (3.45) is obtained from (6.44) with (6.115) by

$$\hat{V} = n_0/n_1 \tag{6.119}$$

where n_0 denotes the number of $\boldsymbol{x}_i \in \mathcal{X}_0$ and n_1 the number of $\boldsymbol{x}_i \in \mathcal{X}_1$. To test a point null hypothesis by means of a confidence region according to (3.82), the density value p_B at the boundary of the confidence region is determined by (6.118).

If instead of the posterior density function $p(\boldsymbol{x}|\boldsymbol{y}, C)$ the posterior density function $\bar{p}(\boldsymbol{x}|\boldsymbol{y}, C)$ from (6.37) and instead of (6.106) the conditional density function

$$\bar{p}(x_k|x_1, \ldots, x_{k-1}, x_{k+1}, \ldots, x_u, \boldsymbol{y}, C) \quad \text{for} \quad k \in \{1, \ldots, u\} \tag{6.120}$$

are available which both are not normalized, random values for x_k may be generated nevertheless by the rejection method (6.10) or the SIR the algorithm (6.31), as already mentioned in connection with (6.106). The normalization constant c in (6.37) for $\bar{p}(\boldsymbol{x}|\boldsymbol{y}, C)$, however, cannot be directly computed by the Gibbs sampler. But the results (6.112) to (6.114), (6.117) and (6.119) can be given, since the MAP estimate (6.114), as already mentioned in connection with (6.40), is valid also for the density functions which are not normalized. To determine the index B from (6.117), also density values not normalized may be sorted in decreasing order. By means of the index B the boundary of the confidence region is then determined by the not normalized density values, as described for (6.42). A point null hypothesis may be also tested with (3.82) by density functions which are not normalized.

If one needs instead of $\bar{p}(\boldsymbol{x}|\boldsymbol{y},C)$ the normalized posterior density function $p(\boldsymbol{x}|\boldsymbol{y},C)$, it has to be estimated from the random variates $\boldsymbol{x}_i$ with $i \in \{1,\ldots,m\}$ in (6.109). The parameter space $\mathcal{X}$ with $\boldsymbol{x} \in \mathcal{X}$ is divided as in (6.56) by intervals on the coordinate axes into small subspaces $\Delta\mathcal{X}_j$ with $j \in \{1,\ldots,J\}$, for instance, the plane into small squares by parallels to the coordinate axes. Let $\boldsymbol{x}_j$ represent the space $\Delta\mathcal{X}_j$ either by (6.57) as a mean of the random variates $\boldsymbol{x}_i \in \Delta\mathcal{X}_j$ or as the midpoint of $\Delta\mathcal{X}_j$ and let $p(\boldsymbol{x}_j|\boldsymbol{y},C)$ be the value of the density function of $p(\boldsymbol{x}|\boldsymbol{y},C)$ for $\boldsymbol{x}_j$. It is estimated corresponding to (6.56) by summing over the importance weights w_i of the subspace $\Delta\mathcal{X}_j$ which gives with (6.115)

$$p(\boldsymbol{x}_j|\boldsymbol{y},C) = \sum_{\boldsymbol{x}_i \in \Delta\mathcal{X}_j} w_i/V_{\Delta\mathcal{X}_j} = n_{\Delta\mathcal{X}_j}/(mV_{\Delta\mathcal{X}_j}) \quad \text{for} \quad j \in \{1,\ldots,J\}$$

$$(6.121)$$

where $V_{\Delta\mathcal{X}_j}$ denotes the hypervolume of the subspace $\Delta\mathcal{X}_j$ and $n_{\Delta\mathcal{X}_j}$ the number for which

$$\boldsymbol{x}_i \in \Delta\mathcal{X}_j \quad \text{for} \quad j \in \{1,\ldots,J\} \tag{6.122}$$

is valid. The estimate corresponds to the computation (2.24) of probabilities from the relative frequencies of the generated random variates so that a discrete density function is obtained. Dividing by $V_{\Delta\mathcal{X}_j}$ because of (2.72) gives corresponding to (6.58) approximately the continuous density function $p(\boldsymbol{x}_j|\boldsymbol{y},C)$.

The estimate (6.121) may produce a ragged form of the posterior density function $p(\boldsymbol{x}|\boldsymbol{y},C)$. To improve it, the *kernel method* of estimating density functions can be applied, see for instance SILVERMAN (1986, p.76),

$$p(\boldsymbol{x}_j|\boldsymbol{y},C) = \frac{1}{mh^u}\sum_{i=1}^{m} K\{\frac{1}{h}(\boldsymbol{x}_j - \boldsymbol{x}_i)\} \quad \text{for} \quad j \in \{1,\ldots,J\} \tag{6.123}$$

where h denotes the width of the window, u the number of unknown parameters and K the kernel function for which usually a radially symmetric unimodal density function is chosen, for instance, the density function of the multivariate normal distribution (2.195) for a $u \times 1$ random vector $\boldsymbol{x}$ with $\boldsymbol{\mu} = \boldsymbol{0}$ and $\boldsymbol{\Sigma} = \boldsymbol{I}$. A useful kernel function for $u = 2$ has been introduced for the example of Chapter 6.3.6.

6.3.4 Computation of Marginal Distributions

If not all unknown parameters $\boldsymbol{x}$ have to be estimated but only a subset of $\boldsymbol{x}$, the estimates (6.112) to (6.114) are still valid. If a confidence region needs to be established for a subset of the unknown parameters which is collected in the vector $\boldsymbol{x}_1$ with $\boldsymbol{x} = |\boldsymbol{x}_1', \boldsymbol{x}_2'|'$, the posterior marginal density function $p(\boldsymbol{x}_1|\boldsymbol{y}, C)$ has to be determined from (6.45). We therefore generate by the Gibbs sampler m values $\boldsymbol{x}_i$ which are seperated corresponding to the decomposition of $\boldsymbol{x}$ into

$$\boldsymbol{x}_i = |\boldsymbol{x}_{1i}', \boldsymbol{x}_{2i}'|' \ . \tag{6.124}$$

As in (6.56) we divide the space $\mathcal{X}_1$ with $\boldsymbol{x}_1 \in \mathcal{X}_1$ and $\boldsymbol{x}_2 \in \mathcal{X}_2$ by intervals on the coordinate axes into small subspaces $\Delta\mathcal{X}_{1j}$ with $j \in \{1, \ldots, J\}$ and introduce the point $\boldsymbol{x}_{1j}$ which represents the space $\Delta\mathcal{X}_{1j}$ either as the mean (6.57) or as the midpoint. To estimate the value $p_d(\boldsymbol{x}_{1j}|\boldsymbol{y}, C)$ of the discrete marginal density function, we sum the importance weights w_i for all random variates $\boldsymbol{x}_{1i} \in \Delta\mathcal{X}_1$ and $\boldsymbol{x}_{2i} \in \mathcal{X}_2$ and obtain with (6.56) and (6.115)

$$p_d(\boldsymbol{x}_{1j}|\boldsymbol{y}, C) = \sum_{\boldsymbol{x}_{1i}\in\Delta\mathcal{X}_{1j}, \boldsymbol{x}_{2i}\in\mathcal{X}_2} w_i = n_{\Delta\mathcal{X}_{1j}}/m$$

$$\text{for} \quad j \in \{1, \ldots, J\} \tag{6.125}$$

where $n_{\Delta\mathcal{X}_{1j}}$ denotes the number of random variates $\boldsymbol{x}_i$ for which

$$\boldsymbol{x}_{1i} \in \Delta\mathcal{X}_{1j} \quad \text{and} \quad \boldsymbol{x}_{2i} \in \mathcal{X}_2 \tag{6.126}$$

is valid. The continuous marginal density function $p(\boldsymbol{x}_{1j}|\boldsymbol{y}, C)$ for $\boldsymbol{x}_{1j}$ follows approximately from (6.58) with

$$p(\boldsymbol{x}_{1j}|\boldsymbol{y}, C) = p_d(\boldsymbol{x}_{1j}|\boldsymbol{y}, C)/V_{\Delta\mathcal{X}_{1j}} \tag{6.127}$$

where $V_{\Delta\mathcal{X}_{1j}}$ denotes the hypervolume of the subspace $\Delta\mathcal{X}_{1j}$.

A more accurate method of estimating the marginal density function is the kernel method (6.123). We replace $\boldsymbol{x}_j$ in (6.123) by $\boldsymbol{x}_{1j}$ and $\boldsymbol{x}_i$ by $\boldsymbol{x}_{1i}$ and obtain the estimate of the marginal density function $p(\boldsymbol{x}_{1j}|\boldsymbol{y}, C)$ for $\boldsymbol{x}_{1j}$ by

$$p(\boldsymbol{x}_{1j}|\boldsymbol{y}, C) = \frac{1}{mh^{u_1}} \sum_{i=1}^{m} K\{\frac{1}{h}(\boldsymbol{x}_{1j} - \boldsymbol{x}_{1i})\} \quad \text{for} \quad j \in \{1, \ldots, J\} \tag{6.128}$$

where h denotes the width of the window, u_1 the number of unknown parameters in $\boldsymbol{x}_1$ and K the kernel function discussed for (6.123). This kernel estimate has been applied for the example of Chapter 6.3.6.

If the density values $p_d(\boldsymbol{x}_{1j}|\boldsymbol{y}, C)$ from (6.125) or with (6.127) from (6.128) for the discrete marginal density function are arranged in decreasing order such that the series $p_d(\boldsymbol{x}_{1l}|\boldsymbol{y}, C)$ for $l \in \{1, \ldots, J\}$ is obtained, the index B of a point $\boldsymbol{x}_{1B}$ at the boundary of the $1 - \alpha$ confidence region for $\boldsymbol{x}_1$ follows with (6.60) by

$$\sum_{l=1}^{B} p_d(\boldsymbol{x}_{1l}|\boldsymbol{y}, C) = 1 - \alpha \qquad (6.129)$$

and the density value p_B of the point $\boldsymbol{x}_{1B}$ from (6.127) or (6.128) by

$$p_B = p(\boldsymbol{x}_{1B}|\boldsymbol{y}, C) \ . \qquad (6.130)$$

If the density values $p_d(\boldsymbol{x}_{1j}|\boldsymbol{y}, C)$ are computed by (6.125), it may happen that identical density values are obtained for more than one subspace $\Delta \mathcal{X}_{1j}$. If a point $\boldsymbol{x}_{1l}$ with one of these identical density values obtains the index B according to (6.129), the boundary of the confidence region cannot be uniquely determined by the interpolation. To avoid such a situation the kernel method (6.128) needs to be applied.

The ratio (3.62) of the two integrals to test the composite hypothesis (3.45) for $\boldsymbol{x}_1$ is found corresponding to (6.119) by

$$\hat{V} = n_0/n_1 \qquad (6.131)$$

where n_0 denotes the number of $\boldsymbol{x}_{1i} \in \mathcal{X}_{10}$, n_1 the number of $\boldsymbol{x}_{1i} \in \mathcal{X}_{11}$ and $\mathcal{X}_{10}$ and $\mathcal{X}_{11}$ the domains for the integration of $\boldsymbol{x}_1$. The reason is that the random values $\boldsymbol{x}_{1i}$ generated for $\boldsymbol{x}_1$ have the marginal density function $p(\boldsymbol{x}_1|\boldsymbol{y}, C)$ of the joint density function $p(\boldsymbol{x}|\boldsymbol{y}, C)$.

To use the conditional density functions which are obtained by the posterior density function $p(\boldsymbol{x}|\boldsymbol{y}, C)$, GELFAND and SMITH (1990) propose, see also GELFAND et al. (1992), the following computation of the marginal density function $p(x_k|\boldsymbol{y}, C)$ for x_k based on (6.106)

$$p(x_k|\boldsymbol{y}, C) = \frac{1}{m} \sum_{i=1}^{m} p(x_k|x_{1i}, \ldots, x_{k-1,i}, x_{k+1,i}, \ldots, x_{ui}, \boldsymbol{y}, C) \quad (6.132)$$

where the density values are summed, which are obtained except for x_{ki} by the m generated random variates from (6.109). By partitioning $\boldsymbol{x}$ into $\boldsymbol{x} = |\boldsymbol{x}_1', \boldsymbol{x}_2'|'$ the posterior marginal density function $p(\boldsymbol{x}_1|\boldsymbol{y}, C)$ for $\boldsymbol{x}_1$ follows correspondingly by

$$p(\boldsymbol{x}_1|\boldsymbol{y}, C) = \frac{1}{m} \sum_{i=1}^{m} p(\boldsymbol{x}_1|\boldsymbol{x}_{2i}, \boldsymbol{y}, C) \qquad (6.133)$$

where $\boldsymbol{x}_{2i}$ with $i \in \{1, \ldots, m\}$ denote the vectors of random variates generated for $\boldsymbol{x}_2$, which are contained in the set (6.109) of generated values. However, computing the marginal density functions by (6.132) or (6.133) needs the normalization constants for $p(x_k|x_1, \ldots, x_u, \boldsymbol{y}, C)$ or $p(\boldsymbol{x}_1|\boldsymbol{x}_2, \boldsymbol{y}, C)$ which generally are not available. If a normalized conditional density function is known which approximates $p(\boldsymbol{x}_1|\boldsymbol{x}_2, \boldsymbol{y}, C)$, an importance weighted marginal density estimation may be used, as proposed by CHEN et al. (2000, p.98).

The method of computing the marginal density function from (6.133) can be justified by the Monte Carlo integration. We obtain with (2.102) instead of (3.5)

$$p(\boldsymbol{x}_1|\boldsymbol{y}, C) = \int_{\mathcal{X}_2} p(\boldsymbol{x}_1|\boldsymbol{x}_2, \boldsymbol{y}, C)p(\boldsymbol{x}_2|\boldsymbol{y}, C)d\boldsymbol{x}_2 \ .$$

The integral expresses the expected value of $p(\boldsymbol{x}_1|\boldsymbol{x}_2, \boldsymbol{y}, C)$ which is computed by the density function $p(\boldsymbol{x}_2|\boldsymbol{y}, C)$. This expected value follows from (6.133) by the Monte Carlo integration (6.16), since the values $\boldsymbol{x}_{2i}$ generated for the vector $\boldsymbol{x}_2$ by the Gibbs sampler have the density function $p(\boldsymbol{x}_2|\boldsymbol{y}, C)$.

6.3.5 Gibbs Sampler for Computing and Propagating Large Covariance Matrices

When estimating unknown parameters, their covariance matrix (3.11) is needed to judge the accurcy of the estimates. Further quantities are often derived from the unknown parameters and their covariance matrices might be of greater interest than the covariance matrix of the unknown parameters. The covariance matrices of the derived quantities are obtained according to (2.158) by propagating the covarince matrix of the unknown parameters. For instance, when determining the gravity field of the earth from satellite observations, the geopotential is generally expanded into spherical harmonics whose coefficients are the unknown parameters. Several ten thousands of harmonic coefficients are estimated in linear models so that it takes a considerable computational effort to compute the covariance matrix of the harmonic coefficients by inverting according to (4.16), (4.88), (4.125) or (4.170) the matrix of normal equations. The harmonic coefficients are transformed into gridded gravity anomalies, geoid undulations or geostrophic velocities. By orbit integration the positions of satellites result from the harmonic coefficients. The covariance matrices of these derived quantities are obtained in case of linear transformations by multiplying the covariance matrix of the harmonic coefficients from the left by the matrix of the linear transformations and from the right by its transpose, see (2.158). For nonlinear transformations the matrix of transformations contains the derivatives of the transformations, as mentioned in connection with (2.158). These propagations of covariance matrices lead in case of many unknown parameters to cumbersome computations which are simplified by applying the Gibbs sampler. The covariance matrix of nonlinear tranformations of the unknown parameters can be directly computed by the Gibbs sampler, thus avoiding to determine the derivatives of the nonlinear transformations. In addition, only a few significant figures are needed when specifying the accuracy by variances and covariances. Estimating these quantities by the Gibbs sampler can take care of this advantage.

Let the $u \times 1$ vector $\boldsymbol{\beta}$ of unknown parameters be defined in the linear model (4.1). In case of a variance factor σ^2 known by $\sigma^2 = 1$, the posterior density function for $\boldsymbol{\beta}$ is given by the normal distribution (4.14)

$$\boldsymbol{\beta}|\boldsymbol{y} \sim N(\hat{\boldsymbol{\beta}}, D(\boldsymbol{\beta}|\boldsymbol{y})) \tag{6.134}$$

with the Bayes estimate $\hat{\boldsymbol{\beta}}$ of $\boldsymbol{\beta}$ from (4.15)

$$E(\boldsymbol{\beta}|\boldsymbol{y}) = \hat{\boldsymbol{\beta}} = (\boldsymbol{X}'\boldsymbol{P}\boldsymbol{X})^{-1}\boldsymbol{X}'\boldsymbol{P}\boldsymbol{y} \tag{6.135}$$

and the covariance matrix $D(\boldsymbol{\beta}|\boldsymbol{y})$ of $\boldsymbol{\beta}$ from (4.16)

$$D(\boldsymbol{\beta}|\boldsymbol{y}) = \boldsymbol{N}^{-1} = (\boldsymbol{X}'\boldsymbol{P}\boldsymbol{X})^{-1} = \boldsymbol{V} \tag{6.136}$$

with $\boldsymbol{N} = \boldsymbol{X}'\boldsymbol{P}\boldsymbol{X}$ being the matrix of normal equations for $\boldsymbol{\beta}$. The $u \times 1$ vector $\boldsymbol{e}$ of errors, the difference between $\boldsymbol{\beta}$ and $E(\boldsymbol{\beta}|\boldsymbol{y})$, is introduced by

$$\boldsymbol{e} = \boldsymbol{\beta} - E(\boldsymbol{\beta}|\boldsymbol{y}) = \boldsymbol{\beta} - \hat{\boldsymbol{\beta}} \; . \tag{6.137}$$

Its distribution follows from (2.202) and (6.134) by

$$\boldsymbol{e} \sim N(\boldsymbol{0}, \boldsymbol{V}) \tag{6.138}$$

with expectation and covariance matrix

$$E(\boldsymbol{e}) = \boldsymbol{0} \quad \text{and} \quad D(\boldsymbol{e}) = D(\boldsymbol{\beta}|\boldsymbol{y}) = \boldsymbol{V} \; . \tag{6.139}$$

We will use the Gibbs sampler to generate random variates $\boldsymbol{e}_i$ with $i \in \{1, \ldots, m\}$ for the error vector $\boldsymbol{e}$. They lead with (6.137) to random variates $\boldsymbol{\beta}_i$ for the vector $\boldsymbol{\beta}$ of unknown parameters

$$\boldsymbol{\beta}_i = \hat{\boldsymbol{\beta}} + \boldsymbol{e}_i \quad \text{for} \quad i \in \{1, \ldots, m\} \; . \tag{6.140}$$

Let $\boldsymbol{f}(\boldsymbol{\beta})$ be a vector of nonlinear or linear transformations of $\boldsymbol{\beta}$. Its covariance matrix $D(\boldsymbol{f}(\boldsymbol{\beta}))$, where the conditioning on $\boldsymbol{y}$ is omitted for simpler notation, follows with (2.144) and (3.11) by

$$D(\boldsymbol{f}(\boldsymbol{\beta})) = \int_{\mathcal{B}} \Big(\boldsymbol{f}(\boldsymbol{\beta}) - E(\boldsymbol{f}(\boldsymbol{\beta}))\Big)\Big(\boldsymbol{f}(\boldsymbol{\beta}) - E(\boldsymbol{f}(\boldsymbol{\beta}))\Big)' p(\boldsymbol{\beta})d\boldsymbol{\beta} \tag{6.141}$$

where $\mathcal{B}$ denotes the parameter space of $\boldsymbol{\beta}$. The Monte Carlo estimate $\hat{D}(\boldsymbol{f}(\boldsymbol{\beta}))$ of $D(\boldsymbol{f}(\boldsymbol{\beta}))$ is obtained with (6.116) and (6.113) by means of the random variates $\boldsymbol{\beta}_i$ from (6.140) by

$$\hat{D}(\boldsymbol{f}(\boldsymbol{\beta})) = \frac{1}{m} \sum_{i=1}^{m} \Big(\boldsymbol{f}(\boldsymbol{\beta}_i) - \hat{E}(\boldsymbol{f}(\boldsymbol{\beta}))\Big)\Big(\boldsymbol{f}(\boldsymbol{\beta}_i) - \hat{E}(\boldsymbol{f}(\boldsymbol{\beta}))\Big)' \tag{6.142}$$

with

$$\hat{E}(\boldsymbol{f}(\boldsymbol{\beta})) = \frac{1}{m}\sum_{i=1}^{m}\boldsymbol{f}(\boldsymbol{\beta}_i) \ . \tag{6.143}$$

If $\boldsymbol{f}(\boldsymbol{\beta})$ represents a linear transformation of the unknown parameters $\boldsymbol{\beta}$, we obtain with the matrix $\boldsymbol{F}$ which has u columns and with (6.137)

$$\boldsymbol{f}(\boldsymbol{\beta}) = \boldsymbol{F}\boldsymbol{\beta} = \boldsymbol{F}(\hat{\boldsymbol{\beta}} + \boldsymbol{e}) \tag{6.144}$$

so that with (6.139)

$$E(\boldsymbol{f}(\boldsymbol{\beta})) = \boldsymbol{F}\hat{\boldsymbol{\beta}} \tag{6.145}$$

follows and finally

$$\boldsymbol{f}(\boldsymbol{\beta}) - E(\boldsymbol{f}(\boldsymbol{\beta})) = \boldsymbol{F}\boldsymbol{e} \ . \tag{6.146}$$

The estimated covariance matrix $\hat{D}(\boldsymbol{F}\boldsymbol{\beta})$ of the covariance matrix of the linear transformation $\boldsymbol{F}\boldsymbol{\beta}$ results therefore from (6.142) by

$$\hat{D}(\boldsymbol{F}\boldsymbol{\beta}) = \frac{1}{m}\sum_{i=1}^{m}(\boldsymbol{F}\boldsymbol{e}_i)(\boldsymbol{F}\boldsymbol{e}_i)' \ . \tag{6.147}$$

With $\boldsymbol{F} = \boldsymbol{I}$ we find the estimate $\hat{D}(\boldsymbol{\beta}|\boldsymbol{y})$ of the covariance matrix of the unknown parameters $\boldsymbol{\beta}$, see also (6.116), by

$$\hat{D}(\boldsymbol{\beta}|\boldsymbol{y}) = \hat{\boldsymbol{V}} = \frac{1}{m}\sum_{i=1}^{m}\boldsymbol{e}_i\boldsymbol{e}_i' \ . \tag{6.148}$$

If one uses (6.142), to compute the covariance matrix of quantities derived by nonlinear transformations of the vector $\boldsymbol{\beta}$ of unknown parameters, or if (6.147) is applied, to obtain the covariance matrix of linearly transformed quantities, the covariance matrix of $\boldsymbol{\beta}$ estimated by (6.148) is obviously not needed. This reduces the number of computations especially, if the vectors $\boldsymbol{f}(\boldsymbol{\beta}))$ or $\boldsymbol{F}\boldsymbol{\beta}$ are much shorter than $\boldsymbol{\beta}$. In contrast, the covariance matrix $D(\boldsymbol{\beta}|\boldsymbol{y})$ of the unknown parameters is always needed, if we propagate the covariance matrices by (2.158).

To sample random variates $\boldsymbol{e}_i$ from (6.138), we need the covariance matrix $\boldsymbol{V}$ of the unknown parameters which, however, is not known. If the Gibbs sampler is applied, the samples are drawn from conditional density functions which can be expressed by the elements of the matrix $\boldsymbol{N}$ of normal equations, as shown by HARVILLE (1999) who used the Gibbs sampler (6.108). To take care of strongly correlated unknown parameters, the Gibbs sampler (6.110) for grouped unknown parameters is applied here, as proposed by GUNDLICH et al. (2003). The vector $\boldsymbol{e}$ of errors is therefore divided into two subvectors $\boldsymbol{e}_l$ and $\boldsymbol{e}_t$ and the matrices $\boldsymbol{N}$ and $\boldsymbol{V}$ accordingly

$$
\boldsymbol{e} = \begin{vmatrix} \boldsymbol{e}_l \\ \boldsymbol{e}_t \end{vmatrix}, \ \boldsymbol{N} = \begin{vmatrix} \boldsymbol{N}_{ll} & \boldsymbol{N}_{lt} \\ \boldsymbol{N}_{tl} & \boldsymbol{N}_{tt} \end{vmatrix}, \ \boldsymbol{N}^{-1} = \boldsymbol{V} = \begin{vmatrix} \boldsymbol{V}_{ll} & \boldsymbol{V}_{lt} \\ \boldsymbol{V}_{tl} & \boldsymbol{V}_{tt} \end{vmatrix}. \quad (6.149)
$$

The conditional distribution for $\boldsymbol{e}_l$ given $\boldsymbol{e}_t$ is defined because of (6.138) by the normal distribution (2.198)

$$
\boldsymbol{e}_l | \boldsymbol{e}_t \sim N(\boldsymbol{V}_{lt}\boldsymbol{V}_{tt}^{-1}\boldsymbol{e}_t, \boldsymbol{V}_{ll} - \boldsymbol{V}_{lt}\boldsymbol{V}_{tt}^{-1}\boldsymbol{V}_{tl}) \ .
$$

We compute $\boldsymbol{V}^{-1}$ by (4.46) and apply the matrix identity (4.47) to obtain $\boldsymbol{N}_{ll} = (\boldsymbol{V}_{ll} - \boldsymbol{V}_{lt}\boldsymbol{V}_{tt}^{-1}\boldsymbol{V}_{tl})^{-1}$ and the identity (4.48) to find $\boldsymbol{N}_{lt} = -\boldsymbol{N}_{ll}\boldsymbol{V}_{lt}\boldsymbol{V}_{tt}^{-1}$ which gives $-\boldsymbol{N}_{ll}^{-1}\boldsymbol{N}_{lt} = \boldsymbol{V}_{lt}\boldsymbol{V}_{tt}^{-1}$ so that the conditional distribution is defined by the subblocks of the matrix $\boldsymbol{N}$ of normal equations

$$
\boldsymbol{e}_l | \boldsymbol{e}_t \sim N(-\boldsymbol{N}_{ll}^{-1}\boldsymbol{N}_{lt}\boldsymbol{e}_t, \boldsymbol{N}_{ll}^{-1}) \ . \quad (6.150)
$$

The Gibbs sampler (6.110) for the grouping technique shall be applied to take care of correlated unknown parameters. This is necessary for the example of determining the gravity field of the earth from satellite observations mentioned above, because the matrix of normal equations tend to be ill-conditioned. A reordering of the harmonic coefficients by the order of their expansion gives an approximate block diagonal structure of the matrix of normal equations and its inverse so that correlated unknown parameters can be grouped, see GUNDLICH et al. (2003).

As in (6.110) the error vector $\boldsymbol{e}$ is now subdivided into r subvectors $\boldsymbol{e}_1$ to $\boldsymbol{e}_r$ and the matrices $\boldsymbol{N}$ and $\boldsymbol{V}$ accordingly

$$
\boldsymbol{e} = \begin{vmatrix} \boldsymbol{e}_1 \\ \ldots \\ \boldsymbol{e}_r \end{vmatrix}, \boldsymbol{N} = \begin{vmatrix} \boldsymbol{N}_{11} & \ldots & \boldsymbol{N}_{1r} \\ \ldots\ldots\ldots\ldots\ldots \\ \boldsymbol{N}_{r1} & \ldots & \boldsymbol{N}_{rr} \end{vmatrix}, \boldsymbol{V} = \begin{vmatrix} \boldsymbol{V}_{11} & \ldots & \boldsymbol{V}_{1r} \\ \ldots\ldots\ldots\ldots\ldots \\ \boldsymbol{V}_{r1} & \ldots & \boldsymbol{V}_{rr} \end{vmatrix}. (6.151)
$$

At the qth iteration of the Gibbs sampler (6.110) we draw $\boldsymbol{e}_l$ given $\boldsymbol{e}_1^{(q)}, \ldots,$ $\boldsymbol{e}_{l-1}^{(q)}, \boldsymbol{e}_{l+1}^{(q-1)}, \ldots, \boldsymbol{e}_r^{(q-1)}$ from the conditional distribution obtained from (6.150)

$$
\boldsymbol{e}_l | \boldsymbol{e}_1^{(q)}, \ldots, \boldsymbol{e}_{l-1}^{(q)}, \boldsymbol{e}_{l+1}^{(q-1)}, \ldots, \boldsymbol{e}_r^{(q-1)}
$$
$$
\sim N\Big(-\boldsymbol{N}_{ll}^{-1}\Big(\sum_{j<l}\boldsymbol{N}_{lj}\boldsymbol{e}_j^{(q)} + \sum_{j>l}\boldsymbol{N}_{lj}\boldsymbol{e}_j^{(q-1)}\Big), \boldsymbol{N}_{ll}^{-1}\Big) \ . \quad (6.152)
$$

To sample from this normal distribution, (6.13) may be applied. Collecting the m samples of each sth iteration after a burn-in phase of o iterations gives the vectors $\boldsymbol{e}_i$ of random variates

$$
\boldsymbol{e}_i = |\boldsymbol{e}_{1i}', \ldots, \boldsymbol{e}_{ri}'|' \quad \text{for} \quad i \in \{1, \ldots, m\} \ . \quad (6.153)
$$

The distance s between the collected samples $\boldsymbol{e}_i$ depends on the correlation between the samples which can be computed, as shown by GUNDLICH et al.

(2003). By means of the random variates e_i the covariance matrix for $f(\beta)$, $F\beta$ and β can now be estimated by (6.142), (6.147) and (6.148).

The estimate $\hat{V}$ in (6.148) can be improved. The inverse N^{-1} in (6.149) is computed by (4.46) to obtain $V_{ll} = N_{ll}^{-1} + N_{ll}^{-1}N_{lt}V_{tt}N_{tl}N_{ll}^{-1}$ which gives instead of (6.148) the estimate $\bar{V}_{ll}$

$$\bar{V}_{ll} = N_{ll}^{-1} + \frac{1}{m}\sum_{i=1}^{m} N_{ll}^{-1}N_{lt}e_{ti}e_{ti}'N_{tl}N_{ll}^{-1} \ . \tag{6.154}$$

The second identity $N_{lt} = -N_{ll}V_{lt}V_{tt}^{-1}$, which leads to (6.150), gives $V_{lt} = -N_{ll}^{-1}N_{lt}V_{tt}$ and instead of (6.148) the estimate

$$\bar{V}_{lt} = \frac{1}{m}\sum_{i=1}^{m} -N_{ll}^{-1}N_{lt}e_{ti}e_{ti}' \ . \tag{6.155}$$

An improvement with respect to computational efficiency results from the estimates

$$\bar{V}_{ll} = N_{ll}^{-1} + \frac{1}{m}\sum_{i=1}^{m}\boldsymbol{\mu}_{li}\boldsymbol{\mu}_{li}' \quad \text{and} \quad \bar{V}_{lt} = \frac{1}{m}\sum_{i=1}^{m}\boldsymbol{\mu}_{li}e_{ji}'$$

$$\text{for} \quad l,j \in \{1,\ldots,r\}, \ l \neq j \tag{6.156}$$

where the vector $\boldsymbol{\mu}_{li}$ is obtained at the end of each sth iteration from the vector

$$-N_{ll}^{-1}\left(\sum_{j<l} N_{lj}e_j^{(q)} + \sum_{j>l} N_{lj}e_j^{(q-1)}\right)$$

computed for generating random variates from (6.152) for the Gibbs sampler. The estimate (6.156) is generally nonsymmetric which can be avoided, if only the elements on the diagonal and above the diagonal are estimated. Monte Carlo methods and also the Gibbs sampler are well suited for parallel computing. KOCH et al. (2004) therefore implemented the Gibbs sampler using (6.152) for the estimate (6.156) on a parallel computer formed by a cluster of PCs.

It is obvious that the estimate (6.156) of the covariance matrix V improves the estimate (6.148), if the matrix N of normal equations has approximately a block diagonal structure. The inverse N_{ll}^{-1} of the block diagonal N_{ll} for $l \in \{1,\ldots,r\}$ then gives a good approximation for V_{ll} so that the Gibbs sampler improves by (6.156) the approximation. This is also generally true, because the variance of the estimate (6.156) is smaller than the one of (6.148), as shown by GUNDLICH et al. (2003). The estimate (6.156) results from an estimation by conditioning, also called Rao-Blackwellizaton, see for instance LIU (2001, p.27). It means that by intoducing a conditional density function a part of the integration can be solved analytically so that the variance of the estimate is reduced.

Only a few significant figures are needed for specifying the accuracy by variances and covariances. To determine the accuracy of the estimate (6.156) GUNDLICH et al. (2003) used the scaled Frobenius norm

$$d = \Big(\frac{1}{u^2\max(v_{ii})^2}\sum_{i=1}^{u}\sum_{j=1}^{u}(\bar{v}_{ij}-v_{ij})^2\Big)^{1/2} \tag{6.157}$$

where u denotes the number of unknown parameters $\boldsymbol{\beta}$, $\bar{v}_{ij}$ and v_{ij} the elements of $\bar{\boldsymbol{V}}$ and $\boldsymbol{V}$. The norm d gives approximately the variance of the estimate $\bar{v}_{ij}$ averaged over all elements of $\bar{\boldsymbol{V}}$ and scaled by the maximum variance v_{ii} because of $|v_{ij}| \leq \max(v_{ii})$. By taking the square root the scaled averaged standard deviation is obtained. Thus, d indicates the number of significant digits not distorted by errors of the estimate. For instance, $d = 1 \times 10^{-3}$ means on the average three significant digits in the estimated elements $\bar{v}_{ij}$ of $\bar{\boldsymbol{V}}$.

For computing d from (6.157) the covariance matrix $\boldsymbol{V}$ is needed which is only available for test computations. However, the estimated covariance matrix $\bar{\boldsymbol{V}}$ follows from (6.156) as the mean value of random variates. The variance of this mean can be easily obtained, if the samples are independent. This can be approximately assumed because of the grouping technique and because only each sth generated sample is selected as random variate. The variances thus obtained can be used to compute the Frobenius norm (6.157) with a good approximation. Applying a parallel computer these approximations are available during the sampling process, as shown by KOCH et al. (2004).

To determine by hypothesis tests the maximum degree of harmonic coefficients in a geopotential model, parallel Gibbs sampling was applied by KOCH (2005B) to compute random variates for the harmonic coefficients. They were nonlinearly transformed to random variates for quantities whose density functions were computed by these random values for the hypothesis tests.

6.3.6   Continuation of the Example: Confidence Region for Robust Estimation of Parameters

As in Chapter 6.2.5 confidence regions shall be determined with (3.35) for the unknown parameters $\boldsymbol{\beta}$ whose estimates are determined by the robust estimation presented in Chapter 4.2.5. Since the integration of the posterior density function $p(\boldsymbol{\beta}|\boldsymbol{y})$ for $\boldsymbol{\beta}$ from (4.62) could not be solved analytically, the Gibbs sampler is applied.

The posterior density function, which is not normalized and again is called $\bar{p}(\boldsymbol{\beta}_T|\boldsymbol{y})$, for the vector $\boldsymbol{\beta}_T = \boldsymbol{\beta} - \hat{\boldsymbol{\beta}}$ of the unknown parameters transformed by (6.65) is obtained by (6.88) with $\bar{e}_{Tv}$ from (6.87)

$$\bar{p}(\boldsymbol{\beta}_T|\boldsymbol{y}) = \prod_{v=1}^{n} \bar{p}(\bar{e}_{Tv}|\boldsymbol{\beta}_T) \ . \tag{6.158}$$

The conditional density function for β_k with $\boldsymbol{\beta}_T = (\beta_k)$ given the remaining components of $\boldsymbol{\beta}_T$ which are collected in the vector $\boldsymbol{\beta}_t$ follows with (6.158) corresponding to (6.106), where the denominater does not need to be considered, because the density function is not normalized,

$$\bar{p}(\beta_k|\beta_1,\ldots,\beta_{k-1},\beta_{k+1},\ldots,\beta_u) = \prod_{v=1}^{n} \bar{p}(\bar{e}_{Tv}|\beta_k,\boldsymbol{\beta}_t)$$

$$\text{and} \quad k \in \{1,\ldots,u\} \ . \tag{6.159}$$

Random variates for β_k have to be generated by the Gibbs sampler with (6.108). The rejection method (6.10) is applied which gives the random values we are looking for, even if the density function used for the generation is only available in the form (6.159) which is not normalized. However, an envelope must be found for the density function (6.159).

As was shown in Chapter 6.2.5, the conditional normal distribution (6.75) is an approximate distribution for (6.159). We obtain, since $\boldsymbol{\beta}_t$ contains the components of $\boldsymbol{\beta}_T$ without β_k,

$$\beta_k|\boldsymbol{\beta}_t,\boldsymbol{y} \sim N(-\hat{\beta}_k,\sigma_k^2) \tag{6.160}$$

with

$$\begin{aligned}
\hat{\beta}_k &= (\boldsymbol{x}_k'\boldsymbol{P}\boldsymbol{x}_k)^{-1}\boldsymbol{x}_k'\boldsymbol{P}\boldsymbol{X}_t\boldsymbol{\beta}_t \\
\sigma_k^2 &= \sigma^2(\boldsymbol{x}_k'\boldsymbol{P}\boldsymbol{x}_k)^{-1}
\end{aligned} \tag{6.161}$$

from (6.73) and $\boldsymbol{X} = (\boldsymbol{x}_k')$ as in (4.59). An envelope of the standard normal distribution is formed by the Cauchy distribution (2.192) with the translation parameter $\theta = 0$ and the scale parameter $\lambda = 1$ (DEVROYE 1986, p.46). An envelope of the normal distribution (6.160) is obtained with $\theta = -\hat{\beta}_k$ and in addition with $\lambda = \sigma_k$, as will be shown in the following.

First we set $\theta = 0$ and $\hat{\beta}_k = 0$ and determine the minimum of the constant C from (6.9) depending on β_k. We obtain with (2.166) and (2.192)

$$C = \frac{1}{\sqrt{2\pi}\sigma_k} e^{-\frac{\beta_k^2}{2\sigma_k^2}} \Big/ \Big(\frac{\lambda}{\pi(\lambda^2 + \beta_k^2)}\Big) \ . \tag{6.162}$$

Extremal values for C follow after taking the logarithm of (6.162) from

$$\frac{d}{d\beta_k}\Big[\ln\frac{1}{\sqrt{2\pi}\sigma_k} - \ln\frac{\lambda}{\pi} - \frac{\beta_k^2}{2\sigma_k^2} + \ln(\lambda^2 + \beta_k^2)\Big] = 0$$

or

$$-\frac{\beta_k}{\sigma_k^2} + \frac{2\beta_k}{\lambda^2 + \beta_k^2} = 0 \ . \tag{6.163}$$

The first extremal value C_1 is obtained for $\beta_k = 0$, thus from (6.162)

$$C_1 = \frac{\lambda}{\sigma_k}\sqrt{\frac{\pi}{2}} \ . \tag{6.164}$$

The second extremal value C_2 is found from (6.163) for

$$\beta_k^2 = 2\sigma_k^2 - \lambda^2 \quad \text{with} \quad \lambda^2 < 2\sigma_k^2 \quad \text{or} \quad \lambda/\sigma_k < \sqrt{2} \ ,$$

hence from (6.162)

$$C_2 = \frac{\sqrt{2\pi}\sigma_k}{e\lambda} e^{\frac{\lambda^2}{2\sigma_k^2}} \quad \text{for} \quad \lambda/\sigma_k < \sqrt{2} \ . \tag{6.165}$$

For $\lambda/\sigma_k < \sqrt{2}$ the constant C attains the maximum (6.165) with $\beta_k^2 = 2\sigma_k^2 - \lambda^2$ and the minimum (6.164) with $\beta_k = 0$. For $\lambda/\sigma_k \geq \sqrt{2}$ the maximum (6.164) of the constant C is reached with $\beta_k = 0$. Thus, the minimum of the constant C has to be looked for $\lambda/\sigma_k < \sqrt{2}$ which is attained at

$$\lambda/\sigma_k = 1 \tag{6.166}$$

and therefore from (6.165)

$$C = \sqrt{\frac{2\pi}{e}} \ . \tag{6.167}$$

To generate random variates with the density function (6.159) by the rejection method, random values have to be generated with the Cauchy distribution (2.192). Because of (6.160) we have $\theta = -\hat{\beta}_k$ and because of (6.166) $\lambda = \sigma_k$. The random values β_{ki} for β_k therefore follow with (6.6) from

$$\beta_{ki} = \sigma_k \tan(\pi u) - \hat{\beta}_k \ . \tag{6.168}$$

For the rejection method (6.10)

$$uCg(x) < \bar{p}(x) \tag{6.169}$$

needs to be fulfilled where $g(x)$ denotes the density function of the Cauchy distribution and $\bar{p}(x)$ the density function (6.159). When computing $g(x)$ one has to keep in mind that the normal distribution (6.160) contains the constants given in (6.73) which depend on the random values β_{ti} given for β_t. If they are taken into account and if as approximate normalization constant for (6.159) the normalization constant of the normal distribution (6.160) is used, one obtains with (2.192), (6.73), (6.161), (6.167) and (6.168) instead of (6.169)

$$u\sqrt{\frac{2\pi}{e}}\,\frac{\sigma_k}{\pi}\,\frac{1}{\sigma_k^2+(\beta_{ki}+\hat{\beta}_k)^2}\,e^{-\frac{1}{2\sigma^2}(c_1+c_2)} < \frac{1}{\sqrt{2\pi}\sigma_k}\prod_{v=1}^{n}\bar{p}(\bar{e}_{Tv}|\beta_{ki},\boldsymbol{\beta}_{ti})$$

with

$$c_1 = \boldsymbol{\beta}_{ti}'\boldsymbol{X}_t'\boldsymbol{P}\boldsymbol{X}_t\boldsymbol{\beta}_{ti}$$

$$c_2 = -\sigma^2\hat{\beta}_k^2/\sigma_k^2 \tag{6.170}$$

or

$$u\frac{2}{\sqrt{e}}\,\frac{\sigma_k^2}{\sigma_k^2+(\beta_{ki}+\hat{\beta}_k)^2}\,e^{-\frac{1}{2\sigma^2}(c_1+c_2)} < \prod_{v=1}^{n}\bar{p}(\bar{e}_{Tv}|\beta_{ki},\boldsymbol{\beta}_{ti})\ . \tag{6.171}$$

By introducing (6.166) and (6.167) the Cauchy distribution becomes the envelope of the normal distribution (6.160), but it has to be checked also, whether the Cauchy distribution is the envelope of the distribution (6.159). Thus, according to (6.9) together with (6.171)

$$\frac{2}{\sqrt{e}}\,\frac{\sigma_k^2}{\sigma_k^2+(\beta_{ki}+\hat{\beta}_k)^2}\,e^{-\frac{1}{2\sigma^2}(c_1+c_2)} \geq \prod_{v=1}^{n}\bar{p}(\bar{e}_{Tv}|\beta_{ki},\boldsymbol{\beta}_{ti}) \tag{6.172}$$

has to be fulfilled. When numerically checking this inequality for the following example, it turns out that it is not fulfilled for large values of β_{ki}. The constant C from (6.167) therefore needs to be increased so that for a largest possible value β_{ki} the inequality becomes true. However, the constant C may not be chosen too large to avoid an inefficiency of the rejection method. Random values β_{ki} not fulfilling (6.172) do not have the distribution (6.159), we are looking for, but a distribution which approximates the Cauchy distribution. This approximation can be accepted, if it happens for large values for β_{ki} and thus for small density values.

By applying the Gibbs sampler (6.108) in connection with the rejection method (6.10) for the density function (6.159), random values $\boldsymbol{\beta}_{Ti}$ with the posterior density function $p(\boldsymbol{\beta}_T|\boldsymbol{y})$ from (4.62) in connection with (6.65) are obtained, although it is only available in the form (6.158) which is not normalized. If confidence regions for a subset from (6.67) of the set $\boldsymbol{\beta}_T$ of unknown parameters have to be computed or hypotheses to be tested, values of the marginal density function are computed from (6.125) or with (6.127) from (6.128), in order to determine with (6.129) a point at the boundary of the confidence region. For testing hypotheses (6.131) is applied.

Example: Again the confidence region of content $1-\alpha = 95\%$ for the unknown parameters β_0 and β_1 in (6.94) of the Example 1 to (6.93) is computed. The marginal distribution for β_0 and β_1 will be estimated by the kernel method (6.128). For $\boldsymbol{\beta}_T = |\beta_0 - \hat{\beta}_0, \beta_1 - \hat{\beta}_1, \beta_2 - \hat{\beta}_2|'$ the random variates $\boldsymbol{\beta}_{Ti}$, $i \in \{1, \ldots, m\}$ with the density function (6.159) were therefore generated by the Gibbs sampler together with the rejection method in $m = 2\,000\,000$ repetitions with a burn-in phase of $o = 200$, as explained for (6.109). It was already mentioned for (6.172) that the density function of the Cauchy distribution is not an envelope of the density function (6.159) so that the fivefold of (6.167) was introduced as constant. For values smaller than 0.0059 on the right-hand side of (6.172) the inequality is not fulfilled, the maximum value on the right-hand side is equal to one. This approximation is sufficiently accurate.

The plane where β_0 and β_1 are defined is subdivided into $J = 20\,000$ surface elements of equal area. With $\boldsymbol{\beta}_{Tc} = |\beta_0 - \hat{\beta}_0, \beta_1 - \hat{\beta}_1|'$ the midpoints of the area elements are denoted by $\boldsymbol{\beta}_{Tcj}$ with $j \in \{1, \ldots, J\}$. The m random variates $\boldsymbol{\beta}_{Ti}$ generated for $\boldsymbol{\beta}_T$ also contain the random values for $\boldsymbol{\beta}_{Tc}$, these values are denoted by $\boldsymbol{\beta}_{Tci}$ with $i \in \{1, \ldots, m\}$. They are used to estimate the marginal density values $p(\boldsymbol{\beta}_{Tcj})$ for the midpoints $\boldsymbol{\beta}_{Tcj}$ of the area elements by the kernel method (6.128)

$$p(\boldsymbol{\beta}_{Tcj}|\boldsymbol{y}, C) = \frac{1}{mh^2} \sum_{k=1}^{m} K\left\{\frac{1}{h}(\boldsymbol{\beta}_{Tcj} - \boldsymbol{\beta}_{Tci})\right\} \quad \text{for} \quad j \in \{1, \ldots, J\} \quad (6.173)$$

with the kernel (SILVERMAN 1986, p.76)

$$K(\boldsymbol{z}) = \begin{cases} 3\pi^{-1}(1 - \boldsymbol{z}'\boldsymbol{z})^2 & \text{for} \quad \boldsymbol{z}'\boldsymbol{z} < 1 \\ 0 & \text{otherwise} \end{cases}$$

and

$$\boldsymbol{z} = \frac{1}{h}(\boldsymbol{\beta}_{Tcj} - \boldsymbol{\beta}_{Tci}) \,.$$

The window width h is chosen to be the fourfold of the side length of a surface element (KOCH 2000). When applying (6.173) the contribution of each generated random vector $\boldsymbol{\beta}_{Tci}$ to the density values of the midpoints $\boldsymbol{\beta}_{Tcj}$ is computed and then the m contributions are added to obtain the density values $p(\boldsymbol{\beta}_{Tcj}|\boldsymbol{y})$. They are ordered to obtain with (6.127) from (6.129) the index B of a point at the boundary of the confidence region. Its density value p_B follows with

$$p_B = 0.0062$$

in good agreement with the values of the Example 1 to (6.93) and of the Example 2 to (6.103). The confidence region is shown in Figure 6.3. It is nearly identical with the ones of Figure 6.1 and 6.2.

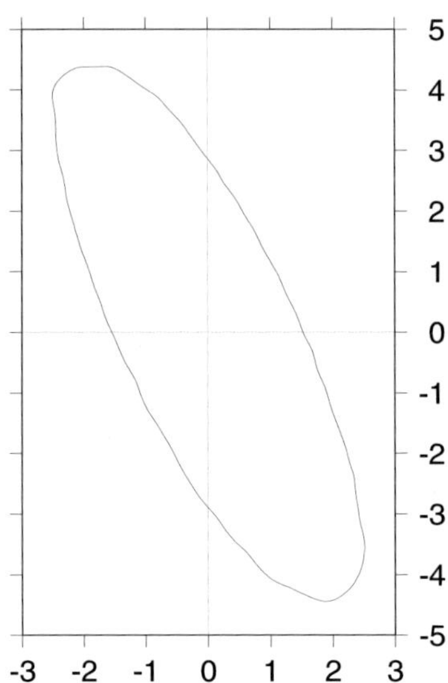

Figure 6.3: Confidence Region for β_0 and β_1 by the Kernel Estimation of
the Marginal Density Function

Δ

References

ALENIUS, S. and U. RUOTSALAINEN (1997) Bayesian image reconstruction for emission tomography based on median root prior. *Eur J Nucl Med*, 24:258–265.

ALKHATIB, H. and W.-D. SCHUH (2007) Integration of the Monte Carlo covariance estimation strategy into tailored solution procedures for large-scale least squares problems. *J Geodesy*, 81:53–66.

ARENT, N., G. HÜCKELHEIM and K.R. KOCH (1992) Method for obtaining geoid undulations from satellite altimetry data by a quasi-geostrophic model of the sea surface topography. *Manuscripta geodaetica*, 17:174–185.

BERGER, J.O. (1985) *Statistical Decision Theory and Bayesian Analysis.* Springer, Berlin.

BERNARDO, J.M. and A.F.M. SMITH (1994) *Bayesian Theory.* Wiley, New York.

BESAG, J.E. (1974) Spatial interaction and the statistical analysis of lattice systems. *J Royal Statist Society*, B 36:192–236.

BESAG, J.E. (1986) On the statistical analysis of dirty pictures. *J Royal Statist Society*, B 48:259–302.

BETTI, B., M. CRESPI and F. SANSO (1993) A geometric illustration of ambiguity resolution in GPS theory and a Bayesian approach. *Manuscripta geodaetica*, 18:317–330.

BETTINARDI, V., E. PAGANI, M.C. GILARDI, S. ALENIUS, K. THIELE-MANS, M. TERAS and F. FAZIO (2002) Implementation and evaluation of a 3D one-step late reconstruction algorithm for 3D positron emission tomography brain studies using median root prior. *Eur J Nucl Med*, 29:7–18.

BISHOP, C.M. (2006) *Pattern Recognition and Machine Learning.* Springer, New York.

BLATTER, C. (1974) *Analysis I, II, III.* Springer, Berlin.

BOX, G.E.P. and M.E. MULLER (1958) A note on the generation of random normal deviates. *Annals Mathematical Statistics*, 29:610–611.

Box, G.E.P. and G.C. Tiao (1973) *Bayesian Inference in Statistical Analysis*. Addison-Wesley, Reading.

Chen, M.-H., Q.-M. Shao and J.G. Ibrahim (2000) *Monte Carlo Methods in Bayesian Computations*. Springer, New York.

Cox, R.T. (1946) Probability, frequency and reasonable expectation. *American Journal of Physics*, 14:1–13.

Cressie, N.A.C. (1991) *Statistics for Spatial Data*. Wiley, New York.

Dagpunar, J. (1988) *Principles of Random Variate Generation*. Clarendon Press, Oxford.

Dean, T.L. and M.P. Wellman (1991) *Planning and Control*. Morgan Kaufmann, San Mateo.

DeGroot, M.H. (1970) *Optimal Statistical Decisions*. McGraw-Hill, New York.

Devroye, L. (1986) *Non-Uniform Random Variate Generation*. Springer, Berlin.

Doucet, A., S. Godsill and C. Andrieu (2000) On sequential Monte Carlo sampling methods for Bayesian filtering. *Statistics and Computing*, 10:197–208.

Fessler, J.A., H. Erdogan and W.B. Wu (2000) Exact distribution of edge-preserving MAP estimators for linear signal models with Gaussian measurement noise. *IEEE Trans Im Proc*, 9(6):1049–56.

Förstner, W. (1979) Ein Verfahren zur Schätzung von Varianz- und Kovarianzkomponenten. *Allgemeine Vermessungs-Nachrichten*, 86:446–453.

Gelfand, A.E. and A.F.M. Smith (1990) Sampling-based approaches to calculating marginal densities. *J American Statistical Association*, 85:398–409.

Gelfand, A.E., A.F.M. Smith and T. Lee (1992) Bayesian analysis of constrained parameter and truncated data problems using Gibbs sampling. *J American Statistical Association*, 87:523–532.

Gelman, A., J.B. Carlin, H.S. Stern and D.B. Rubin (2004) *Bayesian Data Analysis, 2nd Ed.* Chapman and Hall, Boca Raton.

Geman, D., S. Geman and C. Graffigne (1987) Locating texture and object boundaries. In: Devijver, P.A. and J. Kittler (Eds.), *Pattern Recognition Theory and Applications*. Springer, Berlin, 165–177.

GEMAN, S. and D. GEMAN (1984) Stochastic relaxation, Gibbs distributions, and the Bayesian restoration of images. *IEEE Trans Pattern Anal Machine Intell*, PAMI-6:721–741.

GEMAN, S. and D.E. MCCLURE (1987) Statistical methods for tomographic image reconstruction. *Bull Int Statist Inst*, 52-21.1:5–21.

GEORGE, A. and J.W. LIU (1981) *Computer Solution of Large Sparse Positive Definite Systems*. Prentice-Hall, Englewood Cliffs.

GILKS, W.R. (1996) Full conditional distributions. In: GILKS, W.R., S. RICHARDSON and D.J. SPIEGELHALTER (Eds.), *Markov Chain Monte Carlo in Practice*. Chapman and Hall, London, 75–88.

GOLUB, G.H. and U. VON MATT (1997) Generalized cross-validation for large-scale problems. *J Computational and Graphical Statistics*, 6:1–34.

GORDON, N. and D. SALMOND (1995) Bayesian state estimation for tracking and guidance using the bootstrap filter. *J Guidance, Control, and Dynamics*, 18:1434–1443.

GRAFAREND, E.W. and B. SCHAFFRIN (1993) *Ausgleichungsrechnung in linearen Modellen*. B.I. Wissenschaftsverlag, Mannheim.

GREEN, P.J. (1990) Bayesian reconstruction from emission tomography data using a modified EM algorithm. *IEEE Trans Med Imaging*, 9:84–93.

GUI, Q., Y. GONG, G. LI and B. LI (2007) A Bayesian approach to the detection of gross errors based on posterior probability. *J Geodesy*, DOI 10.1007/s00190-006-0132-y.

GUNDLICH, B. (1998) Konfidenzbereiche für robuste Parameterschätzungen. In: FREEDEN, W. (Ed.), *Progress in Geodetic Science at GW 98*. Shaker Verlag, Aachen, 258–265.

GUNDLICH, B. and K.R. KOCH (2002) Confidence regions for GPS baselines by Bayesian statistics. *J Geodesy*, 76:55–62.

GUNDLICH, B., K.R. KOCH and J. KUSCHE (2003) Gibbs sampler for computing and propagating large covariance matrices. *J Geodesy*, 77:514–528.

GUNDLICH, B., P. MUSMAN, S. WEBER, O. NIX and W. SEMMLER (2006) From 2D PET to 3D PET: Issues of data representation and image reconstruction. *Z Med Phys*, 16:31–46.

HAMILTON, A.G. (1988) *Logic for Mathematicians*. Cambridge University Press, Cambridge.

HAMPEL, F.R., E.M. RONCHETTI, P.R. ROUSSEEUW and W.A. STAHEL (1986) *Robust Statistics.* Wiley, New York.

HARVILLE, D.A. (1999) Use of the Gibbs sampler to invert large, possibly sparse, positive definite matrices. *Linear Algebra and its Applications,* 289:203–224.

HEITZ, S. (1968) *Geoidbestimmung durch Interpolation nach kleinsten Quadraten aufgrund gemessener und interpolierter Lotabweichungen.* Reihe C, 124. Deutsche Geodätische Kommission, München.

HÖRMANN, W., J. LEYDOLD and G. DERFLINGER (2004) *Automatic Nonuniform Random Variate Generation.* Springer, Berlin.

HUBER, P.J. (1964) Robust estimation of a location parameter. *Annals Mathematical Statistics,* 35:73–101.

HUBER, P.J. (1981) *Robust Statistics.* Wiley, New York.

HUTCHINSON, M.F. (1990) A stochastic estimator of the trace of the influence matrix for Laplacian smoothing splines. *Comun Statistist-Simula,* 19:433–450.

JAYNES, E.T. (2003) *Probability theory. The logic of science.* Cambridge University Press, Cambridge.

JAZWINSKI, A.H. (1970) *Stochastic Processes and Filtering Theory.* Academic Press, New York.

JEFFREYS, H. (1961) *Theory of Probability.* Clarendon, Oxford.

JENSEN, F.V. (1996) *An Introduction to Bayesian Networks.* UCL Press, London.

JOHNSON, N.L. and S. KOTZ (1970) *Distributions in Statistics: Continuous Univariate Distributions, Vol. 1, 2.* Houghton Mifflin, Boston.

JOHNSON, N.L. and S. KOTZ (1972) *Distributions in Statistics: Continuous Multivariate Distributions.* Wiley, New York.

JUNHUAN, P. (2005) The asymptotic variance-covariance matrix, Baarda test and the reliability of L_1-norm estimates. *J Geodesy,* 78:668–682.

KIRKPATRICK, S., C.D. GELATT and M.P. VECCHI (1983) Optimization by simulated annealing. *Science,* 220:671–680.

KLONOWSKI, J. (1999) *Segmentierung und Interpretation digitaler Bilder mit Markoff-Zufallsfeldern.* Reihe C, 492. Deutsche Geodätische Kommission, München.

KOCH, K.R. (1986) Maximum likelihood estimate of variance components; ideas by A.J. Pope. *Bulletin Géodésique*, 60:329–338.

KOCH, K.R. (1987) Bayesian inference for variance components. *Manuscripta geodaetica*, 12:309–313.

KOCH, K.R. (1990) *Bayesian Inference with Geodetic Applications*. Springer, Berlin.

KOCH, K.R. (1994) Bayessche Inferenz für die Prädiktion und Filterung. *Z Vermessungswesen*, 119:464–470.

KOCH, K.R. (1995A) Bildinterpretation mit Hilfe eines Bayes-Netzes. *Z Vermessungswesen*, 120:277–285.

KOCH, K.R. (1995B) Markov random fields for image interpretation. *Z Photogrammetrie und Fernerkundung*, 63:84–90, 147.

KOCH, K.R. (1996) Robuste Parameterschätzung. *Allgemeine Vermessungs-Nachrichten*, 103:1–18.

KOCH, K.R. (1999) *Parameter Estimation and Hypothesis Testing in Linear Models, 2nd Ed.* Springer, Berlin.

KOCH, K.R. (2000) Numerische Verfahren in der Bayes-Statistik. *Z Vermessungswesen*, 125:408–414.

KOCH, K.R. (2002) Monte-Carlo-Simulation für Regularisierungsparameter. *ZfV–Z Geodäsie, Geoinformation und Landmanagement*, 127:305–309.

KOCH, K.R. (2005A) Bayesian image restoration by Markov Chain Monte Carlo methods. *ZfV–Z Geodäsie, Geoinformation und Landmanagement*, 130:318–324.

KOCH, K.R. (2005B) Determining the maximum degree of harmonic coefficients in geopotential models by Monte Carlo methods. *Studia Geophysica et Geodaetica*, 49:259–275.

KOCH, K.R. (2006) ICM algorithm for the Bayesian reconstruction of tomographic images. *Photogrammetrie, Fernerkundung, Geoinformation*, 2006(3):229–238.

KOCH, K.R. (2007) Gibbs sampler by sampling-importance-resampling. *J Geodesy*, DOI 10.1007/s00190-006-0121-1.

KOCH, K.R. and J. KUSCHE (2002) Regularization of geopotential determination from satellite data by variance components. *J Geodesy*, 76:259–268.

KOCH, K.R. and J. KUSCHE (2007) Comments on Xu et al. (2006) Variance component estimation in linear inverse ill-posed models, J Geod 80(1):69–81. *J Geodesy*, DOI 10.1007/s00190-007-0163-z.

KOCH, K.R. and H. PAPO (2003) The Bayesian approach in two-step modeling of deformations. *Allgemeine Vermessungs-Nachrichten*, 110,111:365–370,208.

KOCH, K.R. and M. SCHMIDT (1994) *Deterministische und stochastische Signale*. Dümmler, Bonn.

KOCH, K.R. and Y. YANG (1998A) Konfidenzbereiche und Hypothesentests für robuste Parameterschätzungen. *Z Vermessungswesen*, 123:20–26.

KOCH, K.R. and Y. YANG (1998B) Robust Kalman filter for rank deficient observation models. *J Geodesy*, 72:436–441.

KOCH, K.R., H. FRÖHLICH and G. BRÖKER (2000) Transformation räumlicher variabler Koordinaten. *Allgemeine Vermessungs-Nachrichten*, 107:293–295.

KOCH, K.R., J. KUSCHE, C. BOXHAMMER and B. GUNDLICH (2004) Parallel Gibbs sampling for computing and propagating large covariance matrices. *ZfV–Z Geodäsie, Geoinformation und Landmanagement*, 129:32–42.

KÖSTER, M. (1995) *Kontextsensitive Bildinterpretation mit Markoff-Zufallsfeldern*. Reihe C, 444. Deutsche Geodätische Kommission, München.

KRARUP, T. (1969) *A contribution to the mathematical foundation of physical geodesy*. Geodaetisk Institut, Meddelelse No.44, Kopenhagen.

KULSCHEWSKI, K. (1999) *Modellierung von Unsicherheiten in dynamischen Bayes-Netzen zur qualitativen Gebäudeerkennung*. Reihe Geodäsie, Band 3. Shaker Verlag, Aachen.

KUSCHE, J. (2003) A Monte-Carlo technique for weight estimation in satellite geodesy. *J Geodesy*, 76:641–652.

LANGE, K. and R. CARSON (1984) EM reconstruction algorithms for emission and transmission tomography. *J Comput Assist Tomogr*, 8:306–316.

LANGE, K., M. BAHN and R. LITTLE (1987) A theoretical study of some maximum likelihood algorithms for emission and transmission tomography. *IEEE Trans Med Imaging*, MI-6:106–114.

LEAHY, R.M. and J. QI (2000) Statistical approaches in quantitative positron emission tomography. *Statistics and Computing*, 10:147–165.

LEONARD, T. and J.S.J. HSU (1999) *Bayesian Methods.* Cambridge University Press, Cambridge.

LINDLEY, D.V. (1957) A statistical paradox. *Biometrika,* 44:187–192.

LIU, J.S. (2001) *Monte Carlo Strategies in Scientific Computing.* Springer, Berlin.

LOREDO, T. J. (1990) From Laplace to Supernova SN 1987A: Bayesian inference in astrophysics. In: FOUGÈRE, P. F. (Ed.), *Maximum Entropy and Bayesian Methods.* Kluwer Academic Publ., Dordrecht, 81–142.

MARSAGLIA, G. and T.A. BRAY (1964) A convenient method for generating normal variables. *SIAM Review,* 6:260–264.

MAYER-GÜRR, T., K.H. ILK, A. EICKER and M. FEUCHTINGER (2005) ITG-CHAMP01: A CHAMP gravity field model from short kinematical arcs of a one-year observation period. *J Geodesy,* 78:462–480.

MEIER, S. and W. KELLER (1990) *Geostatistik.* Springer, Wien.

MENZ, J. and J. PILZ (1994) Kollokation, Universelles Kriging und Bayesscher Zugang. *Markscheidewesen,* 101:62–66.

METROPOLIS, N., A.W. ROSENBLUTH, M.L. ROSENBLUTH, A.H. TELLER and E. TELLER (1953) Equation of state calculations by fast computing machines. *J Chem Phys,* 21:1087–1092.

MODESTINO, J.W. and J. ZHANG (1992) A Markov random field model-based approach to image interpretation. *IEEE Trans Pattern Anal Machine Intell,* 14:606–615.

MORITZ, H. (1969) *A general theory of gravity processing.* Report 122. Department of Geodetic Science, Ohio State University, Columbus, Ohio.

MORITZ, H. (1973) *Least-squares collocation.* Reihe A, 75. Deutsche Geodätische Kommission, München.

MORITZ, H. (1980) *Advanced Physical Geodesy.* Wichmann, Karlsruhe.

NEAPOLITAN, R.E. (1990) *Probabilistic Reasoning in Expert Systems.* Wiley, New York.

NIEMANN, H. (1990) *Pattern Analysis and Understanding.* Springer, Berlin.

NOVIKOV, P.S. (1973) *Grundzüge der mathematischen Logik.* Vieweg, Braunschweig.

O'HAGAN, A. (1994) *Bayesian Inference, Kendall's Advanced Theory of Statistics, Vol. 2B.* Wiley, New York.

OLIVER, R.M. and J.R. SMITH (Eds.) (1990) *Influence Diagrams, Belief Nets and Decision Analysis.* Wiley, New York.

O'SULLIVAN, F. (1986) A statistical perspective on ill-posed inverse problems. *Statistical Science*, 1:502–527.

OU, Z. (1991) Approximate Bayes estimation for variance components. *Manuscripta geodaetica*, 16:168–172.

OU, Z. and K.R. KOCH (1994) Analytical expressions for Bayes estimates of variance components. *Manuscripta geodaetica*, 19:284–293.

PEARL, J. (1986) Fusion, propagation, and structuring in belief networks. *Artificial Intelligence*, 29:241–288.

PEARL, J. (1988) *Probabilistic Reasoning in Intelligent Systems.* Morgan Kaufmann, San Mateo.

PILZ, J. (1983) *Bayesian Estimation and Experimental Design in Linear Regression Models.* Teubner, Leipzig.

PILZ, J. and V. WEBER (1998) Bayessches Kriging zur Erhöhung der Prognosegenauigkeit im Zusammenhang mit der UVP für den Bergbau. *Markscheidewesen*, 105:213–221.

PRESS, S.J. (1989) *Bayesian Statistics: Principles, Models, and Applications.* Wiley, New York.

QI, J., R.M. LEAHY, S.R. CHERRY, A. CHATZIIOANNOU and T.H. FARQUHAR (1998) High-resolution 3D Bayesian image reconstruction using the microPET small-animal scanner. *Phys Med Biol*, 43:1001–1013.

RAIFFA, H. and R. SCHLAIFER (1961) *Applied Statistical Decision Theory.* Graduate School of Business Administration, Harvard University, Boston.

REIGBER, CH., H. JOCHMANN, J. WÜNSCH, S. PETROVIC, P. SCHWINTZER, F. BARTHELMES, K.-H. NEUMAYER, R. KÖNIG, CH. FÖRSTE, G. BALMINO, R. BIANCALE, J.-M. LEMOINE, S. LOYER and F. PEROSANZ (2005) Earth gravity field and seasonal variability from CHAMP. In: REIGBER, CH., H. LÜHR, P. SCHWINTZER and J. WICKERT (Eds.), *Earth Observation with CHAMP–Results from Three Years in Orbit.* Springer, Berlin, 25–30.

RIESMEIER, K. (1984) *Test von Ungleichungshypothesen in linearen Modellen mit Bayes-Verfahren.* Reihe C, 292. Deutsche Geodätische Kommission, München.

RIPLEY, B.D. (1987) *Stochastic Simulation.* Wiley, New York.

RIPLEY, B.D. (1996) *Pattern Recognition and Neural Networks.* University Press, Cambridge.

ROBERT, C.P. (1994) *The Bayesian Choice.* Springer, Berlin.

ROBERTS, G.O and A.F.M. SMITH (1994) Simple conditions for the convergence of the Gibbs sampler and Metropolis-Hastings algorithms. *Stochastic Processes and their Applications*, 49:207–216.

ROUSSEEUW, P.J. (1984) Least median of squares regression. *J American Statistical Association*, 79:871–880.

ROUSSEEUW, P.J. and A.M. LEROY (1987) *Robust Regression and Outlier Detection.* Wiley, New York.

RUBIN, D.B. (1988) Using the SIR algorithm to simulate posterior distributions. In: BERNARDO, J.M., M.H. DEGROOT, D.V. LINDLEY and A.F.M. SMITH (Eds.), *Bayesian Statistics 3.* Oxford University Press, Oxford, 395–402.

RUBINSTEIN, R.Y. (1981) *Simulation and the Monte Carlo Method.* Wiley, New York.

SHEPP, L.A. and Y. VARDI (1982) Maximum likelihood reconstruction for emission tomography. *IEEE Trans Med Imaging*, MI-1:113–122.

SILVERMAN, B.W. (1986) *Density Estimation for Statistics and Data Analysis.* Chapman and Hall, London.

SIVIA, D.S. (1996) *Data Analysis, a Bayesian Tutorial.* Clarendon Press, Oxford.

SKARE, O., E. BOLVIKEN and L. HOLDEN (2003) Improved sampling-importance resampling and reduced bias importance sampling. *Scandinavian Journal of Statistics*, 30:719–737.

SMITH, A.F.M. and A.E. GELFAND (1992) Bayesian statistics without tears: a sampling-resampling perspective. *American Statistician*, 46:84–88.

SMITH, A.F.M. and G.O. ROBERTS (1993) Bayesian computation via the Gibbs sampler and related Markov Chain Monte Carlo methods. *J Royal Statist Society*, B 55:3–23.

SPÄTH, H. (1987) *Mathematische Software zur linearen Regression.* Oldenbourg, München.

STASSOPOULOU, A., M. PETROU and J. KITTLER (1998) Application of a Bayesian network in a GIS based decision making system. *Int J Geographical Information Science*, 12:23–45.

TIKHONOV, A.N. and V.Y. ARSENIN (1977) *Solutions of Ill-Posed Problems*. Wiley, New York.

VARDI, Y., L.A. SHEPP and L. KAUFMAN (1985) A statistical model for positron emission tomography. *J American Statist Ass*, 80:8–37.

VINOD, H.D. and A. ULLAH (1981) *Recent Advances in Regression Methods*. Dekker, New York.

WANG, W. and G. GINDI (1997) Noise analysis of MAP–EM algorithms for emission tomography. *Phys Med Biol*, 42:2215–2232.

WEST, M. and J. HARRISON (1989) *Bayesian Forecasting and Dynamic Models*. Springer, Berlin.

WHITESITT, J.E. (1969) *Boolesche Algebra und ihre Anwendungen*. Vieweg, Braunschweig.

WIENER, N. (1949) *Extrapolation, Interpolation and Smoothing of Stationary Time Series with Engineering Applications*. Wiley, New York.

WOLF, H. (1968) *Ausgleichungsrechnung nach der Methode der kleinsten Quadrate*. Dümmler, Bonn.

WOLF, H. (1975) *Ausgleichungsrechnung, Formeln zur praktischen Anwendung*. Dümmler, Bonn.

WOLF, H. (1979) *Ausgleichungsrechnung II, Aufgaben und Beispiele zur praktischen Anwendung*. Dümmler, Bonn.

XU, P. (2005) Sign-constrained robust least squares, subjective breakdown point and the effect of weights of observations on robustness. *J Geodesy*, 79:146–159.

XU, P., Y. SHEN, Y. FUKUDA and Y. LIU (2006) Variance component estimation in linear inverse ill-posed models. *J Geodesy*, 80:69–81.

YANG, Y. and W. GAO (2006) An optimal adaptive Kalman filter. *J Geodesy*, 80:177–183.

YANG, Y., L. SONG and T. XU (2002) Robust estimator for correlated observations based on bifactor equivalent weights. *J Geodesy*, 76:353–358.

ZELLNER, A. (1971) *An Introduction to Bayesian Inference in Econometrics*. Wiley, New York.

Index

Printing: Krips bv, Meppel
Binding: Stürtz, Würzburg